Social Perception

SOCIAL PERCEPTION

Detection and Interpretation of Animacy, Agency, and Intention

edited by M. D. Rutherford and Valerie A. Kuhlmeier

The MIT Press
Cambridge, Massachusetts
London, England

MIT Press books may be purchased at special quantity discounts for business or sales promotional use. For information, please email special_sales@mitpress.mit.edu or write to Special Sales Department, The MIT Press, 55 Hayward Street, Cambridge, MA 02142.

This book was set in Stone Sans and Stone Serif by Toppan Best-set Premedia Limited, Hong Kong. Printed and bound in the United States of America.

Library of Congress Cataloging-in-Publication Data

Social perception : detection and interpretation of animacy, agency, and intention / edited by M. D. Rutherford and Valerie A. Kuhlmeier.
pages cm
Includes bibliographical references and index.
ISBN 978-0-262-01927-9 (hbk. : alk. paper)
1. Social perception. I. Rutherford, M. D. II. Kuhlmeier, Valerie A.
BF323.S63S643 2013
153.7—dc23
2012049080

10 9 8 7 6 5 4 3 2 1

Contents

1 Introduction: The Interdisciplinary Study of Social Perception

M. D. Rutherford and Valerie A. Kuhlmeier

As we enter a room full of people, we instantly have a number of social perceptions. First, and most fundamentally, we know which objects in the room are animate and which are inanimate. With a mere glance, we can infer that someone has agency, and from a continuous stream of behavior, we extract goals, desires, and dispositions. The aim of this book is to advance our understanding of those perceptual and cognitive processes that are part of our human psychology, allowing us to experience these percepts and understandings quickly and seemingly effortlessly.

The term *social perception* is most closely associated with the field of social psychology, particularly to the theoretical and empirical work of Fritz Heider. Heider was interested in causal attribution and schooled in the nonsocial causal attribution that was of interest in his time, but he made a novel advance by bringing the logic of causal attribution into the social domain (Schönpflug, 2008). He aimed to study the perception, reasoning, and evaluation of social behavior, and he suggested that this "social perception" follows many of the same rules as the perception of physical objects and nonsocial causality (e.g., Heider, 1944). Perhaps most famously, he showed subjects a two-minute video of simple geometric figures moving in relation to one another. He found that despite the "characters" having no morphological features to suggest agency, participants readily perceived agency and described the actions in language rich with intentional and mental-state attributions (Heider & Simmel, 1944). Of particular interest to Heider was the process of perception—that is, the properties of stimuli that produce the phenomenon of social interaction. As stated in that seminal paper, "our aim has not been to determine the correctness of the response, but instead the dependence of the response on stimulus-configurations" (Heider & Simmel, 1944, p. 244).

Further, Heider and Simmel (1944) noted that the movements of the shapes were "organized in terms of acts of persons" (p. 256) and that observers described "what motive or need within that person, is responsible for the movement" (p. 257). Heider expanded on these findings and others in his 1958 book, *The Psychology of Interpersonal Relations*, in which he posited two construals for perceiving and understanding the behavior of others: impersonal causality and personal causality. The former applies to unintentional behavior, such as sneezing, yawning, or holding an emotion such as sadness, while the latter applies to actions that are intentional or purposeful (Heider, 1958). As Malle (2008) points out, however, the distinction that Heider made between intentional and unintentional models faded within social psychological research as the evolving attribution theory instead emphasized inferences of stable dispositions and considered differences between attributions based on traits versus situations (e.g., Jones & Davis, 1965; Kelley, 1967).

Thus, while Heider is often considered one of the major influences in attribution research within social psychology (e.g., Fiske & Taylor, 1991), his emphasis on the perception of intentional action may have had even greater influence within the field of developmental psychology. Research in this field (coupled with work in neuroscience and comparative cognitive psychology) has laid the groundwork for a vast literature on theory of mind, examining the ability to represent others' agency, intentions, desires, and beliefs. Within this literature exist seminal models of the causal understanding of behavior (e.g., Leslie, 1995; Premack & Premack, 1995; see Sperber, Premack, & Premack, 1995, for a collection) and models of early, infant construals of goal-directed action (e.g., Baird & Baldwin, 2001; Gergely et al. 1995; Woodward, 1998).

Consideration of social perception outside of the major thrust of social psychology can also be seen in vision science. In particular, the study of biological motion (often depicted through point-light displays that show dots moving along with major joints of the body) has focused on both the mechanisms of perceptual organization that allow for the integration of the dots into the shape of a body (e.g., Johansson, 1973) and the social significance of the motion (see Troje, this volume, for discussion of both). The latter, through the use of point-light displays, allows for the examination of social perception from movement patterns in the near absence of other morphological features. Relatedly, the study of animate motion (in which whole objects move in space relative to each other) often uses displays similar to those of Heider and Simmel (1944) and examines the perception of social causality and agency (e.g., Scholl & Tremoulet, 2000).

The focus of this book is on these aspects of social perception. That is, the work presented does not follow the traditional routes of attribution theory (though attributions of dispositions and preferences to explain goal-directed action are touched on in chapters by Luo and Choi and by Kuhlmeier); instead, the emphasis is on the visual detection and interpretation of intentional social entities. Three caveats are in order, though. First, this book focuses on movement and action, which means that exciting recent work in face perception is not included (see Calder, Rhodes, Johnson, & Haxby, 2011, for summary). Second, since excellent volumes and review papers about theory-of-mind research into desire and belief representations already exist (see Baron-Cohen, Tager-Flusberg, & Cohen, 1993; Malle, 2004; Wellman, 1990; Wellman & Liu, 2004), that field of research is not covered here. Finally, although this book has some conceptual overlap (and common authors) with an oft-cited volume, *Intentions and Intentionality: Foundations of Social Cognition* (Malle, Moses, & Baldwin, 2001), the present volume places greater emphasis on vision science, developmental science, and neuroscience owing to a recent upswing in research in these areas coupled with technological advances.

Theoretical Frameworks and Methodological Paradigms

This book was developed and organized at a workshop held in June of 2011 at McMaster University in Hamilton, Ontario. Authors from all of the contributed chapters were invited to spend several days together and to present their chapters to one another. The idea behind the meeting was to allow for integration across chapters; the goal was a collection of chapters telling a coherent story about three aspects of social perception: (1) the perception of biological motion, (2) the perception of animacy and the attributions of intentionality made based on animate motion, and (3) The recognition and interpretation of goal directed behavior, based on human action. The meeting provided a venue for conversation among researchers studying these topics; among those included were researchers who mainly see themselves as vision scientists, those who are developmental psychologists, those who are known for their research in autism, and those who rely on imaging technology for their work in neuroscience.

It was evident in the conversations that there are theoretical frameworks and methodological paradigms that cut across four areas:

Developmental science is and has been important to our understanding of the human psychology that allows for social perception. Much of this work has considered the human infant, with research made possible by innovations in methodologies such as the habituation paradigm, the visual-expectation paradigm, eye-tracking (see Woodward & Cannon, this volume), and the dwell-time paradigm (see Baldwin & Sage, this volume), that allows for examination of what preverbal infants expect and how they categorize stimuli. The developmental framework allows for consideration of the foundational cognitive structures of social perception; some cases consider abilities present at birth (see Simion, Bardi, Mascalzoni, & Regolin, this volume), the role of experience such as self-motion (see chapters by Hauf, Woodward & Cannon and Loucks & Summerville, this volume), or both (Luo & Choi, this volume). This new understanding of the interplay between motoric and cognitive systems is but one novel contribution of this research.

Evolutionary Psychology provides a framework that will be evident in several chapters throughout this book, bringing a fitness-focused, functional orientation to the study of social perception. Evolutionary psychology considers all functional complexity in biology, including neural-cognitive complexity, as the result of evolution by natural selection. Like physical organs, human psychological adaptations are viewed as the result of incremental changes accumulating across generations; these adaptations enable individuals to develop solutions, including learning about others' mental states (which guide their actions), for solving specific adaptive problems.

Our evolutionary history bears directly on the social-perceptual aspects of our psychology. Humans have been intensely social throughout evolutionary history. In the Pleistocene hunter-gatherer environment in which we evolved, humans probably lived most of their lives in groups of approximately fifty to two hundred individuals (Dunbar, 1996). Successful membership in these groups depended on one's ability to solicit friendships, monitor allegiances, detect insult and exploitation, avoid offense, and track and fulfill obligations. These abilities rely on facets of human psychology. An individual's failure to succeed in group living might have led to ostracism, and in that hunter-gatherer environment, such ostracism could have been deadly (e.g., "Social Brain Hypothesis," Dunbar, 1998; Brothers, 1990; Barton & Dunbar, 1997; Jolly, 1966). A serious consideration of evolutionary psychology, in this context, leads us to the prediction that the human mind includes a suite of adaptations for solving the social-perceptual problems we discuss herein (see chapters by Rutherford; Kuhlmeier; McAleer & Love; and Troje, this volume). This perspective also predicts that these social-psychological abilities should develop reliably in a species-typical environment (Frankenhuis & Barrett, this volume).

Neuroscience is important to explorations of social perception, and recent advances in neuroimaging techniques allow us to test hypotheses about brain reactivity to social inputs. Not only can such neuroimaging provide information about the locations in the brain involved in processing social information, but in addition, once these areas are located, neuroimaging can then be used to characterize the cognitive mechanisms underlying social perception (see chapters by Grossman; Hamilton & Ramsey; Shiffrar & Thomas, this volume). In particular, this work aims to address key questions relating to what aspects of social perception might be based on domain-specific versus domain-general cognitive mechanisms and what specific roles the mirror neuron system plays in social cognition.

Clinical approaches are also taken in various chapters throughout the book. The clinical group that is of most interest as we work to understand the basis of social perception is people with autism spectrum disorders. It has been long suggested that this group has a specific impairment in social cognitive processing that can present along with unimpaired cognition in nonsocial domains. This dissociation provides an opportunity to (1) probe the kinds of stimuli that are processed socially; (2) consider what alternative strategies can be motivated to process social stimuli, and to what effect (Rutherford, this volume); and (3) consider which social cognitive strategies are developmentally related and which are independent.

Organization of the Book

The frameworks described in the preceding section weave throughout the three sections of this book. The first section presents work on the perception of biological motion, centering on research using point-light walkers as stimuli. The second section considers infants', children's, and adults' perception of animacy, and the understanding of goal-directed motions. The term "animacy perception" in this context refers to research probing observers' ability to discriminate animate from inanimate visual stimuli. In this tradition, the discrimination is usually based on motion cues, most often presented in computerized animations, sometimes a rendering of recorded actual human motion (McAleer & Love, this volume). One major advantage of this use of simplified and computerized stimuli is that it allows the researchers to decouple behavior and morphological features and, thus, to study each independently. Even infants seem to attribute intentions (Luo & Choi, this volume) and dispositions (Kuhlmeier, this volume) to highly stylized, nonhuman characters.

Finally, the third section details the perception and understanding of simple, yet real, human actions. This work examines the role of action experience in the social perception of goal-directed behavior and, additionally, considers the means by which humans parse a continuous stream of behavior to perceive specific intentions.

The organization of this volume (the sections are grouped by the stimuli used to display the social information) reflects the traditional division of research in the broader field of social perception. Indeed, the research presented in sections one and two tends to be published in vision science and neuroscience journals, while research described in the second half of section two and in section three tends to appear in developmental science journals. A major aim of this volume is to foster collaboration between these areas, and the traditional divisions allow for recent work to be presented in concert with related work, enabling readers from outside of the area to gain a more complete picture of a given aspect of social perception. Within each subject area, authors will share common methods and histories, and these common themes and backgrounds will be summarized in section introductions. Across the three sections, readers will likely find new approaches, while recognizing the themes that cut across the areas.

References

Baird, J. A., & Baldwin, D. A. (2001). Making sense of human behavior: Action parsing and intentional inference. In B. F. Malle, L. J. Moses, & D. A. Baldwin (Eds.), *Intentions and intentionality: Foundations of social cognition* (pp. 193–206). Cambridge, MA: MIT Press.

Baron-Cohen, S., Tager-Flusberg, H., & Cohen, D. (Eds.). (1993). *Understanding other minds: Perspectives from autism.* Oxford: Oxford University Press.

Barton, R. A., & Dunbar, R. I. M. (1997). Evolution of the social brain. In A. Whiten & R. Byrne (Eds.), *Machiavellian intelligence* (Vol. II). Cambridge: Cambridge University Press.

Brothers, L. (1990). The social brain: A project for integrating primate behviour and neurophysiology in a new domain. *Concepts in Neuroscience, 1,* 27–51.

Calder, A. J., Rhodes, G., Johnson, M. H., & Haxby, J. V. (Eds.). (2011). *Handbook of face processing.* Oxford: Oxford University Press.

Dunbar, R. I. M. (1996). Determinants of group size in primates: A general model. In W. G. Runciman, J. M. Smith, & R. I. M. Dunbar (Eds.), *Evolution of social behaviour patterns in primates and man* (pp. 33–58). Oxford: Oxford University Press.

Dunbar, R. I. M. (1998). The social brain hypothesis. *Evolutionary Anthropology, 6,* 178–190.

Fiske, S. T., & Taylor, S. E. (1991). *Social cognition* (2nd Ed.). New York: McGraw-Hill.

Gergely, G., Nadasdy, Z., Csibra, G., & Bíró, S. (1995). Taking the intentional stance at 12 months of age. *Cognition, 56,* 165–193.

Heider, F. (1944). Social perception and phenomenal causality. *Psychological Review, 51,* 358–374.

Heider, F. (1958). *The psychology of interpersonal relations.* New York: Wiley.

Heider, F., & Simmel, M. (1944). An experimental study of apparent behavior. *American Journal of Psychology, 57,* 243–259.

Johansson, G. (1973). Visual perception of biological motion and a model for its analysis. *Perception & Psychophysics, 14,* 201–211.

Jolly, A. (1966). Lemur social behavior and primate intelligence. *Science, 153,* 501–506.

Jones, E. E., & Davis, K. E. (1965). From acts to dispositions: The attribution process in social psychology. In L. Berkowitz (Ed.), *Advances in experimental social psychology* (Vol. 2, pp. 219–266). New York: Academic Press.

Kelley, H. H. (1967). Attribution theory in social psychology. In D. Levine (Ed.), *Nebraska symposium on motivation* (Vol. 15, pp. 192–238). Lincoln: University of Nebraska Press.

Leslie, A. M. (1995). A theory of agency. In D. Sperber, D. Premack, & A. J. Premack (Eds.), *Causal cognition: A multidisciplinary debate* (pp. 121–149). Oxford: Clarendon Press.

Malle, B. F. (2004). *How the mind explains behavior: Folk explanations, meaning, and social interaction.* Cambridge, MA: MIT Press.

Malle, B. F. (2008). Fritz Heider's legacy: Celebrated insights, many of them misunderstood. *Social Psychology, 39,* 163–173.

Malle, B. F., Moses, L. J., & Baldwin, D. A. (2001). *Intentions and intentionality: Foundations of social cognition.* Cambridge, MA: MIT Press.

Premack, D., & Premack, A. J. (1995). Intention as psychological cause. In D. Sperber, D. Premack, & A. J. Premack (Eds.), *Causal cognition: A multidisciplinary debate* (pp. 185–199). Oxford: Clarendon Press.

Scholl, B. J., & Tremoulet, P. (2000). Perceptual causality and animacy. *Trends in Cognitive Sciences, 4*(8), 299–309.

Schönpflug, W. (2008). Fritz Heider—my academic teacher and his academic teachers: Heider's "Seminar in Interpersonal Relations" and comments on his European background. *Social Psychology, 39,* 134–140.

Sperber, D., Premack, D., & Premack, A. J. (Eds.). (1995). *Causal cognition: A multidisciplinary debate* (pp. 185–199). Oxford: Clarendon Press.

Wellman, H. M. (1990). *The child's theory of mind.* Cambridge, MA: MIT Press.

Wellman, H. M., & Liu, D. (2004). Scaling of theory-of-mind tasks. *Child Development, 75*(2), 523–541.

Woodward, A. L. (1998). Infants selectively encode the goal object of an actor's reach. *Cognition, 69,* 1–34.

I BIOLOGICAL MOTION PERCEPTION

Section Introduction: Biological Motion Perception

Nikolaus F. Troje

Humans are social animals who spend a large part of their waking hours attending to other people. We are experts in "reading" friends and strangers alike, and we have evolved a multitude of neural-processing tools that are used to infer, based on incoming sensory data, information on who people are and what they are up to. Some of the more sophisticated tools that serve this purpose are relatively young. Speech production and speech comprehension are probably not more than a few hundred thousand years old. Similarly late in evolutionary terms, we lost a large amount of our facial hair. This, in turn, set the stage for the development of a new, complex visual-signaling system that takes advantage of a wealth of different facial muscles capable of creating all sorts of nuances in facial appearance and expression. Coevolving with these traits were the complex visual abilities required to interpret these nuances in other people's faces.

In contrast to face recognition and speech perception, the information channel that forms the topic of this book is much more ancient. The ability to derive information from visual motion is probably as old as vision itself. Very few seeing animals are able to retrieve information from static visual images, but every animal equipped with an eye—no matter how primitive—will respond to visual motion. Among the most basic visual behaviors are responses that are designed to aid in the avoidance of potential predators, such as the shadow response which causes a barnacle to withdraw into its shell in response to a moving shadow.

In general, the movements of other animals are a rich source of information and are discernible over distances at which other visual cues may not be available. The value of the information contained in an animal's movements goes way beyond the domain of explicit communication. The way a sick or injured animal moves reveals its vulnerable state to its predator. The motion of the predator's attack, on the other hand, breaks the camouflage of the lurking hunter. The optic flow generated by the surrounding animals in a flock of birds, a swarm of bees, or a school of fish is essential to coordinate an animal's behavior with that of its peers. Most courtship displays—whether observed in invertebrates, fish, birds, or mammals—are dominated by sophisticated body movements, which in turn demand sophisticated motion processing in the partner's visual system in order for the partner to adequately assess the actor's qualities.

In humans, the role of visual biological motion as a source of information about other people is probably as important as the role played by other sources, such as speech and facial expression. It seems, however, that we are less aware of using biological motion as a means of conveying information as compared to our awareness regarding other information channels. Specifically, compared to speech, the signals conveyed by biological motion are generally less subject to conscious control. They seem to convey who we are rather than who we want to be. Particularly in the context of locomotion and other actions that involve the investment of substantial metabolic energy, the characteristics of such motions contain signals to the expert receiver that are difficult to fake. "Body language" seems to be less of a language in the sense that the "speaker" of a language is generally in conscious control of the emitted signal. Body language is rather the language of the older, somatic parts of ourselves and reveals to the receiver information that the conscious carrier of this information may not have intended to share. The perceptual systems that listen to this language share its longer evolutionary history and also its nonconscious nature—which may explain why we tend to underestimate the important role that body motion plays in visual social perception.

2 What Is Biological Motion? Definition, Stimuli, and Paradigms

Nikolaus F. Troje

Abstract

The study of biological motion perception was introduced into vision research some forty years ago by Swedish psychologist Gunnar Johansson. Since that time, questions and approaches surrounding this topic have changed considerably. While early work on biological motion perception was primarily concerned with the mechanisms of perceptual organization the more recent research emphasizes the social significance of human and animal motion. In tracing the evolution of the term "biological motion" and its use, this chapter attempts to characterize and classify the various questions and experimental paradigms that have dominated research on biological motion perception over the last few decades.

Gunnar Johansson's Biological Motion: Structure from Nonrigid Motion

The term "biological motion" is closely associated with the work of Gunnar Johansson (1911–1998), an experimental psychologist who received his PhD from the University of Stockholm and then taught and conducted research at the University of Uppsala, Sweden, for most of his career. Rarely does a study on biological motion not contain a reference to at least one of the two papers Johansson published on his observation that observers can effortlessly make out a human figure behind a degraded visual stimulus that consists of nothing but a few dots moving along with the major joints of a human body (Johansson, 1973; 1976).

Johansson was an extremely prolific researcher who was well connected and had strong ties to many of the personalities who shaped European postwar experimental psychology. In addition to his two papers on the perception of human motion, he published almost a hundred papers over the course of his career. He was strongly influenced by the school of Gestalt psychology. Wolfgang Köhler had visited him several times in Uppsala, and the two had planned collaborative work shortly before Köhler's unexpected death in 1967. Johansson also maintained a long academic relationship with James J. Gibson, with whom he shared an interest in the question of how the visual system samples and then extracts information about discrete objects and events from the ambient optic array (i.e., the ever-changing, continuous stream of light irradiating the location of the observer from all directions). Particularly, the two friends discussed (and sometimes argued about) the way the visual system distinguishes between the visual motion induced by the observer moving through the environment and the motion that is due to movements of objects, animals, and people relative to the visual environment.

Johansson's approach to this question was his "vector analysis" theory, which he first laid out in his doctoral thesis (Johansson, 1950). His theory was rooted in Gestalt theory, specifically in Duncker's work on induced motion (Duncker, 1928), and it was based on Johansson's own experimental work for his doctoral thesis under the supervision of Gestalt psychologist David Katz at Stockholm University. Johansson's dissertation had set out to demonstrate how the human visual system breaks down the motion of simple objects into "common" motion, encoded in a global reference system, and "relative" motion represented within an object-based reference system. Common motion of the entire visual field with respect to the observer would be indicative of the observer's motion, while motion relative to the global, allocentric reference system has the potential to characterize objects—both in terms of motion parallax induced by the observer and in terms of the object's own movements. On the object level itself, the same distinction seemed to make sense, too. The motion common to all the parts of an object describes its trajectory through space; the relative motion between the parts bears information about changes of orientation or the configuration of the internal structure of a deformable object. "Vector analysis" emphasized the idea that the visual system decomposes motion in the optic array into components of motion occurring within a hierarchy of nested reference systems (Johansson, 1974).

What Johansson later called "biological motion" seemed to be an ideal domain to further explore these ideas and provided a natural continuation of his earlier studies. "The biological motion effect is a most efficient demonstration and verification of the theory of perceptual vector analysis," Johansson said toward the end of his career in an interview conducted by William Epstein in 1992 (Jansson, Bergstroem, & Epstein, 1994). To Johansson, the point-light display technique and the term "biological motion" that he used for it in his first paper on the topic (Johansson, 1973) served a very specific purpose. He was interested in the mechanisms of perceptual organization. Specifically, he was interested in the strategies the visual system uses to integrate the individual moving dots into the percept of the coherent, articulated shape of a human or animal body. He used his new stimuli to demonstrate that the visual system decomposes motion not just into common and relative motion, but into a whole hierarchy of partial motions, each of which is encoded relative to its parent coordinate system. In a way, his work and theories laid the foundation for the interest in visual "structure-from-motion," which soon became a popular research topic within vision science (e.g., Ullman, 1979). A number of early papers in biological motion research elaborated on Johansson's general idea and provided computational methods to retrieve structure from the nonrigid motion of a moving body (Cutting, 1981; Hoffman & Flinchbaugh, 1982; Webb & Aggarwal, 1982).

Today, research on biological motion perception has deviated in many ways from the questions that motivated Johansson in the first place. In fact, in the above-mentioned interview (Jansson, et al., 1994), Johansson expressed some degree of frustration about references to his work in contexts not reflective of his initial interest in the perceptual dissociation between common and relative motion. However, Johansson, in his early work, did not explicitly define what he meant by the term "biological motion" and how he wanted it to be used. Rather than introducing "biological motion" as a technical term, he used it in an informal way to refer to the stimulus domain from which he derived the displays needed to further his studies on the role of motion in perceptual organization: stylized depictions of the articulated motion of the human (or other animal) body. Biological motion in this sense is obviously a rich source of information about conspecifics as well as nonconspecific animals that may serve as prey or provide a predatory threat to the observer. As such, biological motion in this broader sense provides an interesting stimulus class for the vision researcher. Particularly for social animals such as humans, the rich information about the identity, actions, and intentions of an agent contained in the way the agent moves affords adaptations to exploit it. Other information channels, such as facial features, vocalizations, or body posture, may carry similar information. However, motion as a source for socially relevant information might be the evolutionarily oldest resource and certainly still plays a key role in person perception in humans. Its social significance motivates and justifies the wide interest that biological motion research has experienced in recent years.

As interesting as they are, these newer avenues of research into biological motion have little to do with what Gunnar Johansson was initially concerned with. The connections between Johansson's work and the emerging field of social perception are only superficial. One such connection is that point-light displays seem to provide a good solution to a technical problem. Studies on social perception from movement patterns often attempt to isolate motion from other sources of information, and point-light displays certainly provide that dissociation to a reasonable extent. Specifically, they are deprived of facial information, hair, clothing, and at least some aspects of the morphology of the body. Note, however, that information about the general proportions of the body is preserved. The motion of the body is still carried by a substrate that in itself contains some information about actor and action.

The other connection might be the term itself. "Biological motion" has become tightly associated with the point-light displays of Johansson's work. In a more colloquial sense, however, it has a much wider meaning. What other label but "biological motion" could we use when we talk about the movements of animals and humans and the information they convey? "Animate motion" (not to be confused with "animated motion") would be a good term to use if we wanted to avoid the term "biological motion" and its association with Johansson's question about structure from motion. However, the term "animate motion" suffers a somewhat similar fate as "biological motion." It has been adopted by another interesting domain within the field of social perception (see the second section of this volume)—a domain that is in many respects complementary to what is commonly associated with biological motion. Biological motion covers intrinsic motion—that is, the motion of a deforming body relative to reference systems centered on the body or its parts. The term "animate motion," on the other hand, gets commonly associated with the motion of whole objects through space and relative to each other. In order to isolate this aspect, rigid shapes such as triangles, circles, or squares are often used to study animate motion. Under certain conditions, as was first demonstrated by Heider and Simmel (1944), such rigid object movement is readily interpreted by a human observer as intentional and thus animate. For a review of more recent studies on animate motion, see Scholl and Tremoulet (2000; see also Scholl & Gao, this volume). With respect to Johansson's initial terminology, biological motion is primarily concerned with "relative" motion, while animate motion is all about the "common motion" of the entire object through space.

With both "biological motion" and "animate motion" being already reserved for more specific aspects, I will use in this chapter the more general term "life motion" for visual motion that expresses any sort of aspect characteristic for the motion of living beings. "Life motion" includes any kind of visual stimulus that elicits the percept of something or someone being alive—no matter what specific depiction is being used and what aspect of the motion is being emphasized. I am aware that the term "life motion" is at risk to suffer a similar fate as "biological motion" and "animate motion" as it has recently been used to label a specific aspect contained in the local motion of distal body parts (e.g., Wang, Zhang, He, & Jiang, 2010). At this point, however, it appears still a reasonable term to be used for the most general of the potential interpretations of biological motion.

This allows me—at least for the purpose of the current chapter—to reserve the term "biological motion" for stimuli and studies concerned with the intrinsic, nonrigid motion of the deforming body, rather than the motion path that its center of gravity subscribes through space. This usage includes Johansson's question about the retrieval of articulated structure, but does not restrict it to this topic. Staying in line with the predominant way the term was used in the literature, it also includes a wealth of work interested in what the visual system can retrieve from the motion of animals and people once the structure-from-motion problem is solved. Also note that this definition does not restrict us to point-light displays. Any stimulus that attempts to isolate the intrinsic motion component from other sources of information about actor and action is suitable for portraying biological motion and studying its perception. If not specifically interested in the structure-from-motion problem, stick figures may be as good as—or even better than—point-light displays to study the role of intrinsic motion without the influence of facial features or detailed information about the morphology of a body. The amount of additional information and the degree of realism of a depiction, however, is an important dimension when talking about stimuli used in life motion research and will be further discussed at the end of this chapter.

Experimental Paradigms: Detection and Direction

Restricting the term "biological motion" to the study of the perception of the "relative," intrinsic deformation of a body in action still encompasses a large variety of questions and methods. In an attempt to further investigate what is meant by "biological motion perception," I will try to categorize and critically discuss the different experimental paradigms and perceptual tasks that investigators have used to assess biological motion perception in human observers.

In trying to bring some order to the large host of literature, a reasonable distinction is the one between studies that aim to assess biological motion perception in general and the studies that ask specific questions about what kind of information can be retrieved from biological motion stimuli.

Assessing biological motion perception in general might be interesting with respect to varied context conditions or observer populations. For instance, after having demonstrated the general phenomenon of biological motion, Johansson varied display durations in order to measure how much time it takes to perceptually organize the dots into the shape of a person (Johansson, 1976). Subjects were shown point-light displays of a walker in sagittal view at different exposure times and were then asked to verbally report what they saw. With sufficient exposure, they generally reported seeing a human walker, and only for exposure times shorter than 200 ms did their performance decrease. James Cutting, who was one of the first researchers to advance Johansson's early work, set out to measure the robustness and salience of biological motion point-light displays presented in visual noise (Cutting, Moore, & Morrison, 1988). He showed stationary (as if on a treadmill) point-light walkers from a sagittal view and asked his observers to indicate whether the point-light walkers were facing left or right while manipulating the nature of the mask. He showed that a so-called scrambled mask, that is, a mask made of dots that move in the same way as the dots that constitute the walker, is much more efficient than any other kind of dot mask. Bertenthal and Pinto (1994) used a similar stimulus, but a different task, to investigate the role of global structure versus local features for biological motion perception; they also showed a mask made of scrambled walkers and embedded a coherent walker into it. Only half of the trials really contained the coherent walker while the other half showed the mask alone. The observer's task was to identify which of two displays contained the walker. Comparing responses to upright and inverted versions of this detection task, and finding a significant inversion effect, they concluded that biological motion perception is driven by global, configural processes and does not require the prior detection of individual features or local relations. This conclusion was based on earlier findings that the face-inversion effect is mainly caused by orientation sensitivity of configural, but not featural, processing.

The three examples just described are based on three different tasks. Johansson's demonstration employed a free-response format in which observers reported their perception of an unmasked point-light display. Performance was in most cases at ceiling, demonstrating the strength of the basic biological motion phenomenon. Masking, as used by Cutting and colleagues (1988) and also in Bertenthal and Pinto's (1994) study, helps to reduce performance and brings it into a range where we can expect to sensitively measure the effects of other experimental manipulations. But it also introduces a confound: Observers might perform poorly because they have problems with the required figure-ground segregation, or they might perform poorly because they have difficulties with the perceptual organization required to turn the individual dots into the coherent, articulated shape of a human body.

The direction-discrimination task used by Cutting adds another uncertainty: After having segregated the target walker from the masking dots, observers then have to determine the facing direction of the walker from whatever cues are available. It turns out there are at least two such cues. One important cue is certainly the motion-mediated structure of the body: Even from a single frame, it would be easy to determine facing direction if the articulated structure of the body was readily available—for instance by drawing explicit connections between the dots, thus turning the point-light display into a stick figure. The motion of the dots might simply help the observer to replace the missing sticks. The cues to direction are then contained in the reconstructed posture of the body and not in its motion.

The effectiveness of a mask made of scrambled walkers already points to a second source of information that observers use to perceive facing direction. It turns out, in fact, that the local motion of the individual dots can also carry sufficient information about facing direction. I will elaborate on that mechanism further below.

The "direction-discrimination task," in which observers are asked to identify facing direction, can thus be solved by the observer in two entirely independent ways: either by means of motion-mediated structure or with the help of local motion cues. Degraded information of one of them might well be compensated by the other and differences in performance between conditions and observer groups are therefore hard to interpret. These along with the ability to segregate the foreground walker from the background mask, mean that three different perceptual abilities are therefore confounded in the popular biological motion direction-discrimination task.

The experimental paradigms discussed above have all been used to assess the ability to perceive biological motion in both normal observers and with special participant populations. Several modifications exist, and new ones have been introduced. For instance, Mather and colleagues (1992) used a task in which coherent, sagittally presented walkers were intermixed with composite walkers in which upper and lower body were facing in opposite directions. Observers had to detect the incoherence.

Gurnsey and colleagues (2010) introduced a modified version of the direction task: Rather than using sagittal views, they reduced the angle between the left-facing and right-facing views to threshold levels—thus avoiding the need to use a mask and therefore eliminating the figure-ground segregation confound. Ahlstrom and colleagues (1997) used a modification of the detection task introduced by Bertenthal and Pinto (1994) in which they employed not just point-light walkers but depictions of all sorts of other actions, too. This particular stimulus set is important because it later became the basis for identifying and localizing the superior temporal sulcus (STS) as a primary area for the processing of biological motion (Grossman et al., 2000; this volume).

Beyond Structure from Motion: Style

While the experimental paradigms discussed in the previous section have been used for general assessment of an observer's ability to derive the articulated structure and kinematics of a point-light depiction, other tasks have been used to further explore the specific nature of the information that can be derived from such displays. The particular style of the movement can reveal properties of the actor that go beyond its mere detection and orientation. Biological motion point-light displays have been used to explore the limits of this ability and to investigate which particular attributes are carrying the information. James Cutting and colleagues were already able to shown that observers can derive the sex of a walker (Kozlowski & Cutting, 1977) from point-light displays and that they were even able to recognize familiar individuals (Cutting & Kozlowski, 1977). A number of studies later used principled manipulations of the point-light displays to investigate the role of motion-mediated structure, on the one hand, and the kinematics of the body, on the other hand, in gender classification and person identification (Mather & Murdoch, 1994; Troje, 2002; Troje, Westhoff, & Lavrov, 2005; Westhoff & Troje, 2007). Both domains definitely play a role, but information encoded in the kinematics of a walker seems to dominate the information obtained from structural cues.

Other properties encoded in the particular style of walking and other movements involve age (Montepare & Zebrowitz-McArthur, 1988), emotional attributes (Atkinson, Dittrich, Gemmell, & Young, 2004; Clarke, Bradshaw, Field, Hampson, & Rose, 2005; Dittrich, Troscianko, Lea, & Morgan, 1996), and personality traits (Troje, 2008). The kinematics of the point-light animations can also reveal sophisticated details about the performed action itself, such as the weight of a lifted box (Runeson & Frykholm, 1981) or the distance of a thrown object (Knoblich & Flach, 2001; Munzert, Hohmann, & Hossner, 2010). Even deception (Runeson & Frykholm, 1983) and intention (Hohmann, Troje, Olmos, & Munzert, 2011) can be derived from stylistic properties of human motion as encoded in point-light displays.

Note that even though all the studies cited in the two previous paragraphs fall into the domain of biological motion perception and often employ point-light displays, they have little in common with the initial question about perceptual organization of the individual dots into a coherent figure. Deriving sophisticated information about actor and action from the stylistic details of a movement most likely requires access to the articulated structure of the body, but how this structure is obtained might not be relevant at all. If it is relevant, it might rather create a confounding variable that hinders the interpretation of the resulting data. Point-light displays are often used because they seem to isolate the kinematics of a movement from all sorts of other sources of information present in more realistic depictions. However, they are not entirely free of structural information which can be reconstructed relatively easily, as Johansson and the early biological motion researchers have shown. If the subject of an investigation is not the structural reconstruction process itself, but rather the semantic contents of movement of the articulated figure, it may be better to present the articulation explicitly to the observer and replace the point-light walker with a stick figure or some other representation that provides the structure of the body directly while still depriving the viewer of morphological details and facial information. That way, stylistic properties of the motion can be investigated without having to worry about potential influences of the processes that mediate the recovery of the articulated structure.

This approach has rarely been taken within visual psychology, but computer-animation researchers employ it regularly (e.g., Brand & Hertzmann, 2000; Calvert, 2007; Hertzmann, O'Sullivan, & Perlin, 2009; Hodgins, O'Brien, & Tumblin, 1998; Unuma & Takeuchi, 1993). Even though this literature did not grow out of the tradition of Johansson's introduction of biological motion point-light displays and does not use the same terminology, it is concerned with the same questions we discussed in the previous paragraphs: How is information about identity, emotion, and personality encoded in the stylistic details of an actor's movements? In computer animation, this question is directly related to an applied one: What is required to generate human or animal motion that convincingly simulates natural human movement while retaining information about the actor? In the case of photorealistic animations in movies and video games, we seek an answer to the question: Which are the artifacts that reveal to the observer the artificial nature of the animation, and how can they be avoided?

Before Structure from Motion: Life Detection

In addition to work on the structure-from-motion problem and the studies exploring stylistic content, there is a third domain that adds to the general phenomenology and the study of biological motion perception. The fact that masking a walker with scrambled versions of the same walker still allows for some degree of recognition has been use to argue for the importance of global processing (e.g., Bertenthal & Pinto, 1994). After all, the additional dots mask the local motion and performance must therefore be carried by global structure. However, the argument can be reversed: Even though observers can still make the walker out if the mask is not too dense, it is clear that the scrambled walker mask is extremely effective when compared to other kinds of dot masks. For instance, Cutting and colleagues (1988) report experiments in which fifty-five dots were sufficient to reduce direction-discrimination performance to chance level if a scrambled mask was used. If the same number of noise dots moved either on linear or circular trajectories, observers responded correctly in about 80 percent of the cases (see also Hiris, Humphrey, & Stout, 2005). This shows that depriving a point-light stimulus of its local motion components seriously impairs recognition.

Kozlowski and Cutting (1977) were probably the first to suggest that the motion of individual dots may contain information entirely independent of their contribution to the retrieval of articulated shape. Their finding that the single dot representing a foot is enough for an observer to discriminate between a male and a female walker was later retracted (Kozlowski & Cutting, 1978), but others were able to confirm that the local motion of individual dots does help to interpret biological motion (Mather, et al., 1992). Troje and Westhoff (2006) used spatially and temporally scrambled versions of a biological motion point-light walker to show that even if deprived of any structural information, observers can still determine the facing direction of a walker. The same study also showed that the ability to do so is orientation dependent and breaks down entirely when the scrambled walker is presented upside down. Given the absence of any sort of structure or configuration in the scrambled walker, this inversion effect challenges the presumed close association between inversion effects and "configural processing," which originated from the face-recognition literature but had also been adopted in biological motion research (e.g., Bertenthal & Pinto, 1994; Farah, 1992). In addition, the inversion effect provides a tool for finding the source of the signatures that communicate facing direction. It turns out that only inverting the trajectory of the feet entirely disrupts the ability to obtain facing direction from scrambled biological motion (Troje & Westhoff, 2006) and that the critical property of that trajectory is the vertical acceleration pattern and whether or not it is consistent with gravitational acceleration (Chang & Troje, 2009a).

Troje and Westhoff's finding also prompted a discussion about a more general purpose for the visual filter that provides the sensitivity to this particular feature. It was suggested that it functions as a general "life detector" that labels a motion stimulus as potentially animate (Johnson, 2006). Evidence that sensitivity to the invariants in foot motion exists in nonhuman species (Vallortigara & Regolin, 2006), and that it might be present in human neonates prior to individual experience (Bardi, Regolin, & Simion, 2011; Simion, Regolin, & Bulf, 2008), points to an evolutionarily ancient origin. Furthermore, it was shown that the same features that convey information about facing direction also contribute to the perceived animacy of a point-light stimulus (Chang & Troje, 2008). A number of recent studies have confirmed the idea that visual invariants contained in the way the feet respond to constraints imposed by gravity and inertial forces are in fact used by the visual system to direct attention and guide orientation responses to people and other terrestrial, legged animals independently of their particular shape (Hirai, Saunders, & Troje, 2011; Jiang & He, 2007; Wang, et al., 2010) (for a review of the literature, see Troje & Chang, 2013). A recent study showed that upright scrambled biological motion displays are perceived to be significantly longer in duration than inverted stimuli (Wang & Jiang, 2012).

All of the studies mentioned in the previous paragraph employed stimuli similar to the ones used in Troje and Westhoff's (2006) initial study. As it turns out, the local cues can be even stronger than suggested by Troje and Westhoff (2006). Hirai and colleagues (Hirai, Chang, Saunders, & Troje, 2011) recently described a point-light stimulus that is entirely free of structural cues to facing direction, but still elicits facing-direction performance at almost the same level as normal intact point-light walkers. Stimuli were derived from a normal sagittal view of a point-light walker by randomly displacing all dots as in the standard scrambled walker—but this time only disrupting the horizontal configuration while leaving the vertical organization intact. Similar to the way that vertical acceleration validates motion as "biological" without providing information about facing direction itself, the vertical location of the dots representing the feet (either above or below the rest of the body) turns out to be an effective validating cue.

Global Shape versus Local Motion: Methodological Implications

For our discussion of the meaning and usage of the term "biological motion," the ability to exploit purely local information has two important implications. I briefly mentioned the first implication above: In general, the frequently used direction-discrimination task ("Which side is the walker facing?") can be solved using two different and, to a large degree, independent sources of information: (1) the motion-mediated articulated structure and (2) the specific, local motion of the feet. The standard direction-discrimination task confounds two very different abilities such that it becomes difficult to arrive at clear conclusions. The same is a potential problem in other paradigms as well. For instance, if the same local cues that indicate facing direction are used to label motion as "biological," as suggested above (Chang & Troje, 2008; Johnson, 2006; Troje & Westhoff, 2006), then the local motion of the feet may as well be used in detection experiments similar to the one introduced by Bertenthal and Pinto (1994).

The second implication is the following: If we accept that the ability of an observer to exploit local invariants in order to derive information about the animacy and orientation of an agent falls into the domain of biological motion perception, then scrambled biological motion cannot be regarded as nonbiological motion. However, it is often used as if it were; for instance, scrambled biological motion is frequently used as a control stimulus in behavioral, physiological, neuropsychological, or imaging studies (e.g., Grossman et al., 2000; Vaina, Solomon, Chowdhury, Sinha, & Belliveau, 2001). In principal, there is no problem with this kind of contrast—so long as the goal, limits, and implications of such experiments are clear. Contrasting the perception of coherent and scrambled biological motion captures an observer's ability to retrieve motion-mediated structure from a point-light display. It does not, however, allow more general conclusions about biological motion perception. The same is true for the processing abilities of a brain area identified by means of functional magnetic resonance imaging (fMRI) responses to that contrast.

The question of which brain areas are involved in biological motion perception is even more complicated. A central area, which frequently comes up in the literature reporting brain-imaging studies on biological motion perception, is the superior temporal sulcus. To what degree STS is really a "biological motion area," and which aspects of biological motion it processes, is not at all clear (Grossmann, this volume). As mentioned already above, an experiment that effectively localizes STS involves the contrast between coherent point-light displays of actions other than walking and scrambled versions of the same displays. The STS therefore seems to respond to the structure-from-motion aspects of biological motion. The problem is that the response is neither specific nor does it generalize. On the one hand, STS responds not just to biological motion but also to many other visual, and even nonvisual, social stimuli (Allison, Puce, & McCarthy, 2000; Kriegstein & Giraud, 2004; Puce & Perrett, 2003). On the other hand, the response to biological motion point-light displays seems to be specific to certain actions but does not generalize to others. For instance, STS does not necessarily respond to the stimulus most widely used in behavioral experiments on biological motion: the display of a walker shown in sagittal view as walking on a treadmill (Jastorff & Orban, 2009).

Figure-Ground Segregation and Biological Motion Perception

As the simultaneous presence of both global motion-mediated form and local motion invariants can make the interpretation of a biological motion experiment complicated, the use of visual dot masks can become problematic, too. Often used to reduce ceiling performance in biological motion experiments or to measure tolerance thresholds, they add to an experiment a task that is, in principle, unrelated to biological motion: They require the observer to efficiently segregate the walker in the foreground from the mask in the background. Figure-ground segregation performance in itself, however, might be affected by experimental manipulations intended to investigate biological motion perception. For instance, Ikeda, Blake, and Watanabe (2005), using a detection task, found that biological motion perception is particularly poor in the visual periphery—much worse than what could be explained by decreasing visual acuity. Scaling the size of the stimulus up to compensate for the latter didn't help. It was later discovered, however, that this phenomenon depended on the walker being masked by noise dots. Using a paradigm that did not require masking noise, Gurnsey and colleagues (Gurnsey, Roddy, Ouhnana, & Troje, 2008) were able to compensate for the decreasing performance at greater eccentricities by increasing the size of the stimulus (see also Thompson, Hansen, Hess, &

Troje, 2007). Only detecting a walker in masking noise was "unscalably poor," not the perception of biological motion itself.

Two other studies also provide data that should make us cautious about using masking noise to reduce ceiling performance and measure thresholds. In a study comparing older and younger adults in a direction discrimination task, performance was equal between the two groups except when masking noise was added, in which case the older adults fared significantly worse than the younger adults (Pilz, Bennett, & Sekuler, 2010). Here, as well as in a study that compared perception of biological motion with the "lazy" eye of amblyopic observers to vision with the normal control eye (Thompson, Troje, Hansen, & Hess, 2008), the results suggested that the deficits observed in older or visually impaired participants were mainly due to a decreased ability to distinguish the walker from the noisy background—much more so, at least, than to structure-from-motion related aspects of biological motion perception or the ability to process local "life detection" invariants.

A Taxonomy of Biological Motion

In this chapter, I have attempted to position the different usages of the term "biological motion" in a space that is spanned by the questions that motivate the research, the experimental paradigms being used, and the stimuli employed to depict human (or animal) motion. Figure 2.1 summarizes some of the considerations made in the previous sections. I consider the scheme depicted here as a work in progress and the basis for further discussion, rather than a final solution to the question I am asking in the title of this chapter.

Realistic versus Stylized

The stimuli used to study visual life motion of humans and animals are often degraded and stylized to some degree in order to isolate the aspects of interest from confounding variables that are correlated with motion signals in more realistic stimuli (e.g., video, full computer animations) or the ones experienced in the real world. Some studies have chosen to use less abstracted depictions, such as more realistic shaded and/or textured 3D avatars whose 3D morphological attributes are kept constant (e.g., Hodgins et al., 1998). A potential problem with this approach is that morphological and kinematic parameters are highly correlated in real-life motion, and the human visual system is probably very sensitive to violations of learned correlations. Rather than studying kinematics in isolation, realistic but normalized morphologies might produce severe consistency

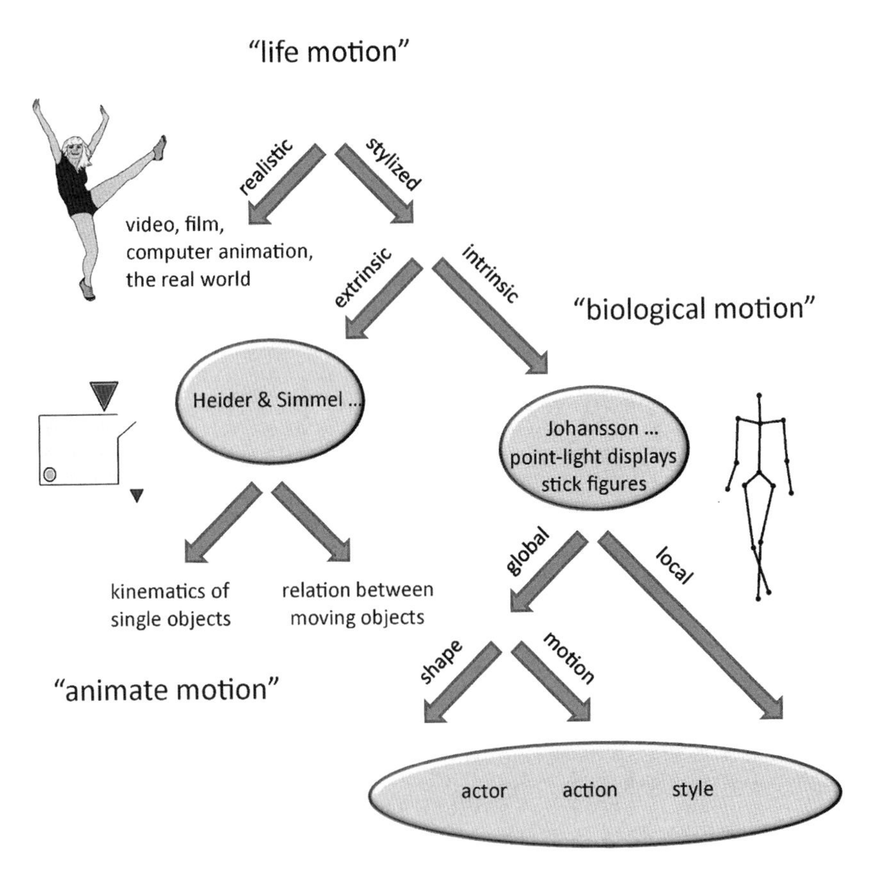

Figure 2.1
The schematic depicts the suggested usage of the term "biological motion" and the context in which the field is positioned.

effects that depend on how well a given motion pattern fits the chosen geometry of the normalized avatar. For instance, using motion data captured from a small, lightweight person to animate a big, heavy avatar will make its movements look unnatural and awkward.

Extrinsic versus Intrinsic

As already discussed above, I suggest restricting the term "biological motion" to the study of the perception of intrinsic, deformable motion of animals and people by means of stimuli that are stylized in order to keep information conveyed by channels other than the kinematics of the body at a minimum. The focus on intrinsic, rather than extrinsic, motion might feel somewhat arbitrary and unnecessary in light of questions surrounding the cognitive mechanisms of social perception, which is a major focus in much of the more recent literature on biological motion perception. However, the distinction makes perfect sense when investigating the perceptual mechanisms that allow us to retrieve the rich information provided by the motion of an agent. Both intrinsic and extrinsic motion may play an important role; however, in unraveling the mechanisms that mediate that information, it is wise to keep the two entirely different visual qualities conceptually distinct.

Global Shape and Global Motion

Within the domain of biological motion perception, one important mechanism is the retrieval of the articulated shape of the body. This process is—almost by definition—global as it is based on the relations between locations and motions of individual parts of the body. An interesting discussion in that context concerns the roles of global motion and global shape. Most of the earlier work emphasized the role of motion for the retrieval of shape (Hoffman & Flinchbaugh, 1982; Johansson, 1973; Webb & Aggarwal, 1982). The movements of small sets of dots that afford a partly rigid interpretation seem to benefit the retrieval of global shape. Cutting (1981) had shown that perception of body shape from point-light displays is much easier when the dots are placed on the joints as compared to displaying off-joint positions. In the latter case, pairs of dots do not retain constant distance in 3D and cannot be interpreted as rigid object parts.

However, the argument can be reversed by putting emphasis on the fact that observers are still able to perceive the articulation of a point-light walker to some degree—even when the dots are placed off joints. More recent work showed that the perception of a coherent shape that undergoes global motion is even possible when the stimulus is entirely deprived of any local motion whatsoever by assigning new dot positions on the body in each frame of the animation (Beintema & Lappe, 2002). This finding motivated the creation of a model in which the incoming sensory data are tested against a prior template that implements knowledge about the articulation and overall shape of a human body (Lange & Lappe, 2006). The findings are interesting because they indicate that the mutual relations between global kinematics and global shape are not necessarily unidirectional. If local motion is not required to retrieve global shape, then the partial rigidity that can be obtained from spatial integration of local motions is also not required. Note, however, that this does not mean that it is not used if available.

Yet another interesting question with respect to the relation between form and motion concerns the role of global motion after the articulated structure has been retrieved. Facing direction, actions, and probably even actor specifications such as gender or emotion can be retrieved from motionless stick figures (Coulson, 2004). Static stick figures can even give rise to the percept of motion, as has been demonstrated by countless cartoonists. In a photograph of a person, it is in most cases easy to determine whether the picture was taken while the person was still or in motion (Kourtzi, Krekelberg, & van Wezel, 2008). So, once motion has served its role in filling in the missing sticks in a point-light animation, it is not clear to what degree motion is still necessary and what kind of information it conveys.

Local Motion

The discussion of the role of local motion for the retrieval of shape-independent information is entirely separate from the dispute detailed in the previous section, which concerns how global coherence of shape and kinematics is obtained. As already discussed earlier in this chapter, local motion by itself can contain interesting information that is independent of global shape and global motion. Because of that independence, it can serve as a generalizing invariant that encodes the presence of another living being—largely irrespective of its particular shape.

We introduced the term "life detector" for the invariant first suggested by Troje and Westhoff (2006) and further characterized by Chang and Troje (2009a; 2009b) in order to express that it is independent of the particular shape of the animal in question. However, the term is not indisputable. The proposed visual filter is sensitive to ballistic properties of the feet which reflect an energetically efficient interaction with gravitational and inertial forces and therefore applies to legged, mid- to large-sized animals. However, it would not respond to fish, flying birds, snakes, or animals so small that their kinematics is dominated by air resistance rather than inertial forces. The invariant that we isolated from the foot motion of articulated animals is certainly not the only life detection mechanism. Others have yet to be identified, maybe in terms of "mid-level features" (Casile & Giese, 2005; Sigala, Serre, Poggio, & Giese, 2005) or "sprites" (Cavanagh, Labianca, & Thornton, 2001).

Actor, Action, and Style

Global aspects of shape and motion, as well as local invariants such as the one discussed above, can contain semantic, socially relevant information about the category of the actor (e.g., human or animal? what kind of animal?) and the action itself. The particular style of its execution can be exploited for further information about both the particular identity of the actor and further details about the action. Again, both local and global information contribute in that respect. For instance, it has been shown that both global shape and global kinematics contribute to person identification and gender classification (Mather & Murdoch, 1994; Troje, 2002). Local aspects can convey emotional contents (e.g., Pollick, Paterson, Bruderlin, & Sanford, 2001) but global shape and motion can also do so (e.g., Dittrich et al., 1996).

Final Thoughts

We can gather, from the studies discussed in this chapter, that the information contained in biological motion is encoded in rather redundant ways. As with other stimuli, this redundancy can be exploited by the visual system to replace missing information in the stimulus and reconstruct attributes that are not directly exposed. If we see only the upper body of a person in motion, we can guess with relative accuracy what is happening with the lower body. A 2D projection of a body in motion results in a strong hypothesis about its 3D structure. Temporal and spatial parameters of biological motion are related in specific ways and it has been shown that the visual system has knowledge about these relations and can exploit them to make predictions about missing parameters. For instance, the size of an animal and the cadence of its gait are related in a similar way as the length of a pendulum is related to the frequency with which it swings. When asked to estimate the size of a walking, quadruped stick-figure animal, observers reproduce this relation—thereby demonstrating that their visual system knows about the relation between temporal and spatial scales under conditions of gravity (Jokisch & Troje, 2003).

Little research has been done so far on the degree to which deviations from learned and expected correlations between kinematics and structure, or between different movement parameters, are detected by the visual system and are perceived to be awkward or artificial. However, the fact that it is still not easy to fit a photorealistic avatar with artificial motion to such a degree that the visual system is seriously fooled into believing it sees a real person is indicative of two things: First, it demonstrates the amazing abilities of the visual system to sense even the slightest inconsistencies within the sophisticated orchestra of the different aspects of shape and motion, which may be a trait that helped our hunting ancestors to identify weak or immature prey or to choose a healthy mate. Second, it showcases our lack of understanding of the biomechanical details that determine these correlations—at least to such a degree that we cannot come up with truly convincing generative models.

The nonlinear relation between the realism achieved by a computer-graphics character and the comfort or discomfort that the observer experiences with it (sometimes termed the "uncanny valley") might be a direct result of the level of consistency between the model that is applied to the stimulus and the stimulus itself. Models applied to aliens, monsters, and cartoon figures probably represent much broader probability distributions as compared to the well-defined, peaked correlations that form the models our visual system applies to real people. Biological motion point-light displays and stick figures are used as a means to study life motion because they abstract the morphology of the figure to such a degree that only few expectations can cause uncontrollable consistency problems. For this reason, biological motion point-light displays as introduced by Gunnar Johansson almost forty years ago will continue to be used in the future; this is true even though much more sophisticated depictions of human motion are available in today's age of computer animations and virtual realities.

References

Ahlstrom, V., Blake, R., & Ahlstrom, U. (1997). Perception of biological motion. *Perception, 26*(12), 1539–1548. doi:10.1068/p261539.

Allison, T., Puce, A., & McCarthy, G. (2000). Social perception from visual cues: Role of the STS region. *Trends in Cognitive Sciences, 4*(7), 267–278. doi:10.1016/S1364-6613(00)01501-1.

Atkinson, A. P., Dittrich, W. H., Gemmell, A. J., & Young, A. W. (2004). Emotion perception from dynamic and static body expressions in point-light and full-light displays. *Perception, 33*, 717–746. doi:10.1068/p5096.

Bardi, L., Regolin, L., & Simion, F. (2011). Biological motion preference in humans at birth: Role of dynamic and configural properties. *Developmental Science, 14*, 353–359.

Beintema, J. A., & Lappe, M. (2002). Perception of biological motion without local image motion. *Proceedings of the National Academy of Sciences of the United States of America, 99*(8), 5661–5663. doi:10.1073/pnas.082483699.

Bertenthal, B. I., & Pinto, J. (1994). Global processing of biological motions. *Psychological Science, 5*(4), 221–225. doi:10.1111/j.1467-9280.1994.tb00504.x.

Brand, M., & Hertzmann, A. (2000). Style machines. Paper presented at the Proceedings of SIGGRAPH 2000.

Calvert, T. (2007). Animating dance. Paper presented at the Proceedings of Graphics Interface.

Casile, A., & Giese, M. A. (2005). Critical features for the recognition of biological motion. *Journal of Vision, 5*(4), 348–360. doi:10.1167/5.4.6.

Cavanagh, P., Labianca, A. T., & Thornton, I. M. (2001). Attention-based visual routines: Sprites. *Cognition, 80*(1–2), 47–60. doi:10.1016/S0010-0277(00)00153-0.

Chang, D. H., & Troje, N. F. (2008). Perception of animacy and direction from local biological motion signals. *Journal of Vision, 8*(5), 3.1–10. doi:10.1167/8.5.3.

Chang, D. H., & Troje, N. F. (2009a). Acceleration carries the local inversion effect in biological motion perception. *Journal of Vision, 9*(1), 19.1–17. doi:10.1167/9.1.19.

Chang, D. H., & Troje, N. F. (2009b). Characterizing global and local mechanisms in biological motion perception. *Journal of Vision, 9*(5), 8.1–10. doi:10.1167/9.5.8.

Clarke, T. J., Bradshaw, M. F., Field, D. T., Hampson, S. E., & Rose, D. (2005). The perception of emotion from body movement in point-light displays of interpersonal dialogue. *Perception, 34*(10), 1171. doi:10.1068/p5203.

Coulson, M. (2004). Attributing emotion to static body postures: Recognition accuracy, confusions, and viewpoint dependence. *Journal of Nonverbal Behavior, 28*(2), 117–139. doi:10.1023/B:JONB.0000023655.25550.be.

Cutting, J. E. (1981). Coding theory adapted to gait perception. *Journal of Experimental Psychology: Human Perception and Performance, 7*(1), 71–87. doi:10.1037/0096-1523.7.1.71.

Cutting, J. E., & Kozlowski, L. T. (1977). Recognizing friends by their walk: Gait perception without familiarity cues. *Bulletin of the Psychonomic Society, 9*(5), 353–356.

Cutting, J. E., Moore, C., & Morrison, R. (1988). Masking the motions of human gait. *Perception & Psychophysics, 44*(4), 339–347. doi:10.3758/BF03210415.

Dittrich, W. H., Troscianko, T., Lea, S. E. G., & Morgan, D. (1996). Perception of emotion from dynamic point-light displays represented in dance. *Perception, 25*(6), 727–738. doi:10.1068/p250727.

Duncker, K. (1928). Über induzierte Bewegung. *Psychologische Forschung, 12*, 180–259. doi:10.1007/BF02409210.

Farah, M. J. (1992). Is an object an object an object? Cognitive and neuropsychological investigations of domain specificity in visual object recognition. *Current Directions in Psychological Science, 1*(5), 164–169. doi:10.1111/1467-8721.ep11510333.

Grossman, E. D., Donnelly, M., Price, R., Pickens, D., Morgan, V., Neighbor, G., Blake, R. (2000). Brain areas involved in perception of biological motion. *Journal of Cognitive Neuroscience, 12*(5), 711–720. doi:10.1162/089892900562417.

Gurnsey, R., Roddy, G., Ouhnana, M., & Troje, N. F. (2008). Stimulus magnification equates identification and discrimination of biological motion across the visual field. *Vision Research, 48*(28), 2827–2834. doi:10.1016/j.visres.2008.09.016.

Gurnsey, R., Roddy, G., & Troje, N. F. (2010). Limits of peripheral direction discrimination of point-light walkers. *Journal of Vision, 10*(2). doi:10.1167/10.2.15.

Heider, F., & Simmel, M. (1944). An experimental study of apparent behavior. *American Journal of Psychology, 57*(2), 243–259. doi:10.2307/1416950.

Hertzmann, A., O'Sullivan, C., & Perlin, K. (2009). Realistic human body movement for emotional expressiveness. SIGGRAPH course material. http://webstaff.itn.liu.se/~jonun/web/teaching/2009-TNCG13/Siggraph09/courses/rhbm.pdf.

Hirai, M., Chang, D. H. F., Saunders, D. R., & Troje, N. F. (2011). Body configuration modulates the usage of local cues to direction in biological motion perception. *Psychological Science, 22,* 1543–1549. doi:10.1177/0956797611417257.

Hirai, M., Saunders, D. R., & Troje, N. F. (2011). Allocation of attention to biological motion: Local motion dominates global shape. *Journal of Vision, 11*(3), 1–11. doi:10.1167/11.3.4.

Hiris, E., Humphrey, D., & Stout, A. (2005). Temporal properties in masking biological motion. *Perception & Psychophysics, 67*(3), 435–443. doi:10.3758/BF03193322.

Hodgins, J. K., O'Brien, J. F., & Tumblin, J. (1998). Perception of human motion with different geometric models. *IEEE Transactions on Visualization and Computer Graphics, 4*(4), 307–316. doi:10.1109/2945.765325.

Hoffman, D. D., & Flinchbaugh, B. E. (1982). The interpretation of biological motion. *Biological Cybernetics, 42*(3), 195–204.

Hohmann, T., Troje, N. F., Olmos, A., & Munzert, J. (2011). The influence of motor expertise and motor experience on action and actor recognition. *Journal of Cognitive Psychology, 23*(4), 403–415. doi:10.1080/20445911.2011.525504.

Ikeda, H., Blake, R., & Watanabe, K. (2005). Eccentric perception of biological motion is unscalably poor. *Vision Research, 45*(15), 1935–1943. doi:10.1016/j.visres.2005.02.001.

Jansson, G., Bergstroem, S. S., & Epstein, W. (1994). *Perceiving events and objects.* Hillsdale, NJ: Lawrence Erlbaum.

Jastorff, J., & Orban, G. A. (2009). Human functional magnetic resonance imaging reveals separation and integration of shape and motion cues in biological motion processing. *Journal of Neuroscience, 29*(22), 7315–7329. doi:10.1523/JNEUROSCI.4870-08.2009.

Jiang, Y., & He, S. (2007). Isolating the neural encoding of the local motion component in biological motion. *Journal of Vision, 7*(9), 551a.

Johansson, G. (1950). *Configurations in event perception.* Uppsala: Almquist & Wiksell.

Johansson, G. (1973). Visual perception of biological motion and a model for its analysis. *Perception & Psychophysics, 14*(2), 201–211. doi:10.3758/BF03212378.

Johansson, G. (1974). Vector analysis in visual perception of rolling motion. *Psychological Research, 36*(4), 311–319. doi:10.1007/BF00424568.

Johansson, G. (1976). Spatio-temporal differentiation and integration in visual motion perception. *Psychological Research, 38*, 379–393. doi:10.1007/BF00309043.

Johnson, M. H. (2006). Biological motion: A perceptual life detector? *Current Biology, 16*(10), R376–R377. doi:10.1016/j.cub.2006.04.008.

Jokisch, D., & Troje, N. F. (2003). Biological motion as a cue for the perception of size. *Journal of Vision, 3*(4), 252–264. doi:10.1167/3.4.1.

Knoblich, G., & Flach, R. (2001). Predicting the effects of actions: Interactions of perception and action. *Psychological Science, 12*(6), 467–472. doi:10.1111/1467-9280.00387.

Kourtzi, Z., Krekelberg, B., & van Wezel, R. J. A. (2008). Linking form and motion in the primate brain. *Trends in Cognitive Sciences, 12*(6), 230–236. doi:10.1016/j.tics.2008.02.013.

Kozlowski, L. T., & Cutting, J. E. (1977). Recognizing the sex of a walker from a dynamic point-light display. *Perception & Psychophysics, 21*(6), 575–580. doi:10.3758/BF03198740.

Kozlowski, L. T., & Cutting, J. E. (1978). Recognizing the sex of a walker from a dynamic point-light display: Some second thoughts. *Perception & Psychophysics, 23*(5), 459. doi:10.3758/BF03204150.

Kriegstein, K. V., & Giraud, A. L. (2004). Distinct functional substrates along the right superior temporal sulcus for the processing of voices. *NeuroImage, 22*(2), 948–955. doi:10.1016/j.neuroimage.2004.02.020.

Lange, J., & Lappe, M. (2006). A model of biological motion perception from configural form cues. *Journal of Neuroscience, 26*(11), 2894–2906. doi:10.1523/JNEUROSCI.4915-05.2006.

Mather, G., & Murdoch, L. (1994). Gender discrimination in biological motion displays based on dynamic cues. *Proceedings of the Royal Society of London, Series B: Biological Sciences, 258*, 273–279. doi:10.1098/rspb.1994.0173.

Mather, G., Radford, K., & West, S. (1992). Low-level visual processing of biological motion. *Proceedings of the Royal Society of London, Series B: Biological Sciences, 249*(1325), 149–155. doi:10.1098/rspb.1992.0097.

Montepare, J. M., & Zebrowitz-McArthur, L. (1988). Impressions of people created by age-related qualities of their gaits. *Journal of Personality and Social Psychology, 55*(4), 547. doi:10.1037/0022-3514.55.4.547.

Munzert, J., Hohmann, T., & Hossner, E. J. (2010). Discriminating throwing distances from point-light displays with masked ball flight. *European Journal of Cognitive Psychology, 22*(2), 247–264. doi:10.1080/09541440902757975.

Pilz, K. S., Bennett, P. J., & Sekuler, A. B. (2010). Effects of aging on biological motion discrimination. *Vision Research, 50*(2), 211–219. doi:10.1016/j.visres.2009.11.014.

Pollick, F. E., Paterson, H. M., Bruderlin, A., & Sanford, A. J. (2001). Perceiving affect from arm movement. *Cognition, 82*(2), B51–B61. doi:10.1016/S0010-0277(01) 00147-0.

Puce, A., & Perrett, D. I. (2003). Electrophysiology and brain imaging of biological motion. *Philosophical Transactions of the Royal Society of London, Series B: Biological Sciences, 358*: 435–445. doi:10.1098/rstb.2002.1221.

Runeson, S., & Frykholm, G. (1981). Visual perception of lifted weight. *Journal of Experimental Psychology: Human Perception and Performance, 7*(4), 733–740. doi:10 .1037/0096-1523.7.4.733.

Runeson, S., & Frykholm, G. (1983). Kinematic specification of dynamics as an informational basis for person-and-action perception: Expectation, gender recognition, and deceptive intention. *Journal of Experimental Psychology: General, 112*(4), 585–615. doi:10.1037/0096-3445.112.4.585.

Scholl, B. J., & Tremoulet, P. D. (2000). Perceptual causality and animacy. *Trends in Cognitive Sciences, 4*(8), 299–309. doi:10.1016/S1364-6613(00)01506-0.

Sigala, R., Serre, T., Poggio, T., & Giese, M. (2005). Learning features of intermediate complexity for the recognition of biological motion. *Artificial Neural Networks: Biological Inspirations–ICANN, 2005*, 241–246. doi:10.1007/11550822_39.

Simion, F., Regolin, L., & Bulf, H. (2008). A predisposition for biological motion in the newborn baby. *Proceedings of the National Academy of Sciences of the United States of America, 105*(2), 809–813. doi:10.1073/pnas.0707021105.

Thompson, B., Hansen, B. C., Hess, R. F., & Troje, N. F. (2007). Peripheral vision: Good for biological motion, bad for signal noise segregation? *Journal of Vision, 7*(10), 1–7. doi:10.1167/7.10.12.

Thompson, B., Troje, N. F., Hansen, B. C., & Hess, R. F. (2008). Amblyopic perception of biological motion. *Journal of Vision, 8*(4), 21–14. doi:10.1167/8.4.22.

Troje, N. F. (2002). Decomposing biological motion: A framework for analysis and synthesis of human gait patterns. *Journal of Vision, 2*(5), 371–387. doi:10.1167/2.5.2.

Troje, N. F. (2008). Retrieving information from human movement patterns. In T. F. Shipley & J. M. Zacks (Eds.), *Understanding events: How humans see, represent, and act on events* (pp. 308–334). Oxford: Oxford University Press.

Troje, N. F., & Chang, D. H. F. (2013). Shape-independent processes in biological motion perception. In M. Shiffrar & K. Johnson (Eds.), *Visual perception of the human body in motion: Findings, theory, and practice*. Oxford: Oxford University Press. pp. 84-102

Troje, N. F., & Westhoff, C. (2006). The inversion effect in biological motion perception: Evidence for a "life detector"? *Current Biology, 16*(8), 821–824. doi:10.1016/j.cub.2006.03.022.

Troje, N. F., Westhoff, C., & Lavrov, M. (2005). Person identification from biological motion: Effects of structural and kinematic cues. *Perception & Psychophysics, 67*(4), 667–675. doi:10.3758/BF03193523.

Ullman, S. (1979). The interpretation of structure from motion. *Proceedings of the Royal Society of London, Series B: Biological Sciences, 203*(1153), 405. doi:10.1098/rspb.1979.0006.

Unuma, M., & Takeuchi, R. (1993). Generation of human motion with emotion. *Computer Animation, 93*, 77–88.

Vaina, L. M., Solomon, J., Chowdhury, S., Sinha, P., & Belliveau, J. W. (2001). Functional neuroanatomy of biological motion perception in humans. *Proceedings of the National Academy of Sciences of the United States of America, 98*(20), 11656–11661. doi:10.1073/pnas.191374198.

Vallortigara, G., & Regolin, L. (2006). Gravity bias in the interpretation of biological motion by inexperienced chicks. *Current Biology, 16*(8), R279–R280. doi:10.1016/j.cub.2006.03.052.

Wang, L., & Jiang, Y. (2012). Life motion signals lengthen perceived temporal duration. *Proceedings of the National Academy of Sciences of the USA*. Published online. doi:10.1073/pnas.1115515109.

Wang, L., Zhang, K., He, S., & Jiang, Y. (2010). Searching for life motion signals. *Psychological Science, 21*(8), 1083. doi:10.1177/0956797610376072.

Webb, J. A., & Aggarwal, J. K. (1982). Structure from motion of rigid and jointed objects. *Artificial Intelligence, 19*(1), 107–130. doi:10.1016/0004-3702(82)90023-6.

Westhoff, C., & Troje, N. F. (2007). Kinematic cues for person identification from biological motion. *Perception & Psychophysics, 69*(2), 241–253. doi:10.3758/BF03193746.

3 From Motion Cues to Social Perception: Innate Predispositions

Francesca Simion, Lara Bardi, Elena Mascalzoni, and Lucia Regolin

Abstract

The present chapter deals with the role of innate predispositions and experience in the origin of sensitivity to social agents, causal relations, and animacy. Experimental evidence will be reviewed to demonstrate that, from birth, motion is the most informative perceptual cue human and nonhuman animals use to identify living objects. The data highlighting innate predispositions at the origin of perception of physical causality, self-propelled motion, and biological motion will be presented and discussed in the light of the experience-dependent or experience-independent origin of such sensitivity.

Introduction

The ability to identify animated objects is thought to be fundamental for survival. Several perceptual cues, including, crucially, the detection of motion cues, allow human and nonhuman animals to identify living objects. The striking ability of humans to attribute causal relations (i.e., appreciation of the interaction between any two events in terms of a cause-effect relationship) and animacy (i.e., the capability of an object to have self-propelled motion) to moving objects on the basis of motion cues is well attested (Heider & Simmel, 1944; Michotte, 1963; Schlottmann, Ray, Mitchell, & Demetriou, 2006). Also, the ability to discriminate the biological vs. nonbiological nature of a motion has been demonstrated in both adults and newborn infants (Johansson, 1973, Simion, Regolin & Bulf, 2008).

According to some authors (e.g., Spelke, Phillips, & Woodward, 1995), some common principles underlie infants' reasoning about both inanimate and animate entities; these principles are cohesion (i.e., the fact that physical objects move as connected bounded wholes), continuity (i.e., the fact that physical objects move on connected unobstructed paths), and gravity (i.e., the fact that physical objects rest and move on supporting surfaces). Moreover, some specific principles concerning the motion of living organisms—such as self-propulsion, biological motion, goal-directed behavior, and reactivity to social contingencies—seem to underlie the detection of animate beings (Spelke & Kinzler, 2007).

Beginning at birth, motion triggers human attention, and attentional biases are mainly stimulus-driven (Valenza, Simion &, Umiltà, 1994; Farroni, Simion, Umiltà, & Dalla Barba, 1999). The neural explanation for the sensitivity to motion relies on the fact that newborns' visual behavior is mainly controlled by the superior colliculus, which belongs to the subcortical visuomotor pathway (Simion,Valenza, Umiltà, & Dalla Barba, 1995, 1998; Johnson, 2005).

Starting from such knowledge, which demonstrates that infants pay attention either to specific features belonging to animate beings, such as the face, eyes, and gaze (Morton & Johnson, 1991, Farroni, Csibra, Simion, & Johnson, 2002), or to the presence of biological motion, the existence has been hypothesized of some mechanisms that precociously bias humans toward socially relevant moving stimuli. Whether the nature of such mechanisms is experience-independent or experience-dependent is still a matter of dispute. The present chapter presents data from developmental studies that highlight the existence of certain innate predispositions to process motion cues underlying the perception of physical causality, self-propelled motion and biological motion.

1 The Perception of Physical Causality

Humans represent the world in terms of a rich cause-effect texture (Carey, 2009). Causal perception and reasoning are, undoubtedly, skills of great value in that they allow people to predict and control their own environment (Cohen, Rundell, Spellman, & Cashon, 1999) and make sense of almost every aspect of the physical world. Psychological research has focused on exploring both higher-level cognitive processing and reasoning (i.e., the dynamics of causal inferences; Sperber, Premack, & Premack, 1995); at the same time, much critical work has concentrated on the role of perceptual processing (Michotte, 1963). Let's focus on this latter perspective.

Imagine you are presented with two objects, A and B, located at some distance from one another. Object A starts to move toward B at a constant speed and comes into contact with it, whereupon A stops moving and B

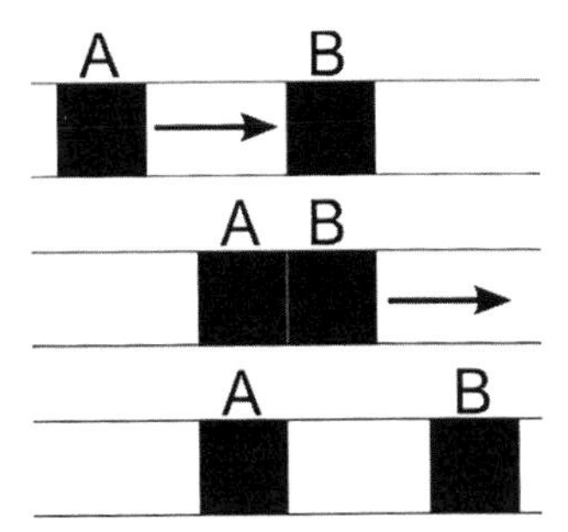

Figure 3.1
The schematic sequence of a launching effect.

starts moving along the same direction, and at the same or slightly lesser speed, as object A (figure 3.1).

You will have probably had the impression that A hit B and caused its movement, even though the objects had moved independently from one another. This type of event is the most well-known example of physical causality; Michotte first described it in 1946 as the "launching effect" (Michotte, 1963). When presented with this type of event, most observers would attribute a causal role to A.

In the case of the launching effect, the impression of causality is immediate and directly experienced; rather than being a conceptually mediated inference, it is the result of a perceptual process. The launching effect occurs when there are certain definite conditions of stimulation and reception, and the effect disappears as a result of appropriate modifications in the stimulus conditions. The leading factors for the causal impression aroused by the launching effect entirely lie in the spatial, temporal, and kinematic features of the motion of the two objects, which particularly favor the integration of the two movements in a unity and are consistent with a single motion transferred from one object to the other (Michotte, 1963).

As for temporal aspects, it is essential to the production of the launching effect that the second movement does rapidly succeed the first (a delay of 50 ms is sufficient to destroy any causal impression). The causal impression is maintained as long as the perception is compatible with the movement of a single object. Also, in the case of the spatial component, there is a better chance to perceive the launching when the paths of the two movements favor the perception of a strong spatial unity. No launching effect is perceived if the trajectory of object B is modified with respect to the trajectory of A; an angle of 25 degrees difference between the two paths is sufficient to weaken the perception of launching, which completely disappears with a 90 degree angle of difference (Michotte, 1963; Kerzel, Bekkering, Wohlschläger, & Prinz, 2000). Finally, with regard to the kinematic aspects, the launching effect is restricted within a critical range of speeds. When A is moderately faster than B, the launching is more reliably perceived. If there is a large difference between the speeds of the two objects, or whenever B is the faster object, the launching effect is not likely to be seen (Michotte, 1963; Schlottmann, Ray, Mitchell, & Demetriou, 2006).

The fact that the perception of causality seems to depend on a highly constrained collection of visual cues (e.g., Michotte, 1963; Choi & Scholl, 2004, 2006; Schlottmann, Ray, Mitchell, & Demetriou, 2006; White, 2006) suggests the possibility of a precocial, ontogenetic origin for such capability. Michotte's work was based on verbal reports from adult observers, but he hypothesized an innate mechanism as the basis for what was observed in adults. A specific perceptual module (i.e., the perceptual input analyzer) gave rise to the perception of physical causality, from which all other types of causal perception and inferences would arise (Michotte, 1963).

The innate or experience-dependent origin of causal representations has been widely investigated by developmental studies (Leslie, 1984; Leslie & Keeble, 1987; Cohen & Amsel, 1998; Cohen & Oakes, 1993; Cohen, Rundell, Spellman, & Cashon, 1999; Oakes, 1994; Oakes & Cohen, 1990, Carey, 2009).

On the one hand, several nativist accounts suggest that the capacity to perceive and represent causal perception might be innate or at least precocial. According to the most prominent nativist proposal (Leslie, 1984; Leslie & Keeble, 1987; Leslie, 1988), causal impressions originate in a low-level visual module with fixed, automatic, and mechanical occurrence, without being influenced at all by higher-level cognitive processes (Leslie & Keeble, 1987). This visual mechanism would be effective from birth and would operate throughout life. It would work at three levels of analysis. At the first level, the system would compute and represent the spatial and temporal relations between the submovements of the event; at the second level, such relations would be combined to produce a redescription of the event in terms of spatiotemporal continuity. Such continuity would allow for an attribution of causal roles and a description of causal structure at level three. The output would subsequently be further processed by the visual system or transferred to central cognitive processes (Leslie & Keeble, 1987). Therefore, infants would automatically perceive causality; this would occur as a function of the spatial and temporal features of the events themselves.

In contrast, according to the modern empiricist-constructivist theories, there may be no innate representation of cause; causal notions in this case would be constructed from sensory primitives (i.e., Cohen, Chaput, & Cashon, 2002).

In the constructivist account, the perception of causality would be the result of perceptual and cognitive development (Oakes & Cohen, 1990) and demonstrably would be neither innate nor domain specific (Cohen & Amsel, 1998; Cohen, Chaput, & Cashon, 2002). Infants' visual information processing would change systematically with development, and causal perception would develop during the first year of life (Cohen & Amsel, 1998). Initially, infants may attend to simple object features, such as shape, color, and type of motion. At a later stage, they may process the relations among those features and attend to the object itself as a unitary entity. Finally, infants may begin to learn about relationships among objects and actions, and integrate them into an event (Cohen, Chaput, & Cashon, 2002).

The debate between nativist and constructivist researchers is difficult to resolve, and both theories still enjoy favor. However, both accounts point out that causal perception arises as a function of the spatiotemporal features of the perceived events. For this reason, developmental studies focus on infants' sensitivity to the distinctive spatiotemporal features of a launching event—this being the most rudimental type of relationship necessary to perceive physical causality. Nevertheless, to be able to perceive the events in terms of a cause-effect relationship, infants' representation of the two stimuli present in the display has to go beyond the mere spatiotemporal analysis; this is because infants would need to infer causality.

The widespread method for obtaining the first type of evidence consists in habituating infants to either a launching event or an event presenting a temporal delay or a spatial gap. Infants are subsequently shown one trial with each of the three test events (e.g., habituation stimulus: launching, test stimuli: launching, a delayed launching and a gap event). Employing this type of paradigm, it has been demonstrated that four-month-old infants successfully discriminate between events on the basis of their spatiotemporal continuity (Leslie, 1982; Cohen, Amsel, Redford, & Casasola, 1998). Nevertheless, such results do not demonstrate that infants are as yet sensitive to physical causality itself (Cohen & Amsel, 1998). In effect, the ability to classify events as causal or noncausal seems to emerge at about six-and-a-half months of age (Leslie & Keeble, 1987; Oakes, 1994; Cohen & Amsel, 1998). Only from this age infants are shown to be able to both discriminate launching events from noncausal control stimuli (i.e., delayed launching and no-collision displays) on the basis of the causal features (Cohen & Amsel, 1998) and assign distinct roles to the agent and the recipient in a launching event (Leslie & Keeble, 1987).

The capability described in the previous paragraph is present only when an event involves simple objects that move along smooth horizontal trajectories. However, starting from ten months of age, infants also demonstrate sensitivity to physical causality in the presence of launching events that involve the movement of complex wheeled objects (Oakes & Cohen, 1990) or objects moving along dissimilar rolling trajectories (Oakes, 1994).

Importantly, the findings so far described are not dependent on a specific type of stimuli; simple computer-generated shapes (Oakes, 1994; Cohen & Amsel, 1998), videotaped movement of objects (Leslie & Keeble, 1987), and real, complex toys (Oakes & Cohen, 1990; Cohen & Oakes, 1993) have been successfully used in such experiments.

Therefore, experimental data suggest that, starting at six months of age, preverbal infants interpret Michotte's launching events as causal. No evidence for representations of causality has been reported as occurring in infants before this age; nor has the ability to process the spatiotemporal cues underlying causal perception been investigated, under conditions of effective control for previous experiences, in our species at birth.

Previous studies demonstrated that, in spite of their poor visual acuity, newborns possess sophisticated perceptual abilities (Macchi Cassia, Simion, Milani, & Umiltà, 2002; Turati, Simion, & Zanon, 2003; Valenza & Bulf, 2007; Simion, Regolin, & Bulf, 2008) that allow them to process and represent several kinds of information.

We investigated the perception of causality in newborns by assessing their spontaneous preferences for either a causal (i.e., launching stimulus) or a noncausal (e.g., delay stimulus) event. Results seem to suggest that two-day-old newborns are able to detect the peculiar spatiotemporal features of physical causality (Mascalzoni, Regolin, Vallortigara, & Simion, in press). These results do not allow us to draw conclusions about newborns' capability to perceive physical causality per se, nor even to determine the domain specificity/generality of the underlying mechanisms. Nevertheless, the results highlight newborns' early predisposition to look at a set of low-level perceptual cues (such as spatiotemporal continuity of motion and linearity of trajectory) that tightly correlate with spatiotemporal parameters associated with physical causality. It seems plausible that this predisposition would prompt attention to causal events and eventually support bootstrapping their perception as causal.

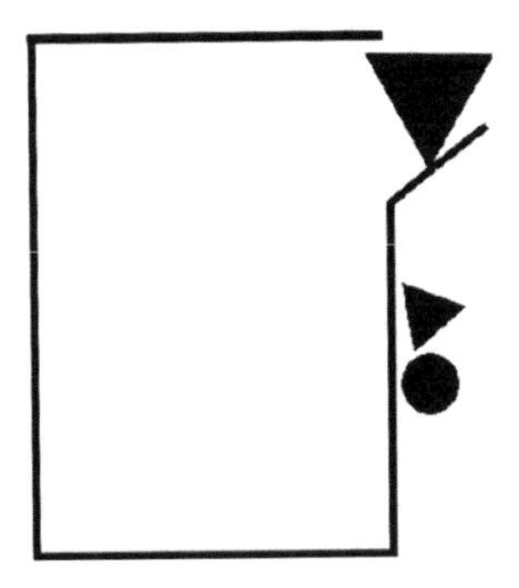

Figure 3.2
Schematic representation of a static frame from the classic Heider and Simmel animation: abstract geometric shapes moving around and inside of a rectangle. (Modified from Heider & Simmel, 1944.)

2 Self-Propelled Motion as a Cue to Animacy

So far we have dealt with humans' early predispositions to extract *physical* causality from well-defined spatiotemporal relationships between the motion of two objects.

Here we will focus on the origin of the ability to process certain specific motion cues so as to perceive animate motion and infer agency.

Humans' striking ability to attribute intentions and motivations to agents, even when these are as rough as geometric figures in motion, is widely demonstrated. The first experimental observations about such phenomena date back to 1944, when the psychologists Fritz Heider and Mary-Ann Simmel devised an animation involving three simple geometric shapes (i.e., a big triangle, a small one, and a little circle; see figure 3.2) that moved at different speeds and accelerations within and around a large rectangle.

Adult humans described the events in terms of an animate motion and inferred intentional action and social interaction (Heider & Simmel, 1944). Naïve observers not only perceived these shapes as "alive," but also attributed to them particular personality traits, beliefs, intentions, and emotions—as if they were animate beings (Heider & Simmel, 1944; Kanizsa & Vicario, 1968; Heberlein & Adolphs, 2004). These anthropomorphic interpretations of the events were automatic and remarkably consistent from one subject to another.

Since Heider and Simmel's work, several studies have tried to reveal the features that characterize an object as an agent. In this context, "agency" has been conceptualized at different levels (Carey, 2009): as mechanical agency, intentional agency, and mentalistic agency. In this paragraph we are going to consider the first, most basic level, the mechanical agency (i.e., the so-called animacy), which implies that an object is capable of (1) self-propelled motion, (2) acting as a mechanical cause, and (3) resisting forces acting upon it. This aspect of agency obviously interacts with the representations of physical causality described in the previous paragraph, since an animate agent is capable of playing the role of causal agent in an interaction and/or resisting a physical contact from another object (Leslie, 1994; Carey, 2009). The concept of mechanical agency, therefore, belongs to the domain of naïve physics; it lies outside the psychological domain of intentions and mental states (Leslie, 1994; Csibra, Gergely, Bíró, Koos, & Brockbank, 1999).

From a naïve view, self-propelled motion seems to provide one of the most powerful cues about what makes an object animate in the sense of its having a distinctive quality not possessed by objects that move only as a result of a physical contact.

This idea dates back at least to Aristotle (*Physics*) and has been incorporated into the developmental psychology doctrine. Several researchers have focused on the origin of the animate-inanimate distinction by investigating infants' ability to relate different types of motion with different kinds of objects (see Rakison & Poulin-Dubois, 2001, for a review). The majority of the theorists considered the physical principles related to the motion of entities as a crucial cue for infants' earliest distinction between animate and inanimate objects. In fact, the most obvious feature that distinguishes animate from inanimate entities seems to be the ability to self-propel motion, as opposed to objects that require external force in order to move. In Leslie's theory of agency (Leslie, 1994, 1995), the self-propelled origin of motion is interpreted in terms of the physical notion of force. It would be detected by the first component of a hierarchically organized system giving rise to the attribution of mechanical agency (i.e., animacy, an agent having an internal source of energy) to the object that it belongs to. Hence, for Leslie and other researchers as well (Premack, 1990; Rakison & Poulin-Dubois, 2001), self-propelled motion is a salient cue that triggers attribution of animacy.

In the last decade, research about the sensitivity to motion-cues that allows human and nonhuman animals to infer animacy has grown enormously (Johnson, 2000; Scholl & Tremoulet, 2000; Rakison & Poulin-Dubois, 2001; for a review see Gelman & Opfer, 2002). Results demonstrate that sensitivity to motion cues of animacy is widespread across cultures (see Morris & Peng, 1994) and subtended by specific brain regions (Castelli,

Happé, Frith, & Frith, 2000; Blakemore, Boyer, Pachot-Clouard, Meltzoff, Segebarth, & Decety, 2003). Humans' attributions of animacy are largely automatic (Scholl & Tremoulet, 2000), so it has been hypothesized that natural selection might have shaped the perceptual system to be particularly sensitive to motion cues of animacy.

Therefore, developmental studies tended to focus on infants' ability to relate self-propelled motion to different kinds of objects (Woodward, Phillips, & Spelke, 1993). Empirical studies have demonstrated that five-month-old infants are able to perceive the self-propelled nature of a movement and that they hold different expectations for physical events involving inert or self-propelled objects when such events possess an internal source of energy (Luo, Kaufman, & Baillargeon, 2009). The results can be readily explained if we assume that infants are capable of reasoning about self-propelled objects, even if in an extremely naïve manner, and perceiving them as endowed with "*an internal source of energy*" that the objects would use directly to control their own motions or indirectly (i.e., by applying force) to control the motion of other objects (Leslie, 1994, 1995). Infants' sensitivity to self-propulsion has been confirmed by further research (such as Kosugi, Ishida, & Fujita, 2003; or Saxe, Tzelnic, & Carey, 2006, 2007). Nevertheless, the current state of research cannot exclude the possibility that prior experience during the first five months of life may shape the responses of infants toward animate objects. To truly test innate predispositions of the animal perceptual system toward motion cues of animacy, we investigated naïve subjects' sensitivity to self-propelled motion as a cue to animacy (Mascalzoni, Regolin, & Vallortigara, 2010). A newly-hatched domestic chick was used as an animal model because it is a precocial species, which allows for the rigorous control of sensory experiences and takes advantage of the phenomenon of filial imprinting (i.e., a process by which the young of some precocial species promptly learns the characteristics of a potential social partner when exposed to it for a short time soon after hatching; Horn, 2004). All subjects came from incubated eggs and, to prevent any visual experience, were hatched in total darkness. On their first day of life, the chicks were exposed to a computer-presented animation sequence picturing two oval-shaped objects of different color. Object motion could be attributed to either a causal agentive role (i.e., the object appeared as self-propelled) or a receptive role (i.e., the object appeared as moved by an external force; see figure 3.3).

After exposure, the chicks were tested for their spontaneous preference between the two objects (Mascalzoni, Regolin, & Vallortigara, 2010). In a first experiment, the chicks saw the two objects involved in the classical

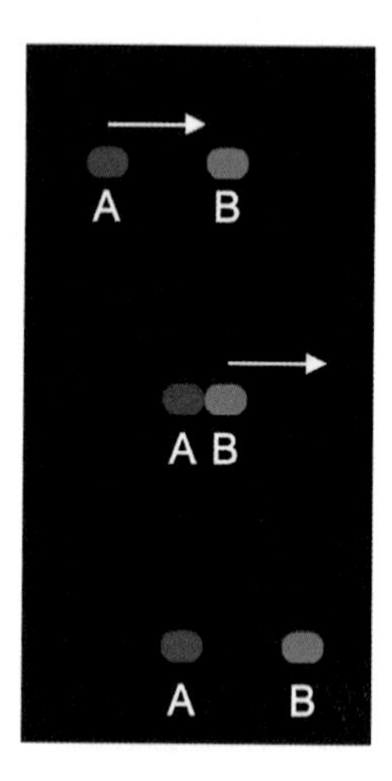

Figure 3.3
Schematic representation of the animation sequence used during chicks' exposure in Experiment 1, the classical launching effect.

example of Michotte's perceived physical causality (i.e., the launching effect described in the first section of this chapter; Michotte, 1963), in which human observers perceive object A as being a "self-propelled causal-agent" launching object B and causing its movement. When tested for their preferences for objects A or B, we found that the chicks showed a clear preference for object A, the self-propelled object playing the causal-agentive role during the exposure phase. The results showed that only when one of the two objects appeared to be self-propelled, and the other not, did a preference emerged: the chicks preferred to watch the self-propelled stimulus. The results thus demonstrate that newly-hatched chicks possess an innate sensitivity allowing them to differentiate and prefer a self-propelled causal agent as a target for imprinting. In fact, physical contact that was not accompanied by physical causation (i.e., when both objects appeared as being self-propelled), or physical causation without any cue about the nature (self/not-self-propelled) of the causal object's motion, sufficed to abolish any preference. Chicks behaved as if they were innately sensitive to self-propulsion as a crucial cue to animacy (Mascalzoni, Regolin, & Vallortigara, 2010). These findings are compatible with the idea that the perceptual system of at least two vertebrate species (domestic chicks and humans) may be designed so that animate entities are inherently appealing and attention is spontaneously triggered toward them. This conclusion is also supported by developmental investigations of the perception of biological motion at birth in humans, which results are summarized in the following section.

3 Biological Motion

The motion of living beings, particularly that of vertebrates, is peculiar. When a vertebrate moves, its limbs and torso undergo a specific pattern of motion constrained by the rigid skeletal structure supporting the body. As a result, spatial relationships among certain body parts are continuously changing, whereas spatial relationships between other parts, such as the distance between certain connected joints of the limbs, remain invariant across the movement. From the original study of Johansson (1973), a range of experimental data has revealed that the human visual system is particularly sensitive to the motion of living creatures. A dozen point-lights, placed on the main joints of a walking person, is sufficient to convey the impression of someone engaged in a coordinated activity. Several animal species are able to discriminate and specifically respond to point-light displays (Omori & Watanabe, 1996; Regolin, Tommasi, & Vallortigara, 2000).

Moreover, to adult observers, point-light biological motion displays provide an important source of socially relevant information, such as gender (Barclay, Cutting, & Kozlowski, 1978; Kozlowski & Cutting, 1977), emotions (Dittrich, Troscianko, Lea, & Morgan, 1996; Pollick, Paterson, Bruderlin, & Sanford, 2001), actions (Dittrich, 1993), and intentions (Blakemore & Decety, 2001). The visual system's impressive ability to organize individual dots into the coherent, articulated shape of a human figure (e.g., Bertenthal & Pinto, 1994) suggests that the visual system rapidly carries out a very complex structure-from-motion analysis from the biological motion animations. This process may rely on (1) the extraction of configural invariants from the relative motion of the elements in the sequence (Bertenthal, Proffitt, & Cutting, 1984) and (2) a global analysis of the array that is probably driven by stored representation of the familiar body form (Bertenthal & Pinto, 1994).

Behavioral findings, along with neuroimaging and electrophysiological evidence, suggest that, in adults, biological motion perception relies on specialized perceptual processes mediated by a dedicated neural system (see Grossman's chapter, this volume). There is evidence that specific brain areas are activated during the perception of point-light walker displays (Grossman, Donnelly, Price, Pickens, Morgan, Neighbor, & Blake, 2000; Puce & Perrett, 2003). Areas identified in neuroimaging studies include the posterior superior temporal gyrus (pSTG) and sulcus (pSTS), motion-sensitive area V5/MT, the ventral temporal cortex, and occasionally the parietal cortex (e.g., Bonda, Petrides, Ostry, & Evans, 1996; Grossman et al., 2000; Grezes, Fonlupt, Bertenthal, Delon-Martin, Segebarth, & Decety, 2001; Vaina & Gross, 2004). The involvement of the pSTG/STS is perhaps the most robust finding (see Puce & Perrett, 2003, for review). Grossman and colleagues (2000) found that biological motion stimuli depicting jumping, kicking, running, and throwing movements produced more right STS activation than did control motion (i.e., random moving dots and scrambled dot displays). This STS brain region is known to be a convergence point for the dorsal and ventral visual streams. Interestingly, contiguous regions along the STS are involved in the perception of other basic social signals, such as facial expression, mouth movement, hand grasp, and eye gaze as well as body movement (see Allison, Puce and McCarthy, 2000).

Different studies investigating the origin of brain specialization in infants suggest that the neural substrates for processing biological motion begin to mature at eight months, as revealed by analysis of the event-related potentials (Reid, Hoehl, & Striano, 2006). However, further works are required to understand how innate predispositions and visual experience interact to determine the neural and functional specialization for biological motion.

Just as holds true with regard to facial perception, accurate perception of biological motion is orientation-dependent; the detection and recognition of point-light walking figures are disrupted when the display is turned upside-down (Pavlova & Sokolov, 2000; Sumi, 1984). A possible explanation for this phenomenon maintains that inversion impairs the configural processing of the familiar shape (Bertenthal & Pinto, 1994). However, recent evidence supports the view that form extraction does not fully explain the processes involved in biological motion perception as well as the related inversion effect. Shipley (2003) showed that when a body was represented walking on its hands instead of walking on foot, subjects were less accurate at detecting a walking-on-hands form when the display was turned upside-down than when it was upright. The result—that the natural direction of the gravity force, and not the familiar orientation of the body form, is crucial for the detection of biological motion—supports accounts of event perception that are based on spatiotemporal patterns of motion associated with the dynamic of the event. Moreover, it has recently been shown that the detection of biological motion, and the discrimination of walking direction in biological motion displays, might be conveyed by local motion cues, such as those provided by the motion of the feet of an animal in locomotion (Chang & Troje, 2008; Troje & Westhoff, 2006). Even when presented with spatially scrambled biological motion displays of humans or animals, which do not contain any structural information, adults correctly judge the direction in which a walker is facing and report seeing a living being even if no animal species can be identified in such displays. However, when either the display or parts of it were inverted, the participants' direction judgment was correct as long as the dots associated with the motion of legs remained intact (Troje & Westhoff, 2006). These results suggest that spatiotemporal motion parameters, based on the dynamic of movement being constrained by the direction of the gravity force acting on the motion of the dots, provide a powerful cue for the detection of a moving, living creature.

Given the adaptive significance of detecting the motion of living beings to survival, and in line with an evolutionary hypothesis, it has been hypothesized that, beginning at birth, humans, as well as some nonhuman animal species, might possess a predisposition to orient toward and preferentially process motion information about living creatures. According to some authors, the existence of a specialized, inborn "life detector" might constitute a primitive and basic mechanism. This mechanism would be shared by vertebrates and driven by motion cues that are not species-specific (Johnson, 2006; Troje & Westhoff, 2006). This *life detection system* should be distinct from an experience-dependent mechanism responsible for the global shape processing required for more specific identification of an agent and its action (Chang & Troje, 2009). According to Chang and Troje (2008), the spatiotemporal parameters of a local motion would drive an evolutionarily ancient mechanism that is available early in life and possibly innate (Troje & Westhoff, 2006). In contrast, an individually acquired global-processing system would be responsible for the sensitivity to global configurations, which subserves the recognition of the agent's identity (Chang & Troje 2009).

The hypothesis about an innate ability of the visual system to extract the most informative features of biological motion was firstly proposed by Johansson (1973) himself.

Developmental studies of infants, using spontaneous visual preference and habituation paradigms, supported this hypothesis. Fox and McDaniel (1982) showed that four- and six-month-old babies manifested a spontaneous preference toward a point-light display depicting a walking person when compared to an array of elements moving in a random fashion. Interestingly, infants were also sensitive to display inversion as they manifested a spontaneous preference for the upright biological motion stimulus when contrasted with the same stimulus upside-down. In a series of studies, Bertenthal and colleagues investigated infants' ability to extract motion-carried configural information. At five months of age, infants discriminate a human point-light walker display from the same stimulus with scrambled spatial relationships (Bertenthal et al., 1984) or with perturbed local rigidity between some joints (Bertenthal, Proffitt, & Kramer, 1987); this suggests that the sensitivity to motion-carried configural information emerges early in life. These results supported the hypothesis that the mechanisms subserving the perception of biological motion may, at least in part, be experience-independent rather than acquired through experience. Nevertheless, these studies cannot exclude the possibility that early experience might have shaped infants' responses toward biological motion displays.

The first convincing evidence of an inborn ability to detect and specifically respond to the dynamic of biological motion came from comparative studies. Newly hatched chicks demonstrated the ability to discriminate between point-light displays depicting biological motion, random dot motion, or the rotating motion of a solid object (Regolin, Tommasi, & Vallortigara, 2000). Inexperienced chicks were able to discriminate a point-light display representing a walking hen from the scrambled version of the same stimulus; apparently the chicks were sensitive to some configural information emerging from the relative motion of the elements in the display. In addition, at their first exposure to point-light displays, chicks preferentially approached biological motion—compared to nonbiological motion—stimuli (Vallortigara et al., 2005). This spontaneous preference emerged irrespective of the species of animals used to produce the stimulus. Chicks did not show any preference for the walking hen as compared to a walking cat or even to a scrambled biological motion display, suggesting that this sensitivity is not species-specific and is due to the dynamic of the motion more than to the configural information present in the display.

These findings seem to be consistent with the idea of an evolutionarily ancient neural mechanism for detecting the movement of other legged vertebrates. This idea has recently been addressed by testing two-day-old infants. On the basis of the above-mentioned results suggesting that the mechanism to detect biological motion is not sensitive to species-specific cues, the same hen-walking animations were used to test both chicks and human babies. Moreover, the use of hen-walking animations ruled out the possibility that newborns may have had any prior experience with the motion depicted in the stimuli. Results from our lab revealed that, at their first exposure, two-day-old babies manifested a spontaneous preference for the biological motion display over a random-motion point-light stimulus (Simion, Regolin, & Bulf, 2008). Interestingly, newborns preferred the upright point-light display depicting a walking hen as compared to the same display upside-down—thus revealing the presence, at birth, of an "orientation effect."

A possible explanation for the orientation effect in newborns is that the human visual system is innately designed to take into account the direction of the gravity force that constraints the movement of terrestrial creatures. The visual system employs a local visual filter that operates independently of structure (see Troje's chapter, this volume).

Overall, newborns' results support the view that humans possess an inborn sensitivity for both the dynamics of biological motion and the gravitational forces acting on motion (Simion et al., 2008). However, alternative interpretations might explain the preference for biological motion displays when contrasted with a random motion display. In effect, the preferred biological motion display depicts a structured stimulus because the movement of the dots allows grouping of individual elements. In contrast, the random display was both nonbiological and nonstructured because the dots moved arbitrarily from one another and in different directions. Thus, the preference might be explained as a general bias toward coherent motion with respect to random, nonstructured stimuli. To test the hypothesis of a preference for a coherent motion, in a recent study a biological motion stimulus depicting a walking hen was compared with a nonbiological, structured motion display representing a hen-like object rotating about its vertical axis (Bardi, Regolin, & Simion, 2011; figure 3.4).

Newborns exhibited a spontaneous preference toward the biological motion stimulus, revealing that preference is triggered by the specific nature of the motion. This result has been explained on the basis of a specific sensitivity to some spatiotemporal parameters present in the dynamic of biological motion stimuli and not shared by the rigid motion of an identical array of dots.

However, these results still leave open the question of newborns' ability to extract the configural or global information revealed by the relative motion of the elements in the biological motion displays. In fact newborns' preference, in the previous experiment, might be based either on the processing of the dynamic of each single dot or on the relationship between the dots; this is because the motion of each single dot was distinctively different in the two arrays. In effect, in the biological motion display, the trajectory of each single dot was either pendular or elliptical. In contrast, in the rigid motion display, each single dot moved with a

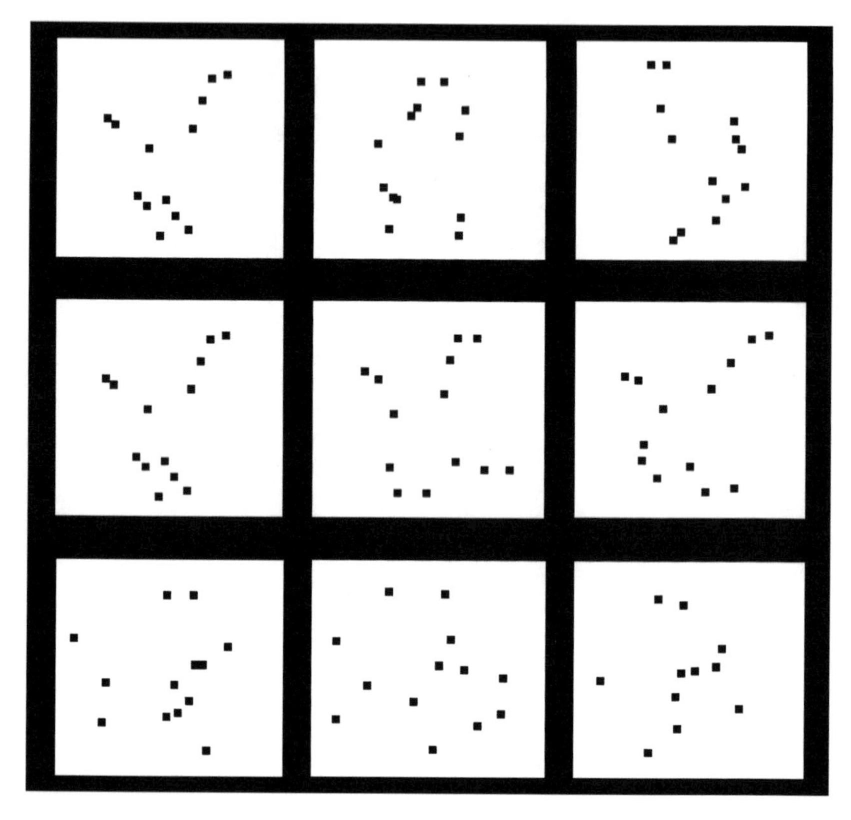

Figure 3.4
Three sample frames of the point-light sequences employed in the study of Bardi et al. (2011). The rigid motion (top), the walking hen (middle), and the scrambled hen (bottom).

continuous translatory motion trajectory. Thus, the discrimination between the two displays, and the preference for one of them, might be explained on the basis of local motion differences.

To investigate whether newborns are able to process the relationships among the moving dots in a condition in which the movement of the single dots is maintained constant, two experiments were carried out by employing a biological display and the spatially scrambled version of it (figure 3.4). In the two displays, the local movement of each dot was kept constant but the global disposition of the dots varied. In other words, the scrambled version was created by shifting the position of each dot and leaving unaltered the local motion. In the first experiment, by use of the habituation technique, newborns were shown to be able to discriminate between the two displays; this revealed that some of the configural information that emerged from the relative motion of the elements in the biological motion stimulus was processed. In the second experiment, we demonstrated the lack of any spontaneous preference for one of the two displays. This outcome revealed that, starting at birth, the human system is selectively triggered by the nature of the motion and turns to any biological motion display—irrespective of the animal form it depicts.

Overall, the information that is crucial in triggering the preference seems to be provided by some spatiotemporal parameters; these parameters define the ballistic-velocity profile, constrained by acceleration owing to the natural direction of the gravity force, that is contained in the dynamic of the dots' motion. The newborns' results support the hypothesis that an inborn predisposition to preferentially process the motion of living creatures is present in humans, as well as in other animal species, prior to any visual experience. Adults' expertise at recognizing the motion of other living beings might originate from this innate predisposition and from attentional biases toward the specific spatiotemporal properties defined by the gravitational acceleration present in the dynamic of biological motion (Johnson, 2006; Simion et al., 2008; Troje & Westhoff, 2006). Even if newborns were shown to be sensitive to some of the configural information revealed by the relative motion of the elements in the biological motion display, such a sensitivity would not explain the spontaneous preference at birth (Bardi et al., 2011). The newborns' preference for the dynamic of biological motion aids in providing babies with the visual experience that supports learning. By means of learning, they become able to extract the fundamental properties of biological motion, properties conveying specific information about conspecifics, such as form information, that supports the identification of agents.

4 Conclusions

Data here discussed support the view that several vertebrate species, including humans, have a primitive bias toward detecting social agents and attending to and preferentially processing sensory information about other living entities (Carey, 2009)—both in terms of causal relations between them and in terms of animacy attributions. The visual system seems to be equipped with mechanisms, evolutionarily well-conserved and developmentally early-emerging, for the detection of the relation between spatial and temporal parameters constrained by gravitational forces embedded in the motion of biological entities. These mechanisms may subserve the crucial abilities for adaptive interaction with other living beings: following the movements of a conspecific, learning by imitation, or directing preferential attention to cues that build on self-propelled and biological motion (such as facial expression and gaze direction). Preferential attention to biological motion, moreover, is seen as a precursor to the capacity for attributing intentions to others (Blakemore & Decety, 2001).

It has recently been found that infants with autism have an impaired perception of physical causality, animacy, and biological motion. As for the perception of causality, a deficit in perceiving the launching effect is present in young children with autism (Ray & Schlottmann, 2007) whereas older children with high-functioning autism showed unimpaired perception of causality (Congiu, Schlottmann, & Ray, 2010). Autistic children, moreover, are impaired in recognizing animacy in artificial stimuli in motion (Congiu, Schlottmann, & Ray, 2010), such as in recognizing point-light displays of biological motion (e.g., Blake, Turner, Smoski, Pozdol, & Stone, 2003). They are instead highly sensitive to the presence of a nonsocial, physical contingency occurring within the stimuli by chance (Klin, Lin, Gorrindo, Ramsay, & Jones, 2009). These observations raise the possibility that the perception of sensory cues about living entities may be impaired in children with autism from a very early age. Therefore, the study of innate predispositions underlying social cognition has significant implications for understanding the altered neurodevelopmental trajectory of brain specialization.

Acknowledgments

This work was supported by grants from the Ministero dell'Università e della Ricerca (No. 2007XFM93B-004) and the University of Padova (No.2007-CPDA075245). We thank Dr. Beatrice Dalla Barba and the nursing staff of the Pediatric Clinic of the University of Padova for their collaboration. We thank Elena LoSterzo for the invaluable help provided in chick testing and care. Many thanks also to the parents and babies who participated in our studies.

References

Allison, T., Puce, A., & McCarthy, G. (2000). Social perception from visual cues: Role of the STS region. *Trends in Cognitive Neuroscience*, *4*, 267–278.

Barclay, C. D., Cutting, J. E., & Kozlowski, L. T. (1978). Temporal and spatial factors in gait perception that influence gender recognition. *Perception & Psychophysics*, *23*, 145–152.

Bardi, L., Regolin, L., & Simion, F. (2011). Biological motion preference in humans at birth: Role of dynamic and configural properties. *Developmental Science, 14,* 353–359.

Bertenthal, B. I., Proffitt, D. R., & Cutting, J. E. (1984). Infant sensitivity to figural coherence in biomechanical motions. *Journal of Experimental Child Psychology, 37,* 213–230.

Bertenthal, B. I., Proffitt, D. R., & Kramer, S. J. (1987). Perception of biomechanical motions by infants: Implementation of various processing constraints. *Journal of Experimental Psychology: Human Perception and Performance, 13,* 577–585.

Bertenthal, B. I., & Pinto, J. (1994). Global processing of biological motions. *Psychological Science, 5,* 221–225.

Blake, R., Turner, L. M., Smoski, M. J., Pozdol, S. L., & Stone, W. L. (2003). Visual recognition of biological motion is impaired in children with autism. *Psychological Science, 14,* 151–157.

Blakemore, S. J., Boyer, P., Pachot-Clouard, M., Meltzoff, A., Segebarth, C., & Decety, J. (2003). The detection of contingency and animacy from simple animations in the human brain. *Cerebral Cortex, 13,* 837–844.

Blakemore, S. J., & Decety, J. (2001). From the perception of action to the understanding of intention. *Nature Reviews: Neuroscience, 2,* 561–667.

Bonda, E., Petrides, M., Ostry, D., & Evans, A. (1996). Specific involvement of human parietal systems and the amygdala in the perception of biological motion. *Journal of Neuroscience, 16,* 3737–3744.

Carey, S. (2009). *The origin of concepts.* Oxford: Oxford University Press.

Castelli, F., Happé, F., Frith, U., & Frith, C. D. (2000). Movement and mind: A functional imaging study of perception and interpretation of complex intentional movement patterns. *NeuroImage, 12,* 314–325.

Chang, D. H. F., & Troje, N. F. (2008). Perception of animacy and direction from local biological motion signals. *Journal of Vision, 8,* 1–10.

Chang, D. H. F., & Troje, N. F. (2009). Characterizing global and local mechanisms in biological motion perception. *Journal of Vision, 9,* 1–10.

Choi, H., & Scholl, B. J. (2004). Effects of grouping and attention on the perception of causality. *Perception & Psychophysics, 66,* 926–942.

Choi, H., & Scholl, B. J. (2006). Measuring causal perception: Connections to representational momentum? *Acta Psychologica, 123,* 91–111.

Cohen, L. B., & Amsel, G. (1998). Precursors to infants' perception of causality. *Infant Behavior and Development, 21,* 713–731.

Cohen, L. B., Amsel, G., Redford, M. A., & Casasola, M. (1998). The development of infant causal perception. In A. Slater (Ed.), *Perceptual development: Visual, auditory, and speech perception in infancy* (pp. 167–209). East Sussex, UK: Psychology Press.

Cohen, L. B., Chaput, H. H., & Cashon, C. H. (2002). A constructivist model of infant cognition. *Cognitive Development, 17,* 1323–1343.

Cohen, L. B., & Oakes, L. M. (1993). How infants perceive a simple causal event. *Developmental Psychology, 29,* 421–433.

Cohen, L. B., Rundell, L. J., Spellman, B. A., & Cashon, C. H. (1999). Infants' perception of causal chains. *Psychological Science, 10,* 412–418.

Congiu, S., Schlottmann, A., & Ray, E. (2010). Unimpaired perception of social and physical causality, but impaired perception of animacy in high functioning children with autism. *Journal of Autism and Developmental Disorders, 40,* 39–53.

Csibra, G., Gergely, G., Bíró, S., Koos, O., & Brockbank, M. (1999). Goal attribution without agency cues: The perception of "pure reason" in infancy. *Cognition, 72,* 237–267.

Dittrich, W. H. (1993). Action categories and recognition of biological motion. *Perception, 22,* 15–23.

Dittrich, W. H., Troscianko, T., Lea, S. E., & Morgan, D. (1996). Perception of emotion from dynamic point-light displays represented in dance. *Perception, 25,* 727–738.

Farroni, T., Csibra, G., Simion, F., & Johnson, M. H. (2002). Eye contact detection in humans from birth. *Proceedings of the National Academy of Sciences of the United States of America, 99,* 9602–9605.

Farroni, T., Simion, F., Umiltà, C., & Dalla Barba, B. (1999). The gap effect in newborns. *Developmental Psychology, 2,* 174–186.

Fox, R., & McDaniel, C. (1982). The perception of biological motion by human infants. *Science, 218,* 486–487.

Gelman, S. A., & Opfer, J. E. (2002). Development of the animate–inanimate distinction. In U. Goswami (Ed.), *Blackwell handbook of childhood cognitive development* (pp. 151–166). Oxford: Blackwell.

Grezes, J., Fonlupt, P., Bertenthal, B., Delon-Martin, C., Segebarth, C., & Decety, J. (2001). Does perception of biological motion rely on specific brain regions? *NeuroImage, 13,* 775–785.

Grossman, E., Donnelly, M., Price, R., Pickens, R., Morgan, V., Neighbor, G., & Blake, R. (2000). Brain areas involved in perception of biological motion. *Journal of Cognitive Neuroscience, 12,* 711–720.

Heberlein, A. S., & Adolphs, R. (2004). Impaired spontaneous anthropomorphizing despite intact perception and social knowledge. *Proceedings of the National Academy of Sciences of the United States of America, 101*, 7487–7491.

Heider, F., & Simmel, M. (1944). An experimental study of apparent behavior. *American Journal of Psychology, 57*, 243–259.

Horn, G. (2004). Pathways of the past: The imprint of memory. *Nature Reviews. Neuroscience, 5*, 108–120.

Johnson, M. H. (2005). *Developmental cognitive neuroscience*. Oxford: Blackwell.

Johnson, M. H. (2006). Biological motion: A perceptual life detector? *Current Biology, 16*, R376–R377.

Johnson, S. C. (2000). The recognition of mentalistic agents in infancy. *Trends in Cognitive Sciences, 4*(1), 22–28.

Johansson, G. (1973). Visual perception of biological motion and a model for its analysis. *Perception & Psychophysics, 14*, 201–211.

Kanizsa, G., & Vicario, G. B. (1968). La percezione della reazione intenzionale. In G. Kanizsa & G. B. Vicario (Eds.), *Ricerche Sperimentali sulla Percezione* (pp. 71–126). Trieste: Edizioni Universitarie.

Kerzel, D., Bekkering, H., Wohlschläger, A., & Prinz, W. (2000). Launching the effect: Representations of causal movements are influenced by what they lead to. *Quarterly Journal of Experimental Psychology, 53A*(4), 1163–1185.

Klin, A., Lin, D. J., Gorrindo, P., Ramsay, G., & Jones, W. (2009). Two-year-olds with autism orient to non-social contingencies rather than biological motion. *Nature, 459*, 257–261.

Kozlowski, L. T., & Cutting, J. E. (1977). Recognizing the sex of a walker from a dynamic point-light display. *Perception & Psychophysics, 21*, 575–580.

Kosugi, D., Ishida, H., & Fujita, K. (2003). 10-month-old infants' inference of invisible agent: Distinction in causality between object motion and human action. *Japanese Psychological Research, 45*, 15–24.

Leslie, A. M. (1982). The perception of causality in infants. *Perception, 11*, 173–186.

Leslie, A. M. (1984). Spatiotemporal continuity and the perception of causality in infants. *Perception, 13*, 287–305.

Leslie, A. M. (1988). The necessity of illusion: Perception and thought in infancy. In L. Weiskrantz (Ed.), *Thought without language* (pp. 185–210). Oxford: Oxford University Press.

Leslie, A. M. (1994). ToMM, ToBY, and agency: Core architecture and domain specificity. In L. A. Hirschfeld & S. A. Gelman (Eds.), *Mapping the mind: Domain specificity in cognition and culture* (pp. 119–148). New York: Cambridge University Press.

Leslie, A. M. (1995). A theory of agency. In D. Sperber, D. Premack, & A. J. Premack (Eds.), *Causal cognition: A multidisciplinary debate* (pp. 121–141). Oxford: Clarendon Press.

Leslie, A. M., & Keeble, S. (1987). Do six-month-olds perceive causality? *Cognition, 25*, 265–288.

Luo, Y., Kaufman, L., & Baillargeon, R. (2009). Young infants' reasoning about physical events involving inert and self-propelled objects. *Cognitive Psychology, 58*, 441–486.

Macchi Cassia, V., Simion, F., Milani, I., & Umiltà, C. (2002). Dominance of global visual properties at birth. *Journal of Experimental Psychology: General, 131*, 398–411.

Mascalzoni, E., Regolin, L., & Vallortigara, G. (2010). Innate sensitivity for self-propelled causal agency in newly hatched chicks. *Proceedings of the National Academy of Sciences of the United States of America, 107*(9), 4483–4485.

Mascalzoni, E., Regolin, L., Vallortigara, G., & Simion, F. (in press). The cradle of causal reasoning: Newborns' preference for physical causality. *Developmental Science.*

Michotte, A. (1963). *The perception of causality*. New York: Basic Books.

Morris, M. W., & Peng, K. (1994). Culture and cause: American and Chinese attributions for social and physical events. *Journal of Personality and Social Psychology, 67*, 949–971.

Morton, J., & Johnson, M. H. (1991). CONSPEC and CONLERN: A two-process theory of infant face recognition. *Psychological Review, 98*, 164–181.

Oakes, L. M. (1994). Development of infants' use of continuity cues in their perception of causality. *Developmental Psychology, 30*, 869–879.

Oakes, L. M., & Cohen, L. B. (1990). Infant perception of a causal event. *Cognitive Development, 5*, 193–207.

Omori, E., & Watanabe, S. (1996). Discrimination of Johansson's stimuli in pigeons. *International Journal of Comparative Psychology, 9*, 92.

Pavlova, M., & Sokolov, A. (2000). Orientation specificity in biological motion perception. *Perception & Psychophysics, 62*, 889–899.

Pollick, F. E., Paterson, H. M., Bruderlin, A., & Sanford, A. J. (2001). Perceiving affect from arm movement. *Cognition, 82*, B51–B61.

Premack, D. (1990). The infant's theory of self-propelled objects. *Cognition, 36*, 1–16.

Puce, A., & Perrett, D. (2003). Electrophysiology and brain imaging of biological motion. *Philosophical Transactions of the Royal Society of London, Series B: Biological Sciences, 35*, 435–445.

Rakison, D. H., & Poulin-Dubois, D. (2001). Developmental origin of the animate–inanimate distinction. *Psychological Bulletin, 127*, 209–228.

Ray, E., & Schlottmann, A. (2007). The perception of social and mechanical causality in young children with autism. *Research in Autism Spectrum Disorders, 1*(3), 266–280.

Regolin, L., Tommasi, L., & Vallortigara, G. (2000). Visual perception of biological motion in newly hatched chicks as revealed by an imprinting procedure. *Animal Cognition, 3*, 53–60.

Reid, V. M., Hoehl, S., & Striano, T. (2006). The perception of biological motion by infants: An event-related potential study. *Neuroscience Letters, 395*, 211–214.

Saxe, R., Tzelnic, T., & Carey, S. (2006). Five-month-old infants know humans are solid, like inanimate objects. *Cognition, 101*, B1–B8.

Saxe, R., Tzelnic, T., & Carey, S. (2007). Knowing who dunnit: Infants identify the causal agent in an unseen causal interaction. *Developmental Psychology, 43*, 149–158.

Schlottmann, A., Ray, E., Mitchell, A., & Demetriou, N. (2006). Perceived physical and social causality in animated motions: Spontaneous reports and ratings. *Acta Psychologica, 123*, 112–143.

Scholl, B. J., & Tremoulet, P. D. (2000). Perceptual causality and animacy. *Trends in Cognitive Sciences, 4*, 299–309.

Shipley, T. F. (2003). The effect of object and event orientation on perception of biological motion. *Psychological Science, 14*, 377–380.

Simion, F., Regolin, L., & Bulf, H. (2008). A predisposition for biological motion in the newborn baby. *Proceedings of the National Academy of Sciences of the United States of America, 105*(2), 809–813.

Simion, F., Valenza, E., Umiltà, C., & Dalla Barba, B. (1995). Inhibition of return is temporo-nasal asymmetrical. *Infant Behavior and Development, 18*, 189–194.

Simion, F., Valenza, E., Umiltà, C., & Dalla Barba, B. (1998). Preferential orienting to faces in newborns: A temporo-nasal asymmetry. *Journal of Experimental Psychology: Human Perception and Performance, 24*(5), 1399–1405.

Spelke, E. S., & Kinzler, K. D. (2007). Core knowledge. *Developmental Science, 10*(1), 89–96.

Spelke, E. S., Phillips, A., & Woodward, A. L. (1995). Infants' knowledge of object motion and human action. In D. Sperber, D. Premack, & A. J. Premack (Eds.), *Causal cognition: A multidisciplinary debate* (pp. 44–78). Oxford: Clarendon Press.

Sperber, D., Premack, D., & Premack, A. J. (1995). *Causal cognition: A multidisciplinary debate*. Oxford: Clarendon Press.

Sumi, S. (1984). Upside-down presentation of the Johansson moving light-spot pattern. *Perception, 13*, 283–286.

Troje, N. F., & Westhoff, C. (2006). The inversion effect in biological motion perception: evidence for a "life detector." *Current Biology, 16*, 821–824.

Turati, C., Simion, F., & Zanon, L. (2003). Newborns' perceptual categorization for closed and open geometric forms. *Infancy, 4*, 309–325.

Vaina, L. M., & Gross, C. G. (2004). Perceptual deficits in patients with impaired recognition of biological motion after temporal lobe lesions. *Proceedings of the National Academy of Sciences of the United States of America, 101*, 16947–16951.

Valenza, E., & Bulf, H. (2007). The role of kinetic information in newborns' perception of illusory contours. *Developmental Science, 10*, 492–501.

Valenza, E., Simion, F., & Umiltà, C. (1994). Inhibition of return in newborn infants. *Infant Behavior and Development, 17*, 293–302.

Vallortigara, G., Regolin, L., & Marconato, F. (2005). Visually inexperienced chicks exhibit spontaneous preference for biological motion patterns. *PLoS Biology, 3*, 1312–1316.

White, P. A. (2006). The role of activity in visual impressions of causality. *Acta Psychologica, 123*, 166–185.

Woodward, A. L., Phillips, A., & Spelke, E. S. (1993). Infants' expectations about the motion of animate versus inanimate objects. In *Proceedings of the fifteenth annual meeting of the Cognitive Science Society* (pp. 1087–1091). Hillsdale, NJ: Erlbaum.

4 Evidence for Functional Specialization in the Human Superior Temporal Sulcus (STS): Consideration of Biological Motion Perception and Social Cognition

Emily D. Grossman

Abstract

The perception of human actions in point-light biological motion animations relies on the coordinated activity of brain regions in the occipital, parietal, and frontal cortex. This chapter discusses neuropsychological, single-unit, and neuroimaging evidence for linking the superior temporal sulcus (STS) to the cognitive demands of perceiving biological motion. Although the evidence accumulated over the past twenty years has suggested STS specialization for the perceptual construction of actions, the relatively recent emergence of findings from studies of social cognition suggest that a domain-specific hypothesis of STS specialization may be misplaced. Some proposals for alternatives theories are discussed.

As inherently social beings, humans engage daily in social interactions requiring swift assessment of people's actions. Most individuals are able to make these assessments quickly and accurately, and it is believed that this ability reflects the coordinated efforts of multiple brain systems. This chapter focuses specifically on those brain mechanisms that reflect the computational demands imposed when visually constructing actions from the shape and kinematics of the body itself.

Historically in vision science, action recognition has been studied using so-called point-light animations (e.g., Johansson, 1973; see introduction to this section). In these sparse sequences, the human body is reduced to the movements of small tokens that replace the joint positions of an actor. Provided that the sequences are in motion, typical observers readily recognize and identify actions in these sparse displays. In contrast, stationary frames are often described as meaningless clouds of dots (Pavlova & Sokolov, 2003). Observers readily extract nuanced details about the actor that are implied by the kinematics; these include gender (Mather & Murdoch, 1994, Troje, Sadr, Geyer, & Nakayama, 2006), identity (Troje, Westhoff, & Lavrov, 2005), emotional state (Dittrich, Troscianko, Lea, &

Morgan, 1996), intentions (Runeson & Frykholm, 1983), and some personality dispositions (Gunns, Johnston, & Hudson, 2002). All of this can be accomplished in just a fraction of one second (e.g., Neri, Morrone, & Burr, 1998).

This chapter outlines the neuropsychological, single-unit, and neuroimaging evidence for specialized neural analysis supporting action recognition in point-light biological motion. In particular, this chapter focuses on the role of a specific brain region—the human superior temporal sulcus (STS). Although there is evidence that occipitotemporal (visual) brain areas and prefrontal (premotor) brain areas support action recognition, the STS appears to play more a specialized, and potentially causal, role in biological motion perception. This chapter will first survey the literature linking the STS to biological motion perception in both nonhuman (macaca) and human primates. The latter part of this chapter discusses a line of research that implicates the human STS in a range of social cognitive tasks quite different from biological motion perception. Thus, the final aspect of this chapter identifies the tacit assumption of domain specificity in much of the biological motion research and discusses domain-general alternatives to the current theories of STS specialization.

The Case for Neural Specialization, from Neuropsychology

The history of research into the isolation of brain structures associated with specific cognitive tasks originates from neuropsychological studies of individuals with brain damage. The earliest reports of deficits in biological motion recognition sought to establish functional dissociations between visual deficits in low-level perceptual computations that could support recognition from point-lights (e.g., form and motion perception) and biological motion perception.

In one such study, Cowey and Vaina (2000) tested a patient who had suffered from two strokes resulting in bilateral damage to her lateral occipital cortex and extending into the posterior parietal cortex. Immediately following her stroke, this patient complained of being unable to recognize friends and family by their face or gait, though she could still identify them by their voices. She soon regained her ability to recognize a familiar face if the individual remained still, but complained that she could not recognize people when they were moving. In testing, researchers found that the patient was unable to discriminate structures defined by moving cues, including biological motion, but could discriminate simpler motion patterns and recognize stationary shapes. The authors concluded that, unlike former motion-blind patients (who could readily recognize point-light biological motion; McLeod, Dittrich, Driver, Perrett, & Zihl, 1996), the pattern of deficits in this patient was evidence of a specific visual motion agnosia.

In a second demonstration of such perceptual dissociations, Schenk and Zihl (1997a,b) identified two individuals who failed to recognize actions in point-light displays when the figures were masked in a stationary field of dots—a manipulation that does not impact recognition for typical observers (Bertenthal & Pinto, 1994). Despite their difficulty with the biological motion sequences, these patients had normal motion perception as measured by coherence thresholds, and they had no difficulty disentangling stationary superimposed shapes. Both patients suffered from damage to the bilateral parietal cortex.

The reverse pattern of results has also been shown in a patient who suffered from bilateral lesions of the temporal-parietal-occipital junction (TPJ); the resulting perceptual impairments to motion sensitivity included difficulty in integrating motion signals into coherent structures. (Vaina, Lemay, Bienfang, Choi, & Nakayama, 1990). Despite these difficulties with moving visual displays, this patient readily recognized actions portrayed in point-light sequences.

Together, the patient findings just outlined would seem to imply that biological motion depends on neural mechanisms that operate independently from, or alongside, the perceptual systems responsible for encoding object shape and motion features. Critically, the lesions associated with impaired biological motion perception in these studies implicate cortical structures in the posterior parietal cortex. Indeed, Pavlova et al. (2003) found a significant negative correlation between the extent of lesioned cortical volume in the posterior parietal cortex and biological motion sensitivity, such that patients with more damage were less able to detect point-light sequences embedded in motion-matched noise.

It may not be surprising, therefore, to learn that on the basis of patient studies, researchers have linked the neural mechanisms required for focused attention, and in particular for the act of selecting dynamic events, to biological motion perception. Patients with lesions to the TPJ regions of the posterior parietal cortex have difficulty individuating events from within a rapid succession; this leads to deficits in a wide range of so-called high-level motion tasks requiring feature-based attention (Battelli, Cavanagh, Martini, & Barton, 2003; Battelli, Pascual-Leone, & Cavanagh, 2007). These patients also have difficulty attending to biological motion; this is evidenced by slower response times when searching for point-targets embedded in arrays of point-light foils and higher error rates (Battelli, Cavanagh, & Thornton, 2003).

It is also worth noting that although most people, young and old, can recognize actions in point-light displays (Carter & Pelphrey, 2006; Freire, Lewis, Maurer, & Blake, 2006; Pavlova, Krageloh-Mann, Sokolov, & Birbaumer, 2001), individuals with certain neurological disorders struggle with biological motion recognition. For example, individuals with schizophrenia fail to properly discriminate point-light biological motion animations from motion-matched nonbiological displays; this is particularly the case when the sequences are masked in noise (Kim, Doop, Blake, & Park, 2005). Some studies have found a similar deficit in children on the autism spectrum (Blake, Turner, Smoski, Pozdol, & Stone, 2003; Saygin, Cook, & Blakemore, 2010) and have identified a correlation between the ability to successfully discriminate masked biological motion and IQ in adults on the autism spectrum (a link not present in typical adults; Rutherford & Troje, 2011). Interestingly, at least one study has shown children with Williams syndrome to be better than typical controls at detecting and discriminating biological motion (Jordan, Reiss, Hoffman, & Landau, 2002). As discussed by Troje in chapter 2 of this volume, the direct implications of these findings are not entirely clear; there are multiple possible strategies for biological motion recognition, and most studies do not attempt to isolate any single one of those perceptual skills. What these studies do tell us, however, is that there are specific visual-cognitive mechanisms that are required for perceptually organizing biological motion, and these are disordered (or perhaps, in the case of Williams syndrome, enhanced) in some specialized populations.

The Human Superior Temporal Sulcus

With the explosion in use of functional magnetic resonance imaging (fMRI) over the past fifteen years, at least thirty studies have measured the neural correlates of biological motion perception in adults and children. These studies have identified a network of brain areas that includes the human superior temporal sulcus (Allison, Puce, & McCarthy, 2000), regions within the occipital-temporal cortex (Michels, Lappe & Vaina, 2005, Vaina, Solomon, Chowdhury, Sinha & Belliveau, 2001), and the premotor cortex (Jastorff, Begliomini, Fabbri-Destro, Rizzolatti, & Orban, 2010, Saygin, Wilson, Hagler, Bates, & Sereno, 2004, Wheaton, Thompson, Syngeniotis, Abbott, & Puce, 2004). Most of these brain areas are not believed to be specialized for action recognition. For example, the neural activity in the motion-sensitive human middle temporal complex (hMT+) is strongly linked to perceiving any kind of visual movement, including the perception of simple translating motion and the experience of motion aftereffects (e.g., Huk, Ress, & Heeger, 2001). Thus the hMT+ is likely involved in the visual analysis of body kinematics (Peuskens, Vanrie, Verfaillie, & Orban, 2005) but not specialized in any way for biological motion perception (Grossman, Donnelly, Price, Pickens, Morgan, Neighbor & Blake, 2000; Peelen, Wiggett, & Downing, 2006). Likewise, the human fusiform body area (FBA) and extrastriate body area (EBA) both have neural signals that dissociate action kinematics from nonbiological motion, but do not appear to be specialized for actions. For example, the EBA and FBA both also support recognition of stationary body postures (Grossman & Blake, 2002; Peelen, Glaser, Vuilleumier, & Eliez, 2009; Taylor, Wiggett, & Downing, 2007; Taylor, Wiggett, & Downing, 2009, Vaina et al., 2001).

Anatomically situated at the center of this large cortical system is the superior temporal sulcus, a long anatomical feature extending from the most anterior aspect of the temporal lobe to the junction of the occipital, temporal, and parietal lobes. It is on the posterior extent of the STS, just at the intersection of the ascending split (Ochiai, Grimault, Scavarda, Roch, Hori, Riviere, Mangin, & Régis, 2004), that studies have identified neural activity as being correlated to the recognition of actions (e.g., Adolphs, 2009; Blake & Shiffrar, 2007).

Many of these studies have directly compared brain activity elicited by biological motion to other types of moving stimuli, and results have consistently revealed stronger STS activation for biological motion. For example, the STS responds more strongly when subjects view point-light sequences compared to when they view coherent translating motion, optic flow, or kinetically defined implied boundaries (Bonda, Petrides, Ostry, & Evans, 1996; Grossman et al., 2000; Howard, Brammer, Wright, Woodruff, Bullmore, & Zeki, 1996). Even when carefully controlling for the unique kinematics of the body by comparing point-light sequences to motion-matched control sequences, researchers reliably localize a focus of activation on the posterior STS when viewing point-light sequences (Grossman & Blake, 2001; Jastorff & Orban, 2009, Saygin, 2007).

However, the STS neural response during biological motion perception is weakened by experimental manipulations that render the point-light sequences more difficult to recognize. Perceptually, biological motion suffers from an inversion effect such that inverting the displays makes them more difficult to recognize (Pavlova & Sokolov, 2000; Sumi, 1984). The same manipulation has the effect of reducing the brain response from the STS as compared to the response when viewing upright displays (Grossman & Blake, 2001; Grezes, et al. 2001). Likewise, embedding point-light sequences in motion-matched noise has the effect of masking the biological motion, with more effective masking using dense fields of motion-matched dynamic noise (Hiris, Humphrey, & Stout, 2005). As might be expected, embedded point-light displays have the effect of reducing the STS neural response (Grezes, Fonlupt, Bertenthal, Delon-Martin, Segebarth, & Decety, 2001), although this reduction can be overcome with training (Grossman, Kim, & Blake, 2004). Lastly, dismembering the body parts to destroy the inherent structure in biological motion has the effect of reducing the STS brain response; this is true even if the body parts move as they normally would when put together as a whole (Pelphrey, Mitchell, McKeown, Goldstein, Allison, & McCarthy, 2003; Thompson, Clarke, Stewart, & Puce, 2005).

Other studies have shown that implying, but not directly showing, biological motion also activates the STS. Kourtzi and Kanwisher (2000) showed subjects images of human bodies in postures, such as rising from a chair, that imply kinetic energy. These implied actions more strongly activated the hMT+ and the STS as compared to images of bodies that failed to imply activity. Even actively engaging in mental imagery of actions without seeing a body in action is sufficient to engage the STS (Grossman & Blake, 2001), likely as a result of reactivating the same neural circuits engaged during perception (e.g. Kosslyn, Ganis, & Thompson, 2001). Interestingly, reading action verbs also engages the STS, but in a region distinct from that involved in biological motion perception (Bedny, Caramazza, Grossman, Pascual-Leone, & Saxe, 2008).

A comparison of the STS response when subjects view other objects, even those with complex movements, consistently reveals stronger activation for biological motion. For example, neural signals in the STS are stronger for biological motion than for kinetically defined shapes such as rotating spheres and cubes (Grezes et al., 2001; Grossman & Blake, 2002). The STS also responds more strongly when observers view biological motion as compared to viewing the pendular and hierarchical movements of a ticking grandfather clock (Pelphrey et al., 2003b). Beauchamp et al. (2003) showed that the STS responds more strongly to the familiar movements of the body than it does when observers view the characteristic movements of commonly used household tools (e.g., hammers and screwdrivers).

In addition to the correlational relationship between the STS and action recognition, two other studies have demonstrated a causal role of the STS in biological motion perception. Repetitive, low frequency (1Hz) transcranial magnetic stimulation (TMS) over the cortex can temporarily interfere with cortical functions associated with the region directly under the stimulation site. When repetitive TMS is applied over the STS, but not the hMT+, observers are significantly less sensitive to point-light displays that are embedded in masking noise (Grossman, Battelli, & Pascual-Leone, 2005). A similar effect has been found using high frequency, theta burst TMS over the STS and premotor cortex (van Kemenade, Muggleton, Walsh, & Saygin, 2010). Both of these studies strongly suggest some specialized computation in the human STS that promotes biological motion perception.

Together, these neuropsychological, neuroimaging, and TMS findings suggest that the human STS has specialized neural machinery for recognizing biological motion, which, in turn, argues for the existence of neurons with neural tuning for biological motion. These putative neurons have never been identified in the human brain, likely because such studies are sufficiently difficult. Researchers have, however, identified a likely homologue to the human STS in the monkey.

Neural Tuning for Biological Motion in the STS

To date, our best understanding of the types of neurons that support visual action recognition comes from single-unit investigations of the macaque superior temporal polysensory area (STPa, sometimes referred to as STSa), which is the likely homologue to human STS (Puce & Perrett, 2003). The STSa, sometimes called the anterior superior temporal polysensory area (STPa) or the temporal-parietal-occipital area (TPO) (Bruce, Desimone, & Gross, 1981; Cusick, Seltzer, Cola, & Griggs, 1995; Pandya & Seltzer, 1982), is a heterogeneous, higher-order brain area that is a convergence zone for the visual, auditory, and somatosensory systems (Cusick, 1996; Jones & Powell, 1970; Seltzer & Pandya, 1978). Although many cells in the STSa are multisensory, the vast majority are visually responsive and tend to have large receptive fields and complex response properties, including a preference for movements of faces, heads, arms, bodies, and body parts (Barraclough, Xiao, Oram, & Perrett, 2006; Bruce et al., 1981; Jellema, Maassen, & Perrett, 2004; Perrett, Smith, Mistlin, Chitty, Head, Potter, Broennimann, Milner, & Jeeves, 1985a; Vangeneugden, Pollick, & Vogels, 2009a).

Among these complex STSa neurons are cells tuned to the postures and kinematics of biological motion, with preferred firing patterns for a wide assortment of actions, including reaching, bending, or walking (Perrett, Smith, Mistlin, Chitty, Head, Potter et al., 1985a). Although many of these neurons respond when the monkey views isolated, animated body parts, most prefer actions that include the whole body of the actor (Oram & Perrett, 1996). And interestingly for the purposes of this chapter, many of these neurons generalize across fully illuminated and point-light biological motion depictions of actions (Oram & Perrett, 1994).

This complex population of action-tuned neurons can be broken down into two subpopulations based on their responses to manipulations commonly associated with changes in viewpoint perspective. Nearly 90 percent of the biological motion-tuned neurons require specific combinations of body postures and movements that render them effectively viewpoint specific (Jellema et al., 2004, Oram & Perrett, 1996). For example, a neuron may fire for a leftward-facing actor that is engaged in walking in the forward direction, but not when that actor bends forward, walks backward, or faces the opposite direction. A smaller population of STSa neurons has a more generalized tuning across changes in viewpoint perspective that render them more viewpoint invariant. For example, these neurons may fire most strongly when the monkey views a person walking backward, whether viewed from the front or the back. Interestingly, both the viewpoint-specific and viewpoint-invariant populations of neurons tend to be position and size invariant (Wachsmuth, Oram, & Perrett, 1994) and are anatomically intermixed (Perrett, Smith, Potter, Mistlin, Head, Milner, & Jeeves, 1985b).

More recent studies have examined the extent to which these biological motion-selective neurons are tuned to the kinematics in actions or to the unique sequence of body poses associated with articulation. The issue of form and motion analysis in biological motion recognition has become a contentious one in computational modeling; some researchers argue that action recognition is inherently a form-based analysis (Lange, Georg, & Lappe, 2006), with others arguing for a more integrated approach (Giese & Poggio, 2003). Psychophysical evidence strongly suggests that humans are sensitive to both posture and kinematic cues and have a preference for using kinematic cues when action sequences are viewed at longer (more natural) durations (Thirkettle, Benton, & Scott-Samuel, 2009; Thurman, Giese, & Grossman, 2010).

In the monkey STSa, there are clearly at least two classes of neurons—those that are tuned for action kinematics and those that respond equally well to stationary images of body patches, with some patchy organizations among these two cell types (Nelissen, Vanduffel, & Orban, 2006; Oram & Perrett, 1994; Perrett et al., 1985a; Vangeneugden, De Maziere, Van Hulle, Jaeggli, Van Gool, & Vogels, 2011). The "snapshot" neurons fire strongly for stationary images of bodies or for short-duration action clips. These neurons may respond equally to all body postures within an action sequence or have tuning for specific postures (Vangeneugden et al., 2009a). Despite the adopted nomenclature, snapshot neurons typically have some dynamic sensitivity; they often fire for dynamic sequences just as strongly as for stationary postures extracted from the same sequence, and they display clear sensitivity to the temporal order of sequentially viewed postures (Singer & Sheinberg, 2010). In contrast, the "action" neurons require body kinematics in order to fire, appear to exist in fewer numbers, are poorly excited by stationary postures, and encode dynamic information across longer durations. Interestingly, while both types of neurons have response properties that would appear to facilitate action recognition, including information that discriminates between actions, it is the action neurons that have firing patterns that more faithfully represent similarity among different types of actions (Vangeneugden et al., 2009a).

Thus, one question that researchers currently grapple with is the extent to which the human STS and monkey STSa are homologous. The evidence for similarity remains mixed. For example, monkeys, even after extensive training, appear to have much more difficulty discriminating biological motion in point-light sequences (Vangeneugden, Vancleef, Jaeggli, VanGool, & Vogels, 2009b). But the human STS appears to have some of the viewpoint-invariant properties observed in the monkey STSa. For example, researchers using fMRI identified the neural signature of action specificity (fMR-adaptation, or repetition suppression, for repeated actions) in observers viewing fully illuminated or point-light actions (Grossman, Jardine, & Pyles, 2010; Kable & Chatterjee, 2006). Moreover, like the monkey STSa, the human STS appears have action specificity that is invariant across changes in position and size (Grossman et al., 2010). However, many questions remain. For example, the relative proportions of viewpoint-specific and object-oriented (e.g., size and position invariant) subpopulations of neurons are not yet clear. Is there evidence for action and snapshot neurons in the human cortex? How might these neural responses relate to perception? And, perhaps the most interesting question—how do we reconcile these findings with those to be discussed in the next section?

The Human STS and Social Cognition

The neurophysiological evidence linking the STS to biological motion recognition is compelling. Across two species, and revealed via multiple physiological techniques, researchers have established a clear correlational relationship, and some studies have offered evidence for a causal relationship. This review, however, would be incomplete without a discussion of a wider range of findings that have, using tasks and stimuli that lack all the features of biological motion, implicated the same region of the STS (e.g., Hein & Knight, 2008).

For example, studies of face recognition were among the first to reveal regions of activation on the posterior STS. The STS is more strongly activated when observers (1) discriminate emotional facial expressions as compared to person identity (Hoffman & Haxby, 2000); (2) attend to meaningful facial movements as compared to nonsensical, or gurning, movements (Campbell, Landis, & Regard, 1986); and (3) attend to changes in eye gaze as compared to other movement patterns (Puce, Allison, Bentin, Gore, & McCarthy, 1998). There are additional, largely overlapping regions on the STS that are activated when subjects view hand and mouth movements, and there is some possibility of a rough somatotopic organization (Pelphrey, Morris, Michelich, Allison, & McCarthy, 2005, Wheaton et al., 2004). That faces, facial movements, and the movements of body parts would drive the STS is not unexpected on the basis of single-unit recordings. Studies of biological motion responses in monkey STSa also report intermixed neurons that have tuning for these types of images (Hietanen & Perrett, 1993; Wachsmuth et al., 1994).

Some researchers have argued, however, that the studies measuring neural responses correlated with viewing bodies and body parts have inadvertently confounded more complex cognitive attributions with action recognition (Saxe & Wexler, 2005). The assertion is that when observers view actors engaged in deliberate actions, the natural consequence is to infer deliberate, goal-directed, or intentional attributes of those actions, and it is this cognitive step that is reflected in the STS activation. There is support for this idea from a number of studies. For example, Saxe et al. (2004) found stronger STS activation when subjects viewed actors who were moving about with clear goals as compared to more innocuous types of actions. Similarly, Morris et al. (2008) found the left hemisphere STS to be more strongly activated when subjects viewed an actor intentionally moving his arm as compared to when the actor's arm was incidentally moved by an external object. This last study is particularly interesting because the actual body movement performed by the actor was identical (or nearly identical) across the different conditions, but the context of the movement was reflected in the unique STS response.

In the previous examples, the STS was more strongly activated when the actor displayed goal-oriented motions as compared to movements that were not goal directed. In the next set of experiments, the STS further differentiates between the contexts of those goals and the congruency between those goals and the actions performed. For example, Shultz et al. (2011) found the STS to be more strongly activated by failed outcome during reaching movements as compared to successful reaching actions. The right hemisphere STS is also more strongly activated when an actor executes an action not in accordance with the expectation established prior to the movement (Wyk, Hudac, Carter, Sobel, & Pelphrey, 2009). Similarly, Jastorff et al. (2010) found the STS to respond more strongly when actors made irrational reaching movements—for example, by reaching high above a short wall to grasp an object (as opposed to a more canonical reaching trajectory).

As we know from the success of Pixar, attribution of goal-directed behavior and intentional states is not limited to portrayals by human actors. Observers readily report their perception of agentive behaviors in robots and other nonhuman objects (e.g., Waytz, Cacioppo, & Epley, 2010). At least three studies have failed to find STS signals that differentiated between human and robot agents (Gobbini, Gentili, Ricciardi, Bellucci, Salvini, Laschi, Guazzelli, & Pietrini, 2010; Pelphrey, Mitchell, McKeown, Goldstein, Allison, & McCarthy, 2003a; Shultz et al., 2011), and one study found a boost in STS activity for mismatches between humanoid features and robot-like behavior (Saygin, Chaminade, Ishiguro, Driver, & Frith, 2011).

One of the more interesting and compelling demonstrations of agentive attribution involves animation sequences of the Heider-Simmel type, in which simple geometric shapes appear to engage in rich social interactions (Heider & Simmel, 1944). Researchers have shown that when these sequences are constructed so as to depict interactive behavior, the STS is activated (Castelli, Happe, Frith, & Frith, 2000; Weisberg, van Turennout, & Martin, 2007), and the more animate and cognizant these geometric shapes appear to be, the stronger the STS response (Schultz, Friston, O'Doherty, Wolpert, & Frith, 2005; Schultz, Imamizu, Kawato, & Frith, 2004). Patients with right hemisphere damage that includes the STS and adjacent regions of the cortex have more difficulty classifying these Heider-Simmel-type movies as depicting a story that includes agents engaged in deliberate behavior (Weed, McGregor, Feldbaek Nielsen, Roepstorff, & Frith, 2010).

With regard to the studies cited in the previous paragraph, it is not yet clear whether the neural mechanisms that support the social cognitive functions described in the studies are the same neural mechanisms that support biological motion recognition; the evidence suggests that the two will be difficult to distinguish using fMRI. Whereas meta-analyses and careful comparisons of localizers have clearly shown a dissociation in the STS among some social cognitive tasks, such as biological motion perception and theory of mind (not discussed in this chapter), there is little evidence separating biological motion from face recognition or perceived interactions in Heider-Simmel sequences (Carrington & Bailey, 2009; Gobbini, Koralek, Bryan, Montgomery, & Haxby, 2007).

In a compelling demonstration of the difficulty ahead for social cognitive neuroscientists, consider an fMRI study in which subjects watched a short sequence involving a two-person interaction (Bahnemann, Dziobek, Prehn, Wolf, & Heekeren, 2009). After viewing the sequence, the subjects made one of three types of judgment: (1) a biological motion judgment (i.e., they judged which direction the person was moving in), (2) a theory-of-mind judgment (i.e., they judged the protagonist's mental state), or (3) a moral judgment. The results showed overlapping activation in the STS for all three tasks; the largest swath of activation occurred during the moral judgments. This large activation had two distinct foci: one overlapped with the activation associated with the biological motion task, and the other overlapped with the theory-of-mind task.

It is worth noting that there is a basis for the idea of multiplexed action analysis in the STS that derives from the single-unit physiology work in monkeys (e.g., Jellema & Perrett, 2006). Recall that the STSa neurons tuned to biological motion in those studies exist intermixed among those that respond to a much wider range of moving and stationary patterns. And within the action-tuned neurons were unique classes of responses—some tuned to motion and form features or precise combinations of these, and others that generalized across viewpoints. From this evidence, it would be entirely reasonable to hypothesize heterogeneous functional specialization, within close anatomical proximity, in the human STS that, depending on the level of analysis, serves both action recognition and social cognition. This remains an open question and is one of the issues at the forefront of current research.

Summary

Point-light biological motion perception, and in particular the perceived richness of the body kinematics from such sparse displays, is a compelling demonstration of the complex cognitive computations and attributions engaged when viewing actions. The engagement of these cognitive processes appears to be largely automated: typical observers lack the ability to perceptually decompose biological motion into its local mechanical parts and are effectively obliged to interpret the displays as the rich and complex actions that they are.

Convergent evidence from neurophysiological measurements has pinpointed the superior temporal sulcus as subserving some important brain computations for either the construction of biological motion or, possibly, the inference of higher level social-cognitive properties inherent in point-light displays. Whereas neural activity in the STS is strongly linked to the brain's ability to detect and discriminate biological motion, new studies have also emerged demonstrating a role for the STS in the attribution of intentional states to the actor.

Thus the current challenge to the cognitive neuroscience community is to develop theoretical proposals of STS functional specialization that encompass a broader range of social cognitive tasks than just biological motion perception. Much research to date has tacitly assumed a domain-specific approach that effectively asserts the sole functional role of the STS as being the extraction of action meanings from biological motion. This is not likely to be true, and a number of alternatives—including theories of multiplexed levels of analysis within this single brain area or domain-general alternative mechanisms—must now be considered. The result will be a better understanding of the critical neural computations taking place in this complex brain area; this understanding can, in turn, help explain why the STS is so important for biological motion recognition.

References

Adolphs, R. (2009). The social brain: Neural basis of social knowledge. *Annual Review of Psychology*, *60*, 693–716.

Allison, T., Puce, A., & McCarthy, G. (2000). Social perception from visual cues: Role of the STS region. *Trends in Cognitive Sciences*, *4*(7), 267–278.

Bahnemann, M., Dziobek, I., Prehn, K., Wolf, I., & Heekeren, H. R. (2009). Sociotopy in the temporoparietal cortex: Common versus distinct processes. *Social Cognitive and Affective Neuroscience*, *5*(1), 48–58.

Barraclough, N. E., Xiao, D., Oram, M. W., & Perrett, D. I. (2006). The sensitivity of primate STS neurons to walking sequences and to the degree of articulation in static images. *Progress in Brain Research*, *154*, 135–148.

Battelli, L., Cavanagh, P., Martini, P., & Barton, J. J. (2003). Bilateral deficits of transient visual attention in right parietal patients. *Brain*, *126*(Pt. 10), 2164–2174.

Battelli, L., Cavanagh, P., & Thornton, I. M. (2003). Perception of biological motion in parietal patients. *Neuropsychologia*, *41*, 1808–1816.

Battelli, L., Pascual-Leone, A., & Cavanagh, P. (2007). The "when" pathway of the right parietal lobe. *Trends in Cognitive Sciences, 11*(5), 204–210.

Beauchamp, M. S., Lee, K. E., Haxby, J. V., & Martin, A. (2003). FMRI responses to video and point-light displays of moving humans and manipulable objects. *Journal of Cognitive Neuroscience, 15*(7), 991–1001.

Bedny, M., Caramazza, A., Grossman, E., Pascual-Leone, A., & Saxe, R. (2008). Concepts are more than percepts: The case of action verbs. *Journal of Neuroscience, 28*(44), 11347–11353.

Bertenthal, B., & Pinto, J. (1994). Global processing of biological motion. *Psychological Science, 5*, 221–225.

Blake, R., & Shiffrar, M. (2007). Perception of human motion. *Annual Review of Psychology, 58*, 47–73.

Blake, R., Turner, L. M., Smoski, M. J., Pozdol, S. L., & Stone, W. L. (2003). Visual recognition of biological motion is impaired in children with autism. *Psychological Science, 14*(2), 151–157.

Bonda, E., Petrides, M., Ostry, D., & Evans, A. (1996). Specific involvement of human parietal systems and the amygdala in the perception of biological motion. *Journal of Neuroscience, 16*(11), 3737–3744.

Bruce, C., Desimone, R., & Gross, C. G. (1981). Visual properties of neurons in a polysensory area in superior temporal sulcus of the macaque. *Journal of Neurophysiology, 46*(2), 369–384.

Campbell, R., Landis, T., & Regard, M. (1986). Face recognition and lipreading: A neurological dissociation. *Brain, 109*, 509–521.

Carrington, S. J., & Bailey, A. J. (2009). Are there theory of mind regions in the brain? A review of the neuroimaging literature. *Human Brain Mapping, 30*(8), 2313–2335.

Carter, E. J., & Pelphrey, K. A. (2006). School-aged children exhibit domain-specific responses to biological motion. *Social Neuroscience, 1*(3–4), 396–411.

Castelli, F., Happe, F., Frith, U., & Frith, C. (2000). Movement and mind: A functional imaging study of perception and interpretation of complex intentional movement patterns. *NeuroImage, 12*(3), 314–325.

Cowey, A., & Vaina, L. M. (2000). Blindness to form from motion despite intact static form perception and motion detection. *Neuropsychologia, 38*(5), 566–578.

Cusick, C. G. (1996). The superior temporal polysensory region in monkeys. In K. S. Rockland, J. H. Kaas, & A. Peters (Eds.), *Extrastriate Cortex in Primates, 14* (pp. 435–468). New York: Plenum Press.

Cusick, C. G., Seltzer, B., Cola, M., & Griggs, E. (1995). Chemoarchitectonics and corticocortical terminations within the superior temporal sulcus of the rhesus monkey: Evidence for subdivisions of superior temporal polysensory cortex. *Journal of Comparative Neurology, 360*(3), 513–535.

Dittrich, W. H., Troscianko, T., Lea, S. E., & Morgan, D. (1996). Perception of emotion from dynamic point-light displays represented in dance. *Perception, 25,* 727–738.

Freire, A., Lewis, T. L., Maurer, D., & Blake, R. (2006). The development of sensitivity to biological motion in noise. *Perception, 35*(5), 647–657.

Giese, M. A., & Poggio, T. (2003). Neural mechanisms for the recognition of biological movements. *Nature Reviews: Neuroscience, 4*(3), 179–192.

Gobbini, M. I., Gentili, C., Ricciardi, E., Bellucci, C., Salvini, P., Laschi, C., Guazzelli, M., & Pietrini, P. (2010). Distinct neural systems involved in agency and animacy detection. *Journal of Cognitive Neuroscience, 23*(8), 1911–1920.

Gobbini, M. I., Koralek, A. C., Bryan, R. E., Montgomery, K. J., & Haxby, J. V. (2007). Two takes on the social brain: A comparison of theory of mind tasks. *Journal of Cognitive Neuroscience, 19*(11), 1803–1814.

Grezes, J., Fonlupt, P., Bertenthal, B., Delon-Martin, C., Segebarth, C., & Decety, J. (2001). Does perception of biological motion rely on specific brain regions? *NeuroImage, 13*(5), 775–785.

Grossman, E., Battelli, L., & Pascual-Leone, A. (2005). TMS over STSp disrupts perception of biological motion. *Vision Research, 45*(22), 2847–2853.

Grossman, E., & Blake, R. (2002). Brain areas active during visual perception of biological motion. *Neuron, 35*(6), 1157–1165.

Grossman, E., Donnelly, M., Price, R., Pickens, D., Morgan, V., Neighbor, G., & Blake, R. (2000). Brain areas involved in perception of biological motion. *Journal of Cognitive Neuroscience, 12*(5), 711–720.

Grossman, E. D., & Blake, R. (2001). Brain activity evoked by inverted and imagined biological motion. *Vision Research, 41,* 1475–1482.

Grossman, E. D., Jardine, N. L., & Pyles, J. A. (2010). fMR-adaptation reveals invariant coding of biological motion on the human STS. *Frontiers in Human Neuroscience, 4,* 15.

Grossman, E. D., Kim, C. Y., & Blake, R. (2004). Learning to see biological motion: Brain activity parallels behavior. *Journal of Cognitive Neuroscience, 16*(9), 1–11.

Gunns, R. E., Johnston, L., & Hudson, S. M. (2002). Victim selection and kinematics: A point-light investigation of vulnerability to attack. *Journal of Nonverbal Behavior, 26*(3), 129–158.

Heider, F., & Simmel, M. (1944). An experimental study of apparent behavior. *American Journal of Psychology, 57*(2), 243–259.

Hein, G., & Knight, R. T. (2008). Superior temporal sulcus—it's my area: Or is it? *Journal of Cognitive Neuroscience, 20*(12), 2125–2136.

Hietanen, J. K., & Perrett, D. I. (1993). Motion sensitive cells in the macaque superior temporal polysensory area. I. Lack of response to the sight of the animal's own limb movement. *Experimental Brain Research, 93*(1), 117–128.

Hiris, E., Humphrey, D., & Stout, A. (2005). Temporal properties in masking biological motion. *Perception & Psychophysics, 67*(3), 435–443.

Hoffman, E. A., & Haxby, J. V. (2000). Distinct representations of eye gaze and identity in the distributed human neural system for face perception. *Nature Neuroscience, 3*(1), 80–84.

Howard, R. J., Brammer, M., Wright, I., Woodruff, P. W., Bullmore, E. T., & Zeki, S. (1996). A direct demonstration of functional specialization within motion-related visual and auditory cortex of the human brain. *Current Biology, 6,* 1015–1019.

Huk, A. C., Ress, D., & Heeger, D. J. (2001). Neuronal basis of the motion aftereffect reconsidered. *Neuron, 32*(1), 161–172.

Jastorff, J., Begliomini, C., Fabbri-Destro, M., Rizzolatti, G., & Orban, G. A. (2010). Coding observed motor acts: Different organizational principles in the parietal and premotor cortex of humans. *Journal of Neurophysiology, 104*(1), 128–140.

Jastorff, J., & Orban, G. A. (2009). Human functional magnetic resonance imaging reveals separation and integration of shape and motion cues in biological motion processing. *Journal of Neuroscience, 29*(22), 7315–7329.

Jellema, T., Maassen, G., & Perrett, D. I. (2004). Single cell integration of animate form, motion, and location in the superior temporal cortex of the macaque monkey. *Cerebral Cortex, 14*(7), 781–790.

Jellema, T., & Perrett, D. I. (2006). Neural representations of perceived bodily actions using a categorical frame of reference. *Neuropsychologia, 44*(9), 1535–1546.

Johansson, G. (1973). Visual perception of biological motion and a model for its analysis. *Perception & Psychophysics, 14,* 195–204.

Jones, E. G., & Powell, T. P. (1970). An anatomical study of converging sensory pathways within the cerebral cortex of the monkey. *Brain, 93*(4), 793–820.

Jordan, H., Reiss, J. E., Hoffman, J. E., & Landau, B. (2002). Intact perception of biological motion in the face of profound spatial deficits: Williams syndrome. *Psychological Science, 13*(2), 162–167.

Kable, J. W., & Chatterjee, A. (2006). Specificity of action representations in the lateral occipitotemporal cortex. *Journal of Cognitive Neuroscience, 18*(9), 1498–1517.

Kim, J., Doop, M. L., Blake, R., & Park, S. (2005). Impaired recognition of biological motion in schizophrenia. *Schizophrenia Research, 77*(2–3), 299–307.

Kosslyn, S. M., Ganis, G., & Thompson, W. L. (2001). Neural foundations of imagery. *Nature Reviews. Neuroscience, 2*(9), 635–642.

Kourtzi, Z., & Kanwisher, N. (2000). Implied motion activates extrastriate motion-processing areas. Response to David and Senior (2000). *Trends in Cognitive Sciences, 4*(8), 295–296.

Lange, J., Georg, K., & Lappe, M. (2006). Visual perception of biological motion by form: A template-matching analysis. *Journal of Vision, 6*(8), 836–849.

Mather, G., & Murdoch, L. (1994). Gender discrimination in biological motion displays based on dynamic cues. *Proceedings of the Royal Society of London, Series B: Biological Sciences, 258*(1353), 273–279.

McLeod, P., Dittrich, W., Driver, J., Perrett, D., & Zihl, J. (1996). Preserved and impaired detection of structure from motion by a "motion-blind" patient. *Visual Cognition, 3*(4), 363–391.

Michels, L., Lappe, M., & Vaina, L. M. (2005). Visual areas involved in the perception of human movement from dynamic form analysis. *Neuroreport, 16*(10), 1037–1041.

Morris, J. P., Pelphrey, K. A., & McCarthy, G. (2008). Perceived causality influences brain activity evoked by biological motion. *Social Neuroscience, 3*(1), 16–25.

Nelissen, K., Vanduffel, W., & Orban, G. A. (2006). Charting the lower superior temporal region, a new motion-sensitive region in monkey superior temporal sulcus. *Journal of Neuroscience, 26*(22), 5929–5947.

Neri, P., Morrone, M. C., & Burr, D. C. (1998). Seeing biological motion. *Nature, 395*(6705), 894–896.

Ochiai, T., Grimault, S., Scavarda, D., Roch, G., Hori, T., Riviere, D., Mangin, J. F., & Régis, J. (2004). Sulcal pattern and morphology of the superior temporal sulcus. *NeuroImage, 22*(2), 706–719.

Oram, M. W., & Perrett, D. I. (1994). Responses of anterior superior temporal polysensory (STPa) neurons to "biological motion" stimuli. *Journal of Cognitive Neuroscience, 6*, 99–116.

Oram, M. W., & Perrett, D. I. (1996). Integration of form and motion in the anterior superior temporal polysensory area (STPa) of the macaque monkey. *Journal of Neurophysiology, 76*(1), 109–129.

Pandya, D. N., & Seltzer, B. (1982). Intrinsic connections and architectonics of posterior parietal cortex in the rhesus monkey. *Journal of Comparative Neurology, 204*(2), 196–210.

Pavlova, M., Krageloh-Mann, I., Sokolov, A., & Birbaumer, N. (2001). Recognition of point-light biological motion displays by young children. *Perception, 30*(8), 925–933.

Pavlova, M., & Sokolov, A. (2000). Orientation specificity in biological motion perception. *Perception & Psychophysics, 62*, 889–898.

Pavlova, M., & Sokolov, A. (2003). Prior knowledge about display inversion in biological motion perception. *Perception, 32*(8), 937–946.

Pavlova, M., Staudt, M., Sokolov, A., Birbaumer, N., & Krageloh-Mann, I. (2003). Perception and production of biological movements in patients with early periventricular brain lesions. *Brain, 126*(Pt 3), 692–701.

Peelen, M. V., Glaser, B., Vuilleumier, P., & Eliez, S. (2009). Differential development of selectivity for faces and bodies in the fusiform gyrus. *Developmental Science, 12*(6), F16–F25.

Peelen, M. V., Wiggett, A. J., & Downing, P. E. (2006). Patterns of fMRI activity dissociate overlapping functional brain areas that respond to biological motion. *Neuron, 49*(6), 815–822.

Pelphrey, K. A., Mitchell, T. V., McKeown, M. J., Goldstein, J., Allison, T., & McCarthy, G. (2003). Brain activity evoked by the perception of human walking: Controlling for meaningful coherent motion. *Journal of Neuroscience, 23*(17), 6819–6825.

Pelphrey, K. A., Morris, J. P., Michelich, C. R., Allison, T., & McCarthy, G. (2005). Functional anatomy of biological motion perception in posterior temporal cortex: An FMRI study of eye, mouth, and hand movements. *Cerebral Cortex, 15*(12), 1866–1876.

Perrett, D. I., Smith, P. A., Mistlin, A. J., Chitty, A. J., Head, A. S., Potter, D. D., Broennimann, R., Milner, A. D., & Jeeves, M. A. (1985a). Visual analysis of body movements by neurons in the temporal cortex of the macaque monkey: A preliminary report. *Behavioural Brain Research, 16*(2–3), 153–170.

Perrett, D. I., Smith, P. A., Potter, D. D., Mistlin, A. J., Head, A. S., Milner, A. D., & Jeeves, M. A. (1985b). Visual cells in the temporal cortex sensitive to face view and gaze direction. *Proceedings of the Royal Society of London, Series B: Biological Sciences, 223*(1232), 293–317.

Peuskens, H., Vanrie, J., Verfaillie, K., & Orban, G. A. (2005). Specificity of regions processing biological motion. *European Journal of Neuroscience, 21*(10), 2864–2875.

Puce, A., Allison, T., Bentin, S., Gore, J. C., & McCarthy, G. (1998). Temporal cortex activation in humans viewing eye and mouth movements. *Journal of Neuroscience, 18*, 2188–2199.

Puce, A., & Perrett, D. (2003). Electrophysiology and brain imaging of biological motion. *Philosophical Transactions of the Royal Society of London, Series B: Biological Sciences, 358*, 435–445.

Runeson, S., & Frykholm, G. (1983). Kinematic specification of dynamics as an informational basis for person-and-action perception: Expectation, gender recognition, and deceptive intention. *Journal of Experimental Psychology: General, 112*, 585–615.

Rutherford, M.D., & Troje, N.F. (2012). IQ predicts biological motion perception in autism spectrum disorders. *Journal of Autism and Developmental Disorders, 42*(4), 557–565.

Saxe, R., & Wexler, A. (2005). Making sense of another mind: The role of the right temporo-parietal junction. *Neuropsychologia, 43*(10), 1391–1399.

Saxe, R., Xiao, D. K., Kovacs, G., Perrett, D. I., & Kanwisher, N. (2004). A region of right posterior superior temporal sulcus responds to observed intentional actions. *Neuropsychologia, 42*(11), 1435–1446.

Saygin, A. P. (2007). Superior temporal and premotor brain areas necessary for biological motion perception. *Brain, 130*(Pt 9), 2452–2461.

Saygin, A. P., Chaminade, T., Ishiguro, H., Driver, J., & Frith, C. (2012). The thing that should not be: Predictive coding and the uncanny valley in perceiving human and humanoid robot actions. *Social Cognitive and Affective Neuroscience, 7*(4), 413–422.

Saygin, A. P., Cook, J., & Blakemore, S. J. (2010). Unaffected perceptual thresholds for biological and non-biological form-from-motion perception in autism spectrum conditions. *PLoS ONE, 5*(10), e13491.

Saygin, A. P., Wilson, S. M., Hagler, D. J., Bates, E., & Sereno, M. I. (2004). Point-light biological motion perception activates human premotor cortex. *Journal of Neuroscience, 24*(27), 6181–6188.

Schenk, T., & Zihl, J. (1997a). Visual motion perception after brain damage I: Deficits in global motion perception. *Neuropsychologia, 35*, 1289–1297.

Schenk, T., & Zihl, J. (1997b). Visual motion perception after brain damage II: Deficits in form-from-motion perception. *Neuropsychologia, 35*, 1299–1310.

Schultz, J., Friston, K. J., O'Doherty, J., Wolpert, D. M., & Frith, C. D. (2005). Activation in posterior superior temporal sulcus parallels parameter inducing the percept of animacy. *Neuron, 45*(4), 625–635.

Schultz, J., Imamizu, H., Kawato, M., & Frith, C. D. (2004). Activation of the human superior temporal gyrus during observation of goal attribution by intentional objects. *Journal of Cognitive Neuroscience, 16*(10), 1695–1705.

Seltzer, B., & Pandya, D. N. (1978). Afferent cortical connections and architectonics of the superior temporal sulcus and surrounding cortex in the rhesus monkey. *Brain Research, 149*(1), 1–24.

Shultz, S., Lee, S. M., Pelphrey, K., & McCarthy, G. (2011). The posterior superior temporal sulcus is sensitive to the outcome of human and non-human goal-directed actions. *Social Cognitive and Affective Neuroscience, 6*(5), 602–611.

Singer, J. M., & Sheinberg, D. L. (2010). Temporal cortex neurons encode articulated actions as slow sequences of integrated poses. *Journal of Neuroscience, 30*(8), 3133–3145.

Sumi, S. (1984). Upside-down presentation of the Johansson moving light-spot pattern. *Perception, 13*(3), 283–286.

Taylor, J. C., Wiggett, A. J., & Downing, P. E. (2007). Functional MRI analysis of body and body part representations in the extrastriate and fusiform body areas. *Journal of Neurophysiology, 98*(3), 1626–1633.

Taylor, J. C., Wiggett, A. J., & Downing, P. E. (2009). fMRI-adaptation studies of viewpoint tuning in the extrastriate and fusiform body areas. *Journal of Neurophysiology, 98*(3), 1626–1633.

Thirkettle, M., Benton, C.P., & Scott-Samuel, N.E. (2009). Contributions of form, motion, and task to biological motion perception. *Journal of Vision, 9*(3), 28.1–11.

Thompson, J. C., Clarke, M., Stewart, T., & Puce, A. (2005). Configural processing of biological motion in human superior temporal sulcus. *Journal of Neuroscience, 25*(39), 9059–9066.

Thurman, S. M., Giese, M. A., & Grossman, E. D. (2010). Perceptual and computational analysis of critical features for biological motion. *Journal of Vision, 10*(12), 15.

Troje, N. F., Sadr, J., Geyer, H., & Nakayama, K. (2006). Adaptation aftereffects in the perception of gender from biological motion. *Journal of Vision, 6*(8), 850–857.

Troje, N. F., Westhoff, C., & Lavrov, M. (2005). Person identification from biological motion: Effects of structural and kinematic cues. *Perception & Psychophysics, 67*(4), 667–675.

Vaina, L. M., Lemay, M., Bienfang, D. C., Choi, A. Y., & Nakayama, K. (1990). Intact "biological motion" and "structure from motion" perception in a patient with impaired motion mechanisms: A case study. *Visual Neuroscience, 5*(4), 353–369.

Vaina, L. M., Solomon, J., Chowdhury, S., Sinha, P., & Belliveau, J. W. (2001). Functional neuroanatomy of biological motion perception in humans. *Proceedings of the National Academy of Sciences of the United States of America, 98*(20), 11656–11661.

van Kemenade, B., Muggleton, N., Walsh, V., & Saygin, A. P. (2010). The effects of TMS over STS and premotor cortex on the perception of biological motion. *Journal of Vision, 10*(7), 785.

Vangeneugden, J., De Maziere, P. A., Van Hulle, M. M., Jaeggli, T., Van Gool, L., & Vogels, R. (2011). Distinct mechanisms for coding of visual actions in macaque temporal cortex. *Journal of Neuroscience, 31*(2), 385–401.

Vangeneugden, J., Pollick, F., & Vogels, R. (2009a). Functional differentiation of macaque visual temporal cortical neurons using a parametric action space. *Cerebral Cortex, 19*(3), 593–611.

Vangeneugden, J., Vancleef, K., Jaeggli, T., VanGool, L., & Vogels, R. (2009b). Discrimination of locomotion direction in impoverished displays of walkers by macaque monkeys. *Journal of Vision, 10*(4), 22.1–19.

Wachsmuth, E., Oram, M. W., & Perrett, D. I. (1994). Recognition of objects and their component parts: Responses of single units in the temporal cortex of the macaque. *Cerebral Cortex, 5*, 509–522.

Waytz, A., Cacioppo, J., & Epley, N. (2010). Who sees human? The stability and importance of individual differences in anthropomorphism. *Perspectives on Psychological Science, 5*, 219–232.

Weed, E., McGregor, W., Feldbaek Nielsen, J., Roepstorff, A., & Frith, U. (2010). Theory of mind in adults with right hemisphere damage: What's the story? *Brain and Language, 113*(2), 65–72.

Weisberg, J., van Turennout, M., & Martin, A. (2007). A neural system for learning about object function. *Cerebral Cortex, 17*(3), 513–521.

Wheaton, K. J., Thompson, J. C., Syngeniotis, A., Abbott, D. F., & Puce, A. (2004). Viewing the motion of human body parts activates different regions of premotor, temporal, and parietal cortex. *NeuroImage, 22*(1), 277–288.

Wyk, B. C., Hudac, C. M., Carter, E. J., Sobel, D. M., & Pelphrey, K. A. (2009). Action understanding in the superior temporal sulcus region. *Psychological Science, 20*(6), 771–777.

5 Beyond the Scientific Objectification of the Human Body: Differentiated Analyses of Human Motion and Object Motion

Maggie Shiffrar and James P. Thomas

Abstract

Vision researchers typically objectify the human body in their studies of body perception by adopting the same theoretical approaches and experimental methodologies used in studies of object perception. Although the human body can be understood as a physical object, it is also much more. Numerous behavioral and neurophysiological studies, with adult and infant observers, have demonstrated important differences in the visual perception of human motion and object motion. At least three factors—motor experience, visual experience, and emotional processes—differentiate the visual perception of animate, or at least observer-like, entities from inanimate entities.

1 Introduction: Is the Human Body Simply an Object?

The differentiation of animate, intentional entities from inanimate ones is fundamentally important for any organism that interacts with its environment. At a basic level, the behavioral interactions that humans have with animate and inanimate beings differ significantly. For example, successful interactions with animate entities involve added complexities, such as the accurate detection and comprehension of intentions. While the class of animate entities is quite large, human observers have by far the most experience interacting with one type of animate entity: people. The goal of this chapter is to review, in both adult and infant observers, differences between the visual analysis of human motion and the motions of inanimate objects. And as recent evidence suggests that the perception of animal motion depends on some of the same mechanisms that underlie the perception of human motion, visual analyses of nonhuman animal motions will also be considered.

While developmental psychologists have long studied differences in the perception of animate and inanimate motions, many vision scientists have assumed that the same processing mechanisms underlie the perception of both categories of motion. For example, let's consider the conclusions of three leading vision researchers of the twentieth century. In his influential book *Vision*, David Marr (1982) argues for a single, hierarchical visual processing system that generates the same types of descriptions for both animate (e.g., humans and animals) and inanimate (e.g., tanks) stimuli. Similarly, Roger N. Shepard (1984) described visual motion perception as dependent on the same processes no matter what the underlying input. As Shepard eloquently stated,

> There evidently is little or no effect of the particular object presented. The motion we involuntarily experience when a picture of an object is presented first in one place and then in another, whether the picture is of a leaf or of a cat, is neither a fluttering drift nor a pounce, it is, in both cases, the same simplest rigid displacement. (Shepard, 1984, p. 426)

The work of Gunnar Johansson provides another example of the scientific objectification of human bodies in action. Johansson's influential vector analysis model of visual motion perception (e.g., Johansson, 1964, 1976) was intended to apply to all categories of dynamic visual stimuli. Indeed, when Johansson first applied the point-light technique to the study of the perception of human movement (Johansson, 1973), he did so with the purpose of determining whether the mechanisms involved in the perception of rolling wheels were sufficient to account for the perception of moving animals. He concluded that they were. When his observers described their visual percepts of point-light displays of human movement as particularly vivid, Johansson (1973) attributed this enhanced vividness to differences in visual experience and most certainly not as a reflection of differentiated visual processes dedicated to the perception of human action.

Of course, the human body is a physical object. And, as will be outlined below, the visual perception of human and object motion share many functional and processing constraints. For example, the perception of a stationary mailbox for the purpose of navigating around it, and the perception of a seated woman for the purpose of navigating around her, likely depend on similar neural mechanisms. Indeed, there are many situations in which human observers perceive human and object motion similarly (e.g., Shiffrar & Freyd, 1990; Funk, Shiffrar, & Brugger, 2005). Consistent with this, magnetoencephalographic (MEG) data indicate that the perception of point-light displays of human and of object motion rely on the same brain regions during the first 200 ms of analysis (Virji-Babul et al.,

2007). Thus, at several levels, the visual perception of people and objects do rely on the same processes.

However, mounting evidence suggests the existence of fundamentally important differences between the visual perception of people and the visual perception of objects. From a functional point of view, we obviously interact differently with people than with objects. When one detects a friend executing a pendular waving motion with her arm, one might wave back or walk over to say hello. But, if one were to see that same pendular motion performed by the branch of a tree, one would neither wave to the tree nor initiate a conversation with it. As an inherently social stimulus, human motion affords behavioral opportunities that are different from those afforded by object motion. Since we interact differently with people than with objects, it makes sense to wonder whether our visual system analyzes people and objects differently. Neurophysiological and psychophysical evidence, summarized in the following sections, overwhelmingly supports the hypothesis that visual analyses of human motion differ in some fundamental ways from visual analyses of object motion.

2 Differences in the Visual Perception of Human and Object Motions: Adult Observers

By definition, motion is a change in spatial location over time. Thus, one simple approach to the question of whether the visual system analyzes human motion and object motion differently is to investigate how these types of stimuli are analyzed over space and time. Because the input to the visual motion perception system comes from retinal receptors with relatively small receptive fields, the movements of real-world stimuli must be inferred from a balanced integration of the luminance changes detected within many tiny regions of the retina. This creates a problem for the visual system that is analogous to John Godfrey Saxe's (1855) rendition of the classic Indian tale of six blind men feeling an elephant. One man feels only the elephant's tail. Another man feels a tusk. Yet another feels a knee, and so forth. Individually, none of the men can identify the animal from the tactile information gathered from one small region of the elephant's body. Instead, the identification of the elephant requires a complex integration of the men's collective haptic experiences. In the same way, no individual neuron in an observer's visual system can identify a real object's motion. Instead, local motion measurements have to be combined across space and time in order to generate global motion percepts.

2.1 Integrating Motion across Space

How does the visual system accurately assess the movements of real-world entities from a multitude of tiny, local measurements? Simply averaging all of the motion measurements across an image would necessarily result in the perception of one big, undifferentiated blob of motion. So how does the visual system resolve this problem? One class of global motion processing models posits that the visual system compares motion measurements across rigidly connected edges of different orientations (e.g., Adelson & Movshon, 1982). Another class of local motion processing models selectively relies on the measurements taken from edge discontinuities, such as endpoints and corners, that indicate where one object ends and the next object begins (e.g., Hildreth, 1984). Over the years, it has become apparent that the human visual system utilizes both global and local motion processes (Shiffrar & Lorenceau, 1996). But, as we'll see below, the balanced interaction of local and global processes depends, in part, on whether the underlying stimulus being analyzed is human motion or object motion.

When observers view objects—such as squares, cars, and scissors—moving behind apertures (an experimental technique that simulates the measurement problem described above), their percepts indicate the use of local motion processes (Shiffrar & Pavel, 1991; Shiffrar & Lorenceau, 1996). Conversely, when observers view human motion through apertures, their percepts indicate the use of global motion processes (Shiffrar, Lichtey, & Heptulla-Chatterjee, 1997). Interestingly, use of these global analyses appears to depend on the physical plausibility of an observed action. When observers view a person walking with impossibly fast or impossibly slow gaits, their percepts indicate a return to a heavy reliance on local motion processes. Conversely, when observers view a person walking with typical, physically possible gaits, their percepts reflect a reliance on global motion processes (Shiffrar et al., 1997).

Point-light displays can also be used to examine how the visual system integrates motion measurements across space. Point-light displays are constructed by attaching small markers, or point-lights, to the major joints of moving actors (see figures 5.1a and 5.1b). These stimuli are filmed so that only the point-lights are visible; this significantly reduces form information while leaving motion kinematics intact. When these highly degraded visual displays are set into motion, observers readily perceive human movement (e.g., Johansson 1973, 1976). Importantly, when a point-light defined walking person appears within a point-light mask, as shown in figures 5.1c and 5.1d, typical adult observers can still reliably detect the person (e.g., Bertenthal & Pinto 1994). Point-light masks are typically constructed by

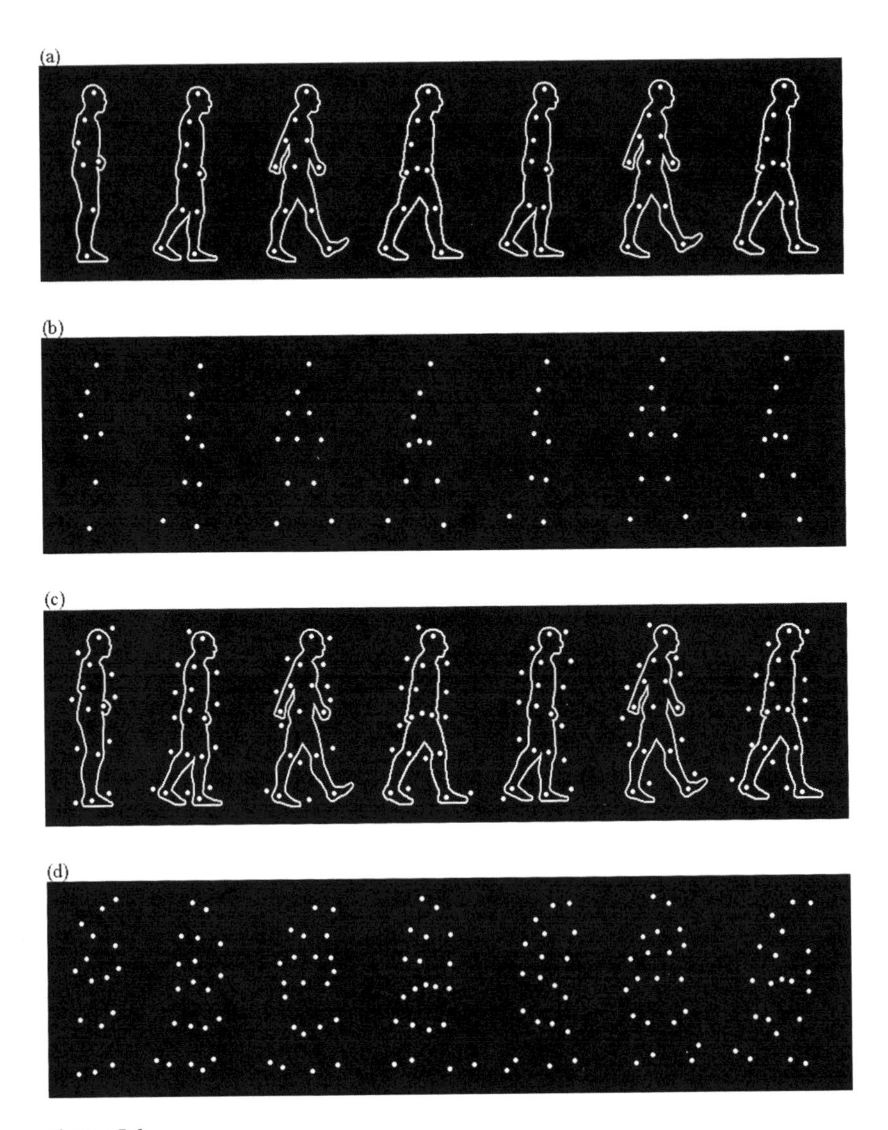

Figure 5.1
(a) A series of static outlines depicting the changing shape of a walking person's body with point-lights positioned on the major joints and head. Not all point-lights are shown. Outlines are not typically shown in experimental stimuli. (b) A point-light walker is constructed by removing everything from each image except the point-lights. When static, these displays are difficult to interpret. However, once set in motion, observers readily detect the presence of a walking person. (c) Static body outline and point-light walker with mask dots added. Mask dots are typically constructed by first duplicating, then randomly scrambling the starting location of each duplicated dot comprising the coherent walker. The body outline does not typically appear in experimental stimuli. (d) A point-light walker presented in a point-light mask. Because the local motion of each mask point matches the motion of a point comprising the coherent walker, walker detection requires a global analysis.

duplicating one or more point-light walkers and then randomizing the starting locations of each of those duplicate points. As a result, the points defining the mask have the same size, luminance, and velocities as the points defining the walker. Since the motions of the points defining the walker and those defining the mask are identical, local analyses of the motions of individual points cannot be used to detect the walker. Because only the global configuration of the locations and motions of multiple points distinguishes the walker from the mask, detection of point-light walkers in a mask must rely on global or large-scale motion processes (Bertenthal & Pinto, 1994). When the same masking technique is used with nonhuman, inanimate point-light motion (e.g., Hiris, Krebeck, Edmonds, & Stout, 2005; Kaiser et al., 2010), typical observers demonstrate marked decrements in their ability to detect these figures relative to their detection of point-light human motion. This pattern of results suggests that observers are better able to integrate human motion across image space than they are able to integrate object motion across space.

2.2 Integrating Motion across Time

Temporal aspects of motion perception have traditionally been examined through the use of apparent motion, which is the illusory perception of movement from rapidly alternating static images. Evidence from this and other time-based research paradigms indicates that the human visual system integrates human and object motions differently over time, at least under some conditions. In the case of apparent motion, when different views of a moving person and different views of a moving object are displayed under conditions that give rise to apparent motion, the resulting percepts of the two classes of stimuli differ. For instance, when static images of a person in two different postures are flashed in rapid alternation (figure 5.2), adult observers generally report the perception of the shortest possible path of motion connecting those postures, even if that path describes a physically impossible motion, such as one hand translating through another. The shortest possible paths of motion are also seen when two different views of an object are presented in rapid alternation. However, when pictures of the human body are presented at slower temporal rates that are consistent with the temporal range of normal human-action production, observers tend to perceive paths of apparent motion that are consistent with the biomechanical constraints on the human body. Conversely, when images of objects are shown at these same slower rates, adult observers continue to report the perception of the shortest possible paths of apparent motion. This pattern of results suggests that human movement

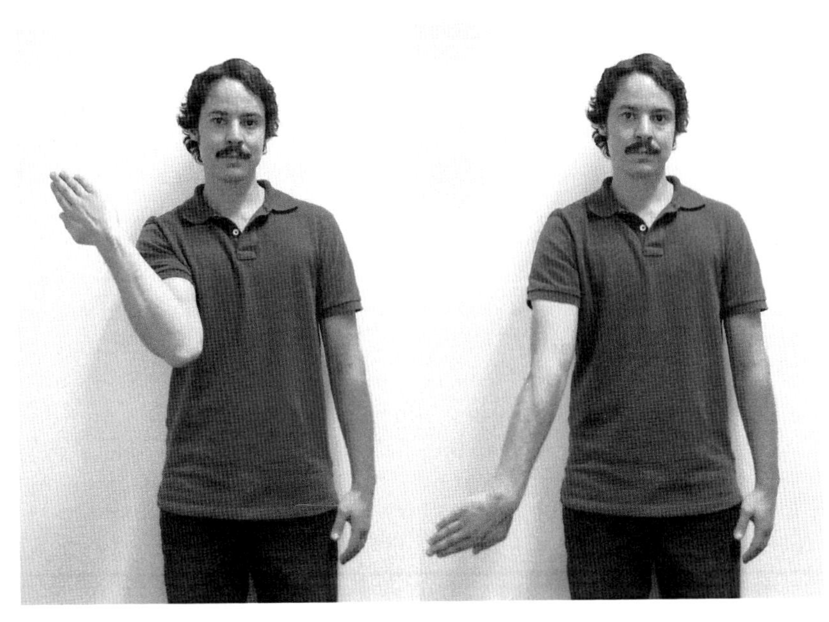

Figure 5.2
An apparent motion display of a person rotating his arm. At long interstimulus intervals, the arm appears to rotate about the elbow toward the body in a biomechanically possible way. With brief interstimulus intervals, the arm appears to move downward following the shortest possible path of apparent motion, which is physically impossible.

is analyzed by motion processes that operate over relatively large temporal windows and that take into account the biomechanical limitations of the human body (e.g., Shiffrar & Freyd, 1990; 1993).

The ability of adult observers to analyze human motion over extended temporal intervals has also been examined with point-light displays. In one such study, a point-light person walked in place while facing either left or right. The walker was then placed, at random locations, within a point-light mask that reduced the utility of local motion processes. A blank display of varying durations separated each of the frames depicting the masked point-light walker. Across trials, observers tried to detect the walker and report whether it faced leftward or rightward (Thornton et al., 1998). Naïve observers were significantly above chance in their detection of the direction in which the point-light walker faced; this was true even when the blank temporal interval between each frame, or the interstimulus interval, was as long as 120 ms. In the domain of visual motion perception,

120 ms represents a long gap that is well outside the range of local motion processes. This is not to say that local analyses are not used during the perception of point-light displays of human motion. Indeed, under most conditions, both local and global motion processes contribute to our perceptions of other people's actions. But, the ability to perceive a coherent point-light walker in a point-light mask across significant separations in both space and time is particularly impressive. This finding adds further support for the hypothesis that the visual perception of human movement is unique in the extent to which observers can tolerate massive amounts of noise (or gaps) in both the spatial and temporal dimensions.

2.3 Are People Simply Animals to the Human Visual System?

The studies described in sections 2.1 and 2.2 compared the visual perception of moving people and moving objects. Thus, it is unclear whether the results of those studies are specific to the perception of human motion, or whether they generalize to the perception of all animate and/or intentional motions. We know that observers can recognize nonhuman animals, including camels, horses, and penguins, from point-light displays (e.g., Mather & West, 1993; Chouchourelou, Golden, & Shiffrar, 2012). But are the same motion processing mechanisms employed during the perception of human movement and animal movement?

A superb, though little known, psychophysical study conducted by Leslie Cohen in her doctoral dissertation (2002) addressed the above question by comparing visual sensitivities to human, dog, and seal locomotion under otherwise identical conditions. Cohen found that observers were better able to detect the presence of coherent human motion than the presence of coherent seal or dog motion in masked point-light displays. One could speculate that this pattern of visual sensitivity simply reflects the differential amounts of real-world visual experience that observers had with these three types of motion. After all, we tend to observe the motion of other humans with significantly more frequency than the motion of seals or dogs. But Cohen's (2002) study provided an elegant control for visual experience. She ran three different groups of observers: seal trainers, dog trainers, and typical undergraduates. If levels of visual sensitivity simply reflected levels of visual experience, then seal trainers should show greater visual sensitivity to seal motion than to dog motion, and dog trainers should show greater visual sensitivity to dog motion than to seal motion. Instead, Cohen (2002) found that all three classes of observers showed exactly the same pattern of visual sensitivity: the greatest visual sensitivity to human motion and the least visual sensitivity to seal motion.

Based on these results, Cohen proposed that visual sensitivity to point-light displays of animal motion might reflect the degree of physical similarity between the observed actions of the animal and the observer's own repertoire of performable actions. Certainly, it is much easier for humans to imitate the gaits of dogs than the undulating whole-body flops of seal locomotion on land.

The pattern of results summarized above raises an interesting question. Let's start with the assumption, which has received extensive support, that the visual perception of human motion relies on the observer's own bodily capabilities (e.g., Wilson & Knoblich, 2005). If visual sensitivity to animal motion reflects the degree of similarity between a seen animal's bodily structure and the bodily structure of the observer, might this suggest that the visual perception of animal motion depends on the same mechanisms employed during the visual perception of human motion? If the perception of animal motion and human motion taps the same mechanisms, then these two classes of motion should share core perceptual characteristics. Studies of visual sensitivity to point-light displays of human movement have demonstrated that the perception of human movement is orientation specific (e.g., Bertenthal & Pinto, 1994) and dependent on global processes (e.g., Thornton et al., 1998). In a series of studies comparing visual sensitivity to point-light displays of human locomotion and horse locomotion, it was found that visual percepts of horse gaits are both orientation specific and spatiotemporally global (Pinto & Shiffrar, 2009). In other words, the visual perception of horse motion shares two fundamental characteristics with the visual perception of human movement. However, detection thresholds in these studies indicated that, overall, human observers are more sensitive to human movement than to horse movement (Pinto & Shiffrar, 2009). Follow-up work further suggested that human observers are better able to recognize the movements of bipedal animals than the movements of apedal animals (Chouchourelou et al., 2012).

Taken together, the behavioral results described above suggest that the analyses of human and animal motions differ in a graded, rather than dichotomous, fashion. Furthermore, visual sensitivity to the movements of nonhuman animals appears to reflect the degree of similarity between an observed animal's body and the body of the observer. Such a conclusion suggests that, when it comes to visual motion perception, human motion may not be simply one exemplar within the perceptual category of animate motion. Instead, human motion may represent the prototypical exemplar at the very center of a graded category of animate motions. Indeed, the graded shape of that category's boundary may even be defined by the physical capabilities and/or the physical shape of the observer's own body. Consistent with this proposition, when the limbs defining a point-light walker are rearranged so that they no longer match the structure of the human body, visual sensitivity to that point-light walker's motion drops significantly (Pinto & Shiffrar, 1999). Furthermore, when wooden beams are sized and arranged so that they simulate the gross structure of the human body, observers perceive those wooden beams as following paths of apparent motion that are consistent with the biomechanical constraints of the human body (Heptulla-Chatterjee et al., 1996). However, it is important to note that a human body is certainly not a prerequisite for the detection of animacy (e.g., Tremoulet & Feldman, 2000; Rutherford, Pennington, & Rogers, 2006; see chapters 6–12 in the current volume).

3 Neurophysiological Differences between Percepts of Human and Object Motion

The psychophysical studies summarized in the previous section suggest that both overlapping and distinctive processes underlie visual percepts of human motion, animal motion, and object motion. Neurophysiological studies also support that conclusion (see Blake & Shiffrar, 2007, for review) and suggest that patterns of activity in at least two neural areas differentiate analyses of human and object motions. In the first such study, David Perrett and his colleagues conducted single-cell recordings in the macaque cortex and discovered that many neurons in the superior temporal sulcus (STS) are selectively responsive to human forms and motions (see Puce & Perrett, 2003 for review). In the human, fMRI studies indicate that the posterior region of the STS (STSp) responds strongly to point-light depictions of upright and coherent, but not inverted or scrambled, human movement (Grossman & Blake, 2001; see also chapter 4 in the current volume). Finally, research with transcranial magnetic stimulation and with lesion patients indicates that normal functionality within area STSp is required for the accurate perception of point-light displays of human movement (Grossman, Battelli, & Pascual-Leone, 2005; Saygin, 2007). Importantly, given the above discussion, STSp responds more strongly to human motion than it does to animal-like motion (Pyles, Garcia, Hoffman, & Grossman, 2007) or object motion (Beauchamp et al., 2003). Studies utilizing MEG indicate that analyses of point-light displays of human and object movement diverge approximately 200 ms after stimulus onset, at which point processes in the right temporal lobe, encompassing the STSp, are triggered during the perception of human movement (Virji-Babul, Cheung, Weeks, Kerns, & Shiffrar, 2007; see also chapter 4 in the current volume).

While it may be tempting to describe the STSp as *the* region responsible for the analysis of human and human-like motion, there is at least one other region that is necessary for the accurate perception of point-light displays of human motion: the motor cortex. Lesion evidence indicates that a functional premotor cortex is also a necessary prerequisite for the differentiation of coherent and scrambled point-light actors and the identification of point-light actions (Saygin, 2007). Electroencephalogram (EEG) data further support the hypothesis that premotor cortex plays a critical role in the perception of coherent, but not scrambled, point-light displays of human motion (Ulloa & Pineda, 2007). Research with hemiplegic patients indicates that lesions to the motor system selectively degrade visual sensitivity to point-light actions corresponding to the observers' paralyzed limbs (Serino et al., 2009). Consistent with this, significantly greater motor system activity is found when observers view actions that they can perform versus actions that they cannot perform (Stevens, Fonlupt, Shiffrar, & Decety, 2000). Finally, observers' motor skills correlate with their visual sensitivity to point-light displays of human movement (Price, Shiffrar, & Kerns, 2012). All of this evidence suggests that the visual perception of human movement relies in part on specialized neural mechanisms.

4 Development of the Visual Perception of Human Movement

The evidence described in the previous section overwhelmingly supports the hypothesis that, at least in adult observers, visual analysis of the movements of humans and human-like animals differs from the visual analysis of object motion. At what developmental point does this differentiation appear?

The ability to perceive point-light displays of human motion arises early in life. The first support for this conclusion came from a classic preferential looking study (Fox & McDaniel, 1982). The researchers found that typical infants between the ages of four and six months could differentiate between point-light depictions of human motion and random motion; they could also differentiate between upright and inverted point-light displays of human motion (Fox & McDaniel, 1982). Robust sensitivity to human motion is also evidenced in 8-month-old infants by differences in the amplitudes of event-related potentials (ERP) during observations of upright and scrambled point-light walkers (Hirai & Hiraki, 2005) and during observations of upright and inverted point-light walkers (Reid, Hoehl, & Striano, 2006). These results are significant because they indicate that infants can differentiate at least one type of human action that they have never performed—that is, walking. On the other hand, when supported, infants exhibit a rhythmic alternation of their legs (Thelen, Fisher, & Ridley-Johnson, 1984) that might be a sufficient approximation of walking. Existence of this spontaneous movement pattern opens up the possibility that the perception of at least some actions, such as walking, may be subserved by innate mechanisms.

More recent work indicates that infants as young as two days old show a preference for unmasked point-light displays of upright biological (in this case, chicken) motion relative to either inverted or scrambled versions of these displays (Simion, Regolin, & Bulf, 2008; see also chapter 3 in this volume). Not surprisingly, the two-day-old human infants who participated in this experiment had no previous visual experience with chicken motion. Thus, this finding adds further support for the hypothesis that at least some aspects of visual sensitivity to biological motion may be innate. Consistent with this, visually inexperienced chicks also exhibit a visual preference for the coherent motions of other animals, especially other chickens (Vallortigara, Regolin, & Marconato, 2005). Such a preference would clearly be beneficial for accurate and rapid imprinting (Vallortigara et al., 2005).

Visual sensitivity to human and/or biological motions may reflect some innate mechanisms, but it is also clear that enhanced visual sensitivity to human motion evolves during an observer's development. For example, three-month-old observers can detect perturbations in the phase relations between the points defining a coherent point-light human walker. But their detection abilities are equivalent for upright and inverted point-light walkers. However, just two months later, at the age of five months, observers demonstrate specialization in their visual sensitivity to upright human motion (Pinto, 2006). Infants also respond to human motion differently from the way they respond to object motion. For example, seven-month-old infants will reproduce the action goals of a person but not the apparent action goals of an object (Mahajan & Woodward, 2009; see also chapter 16 of this volume). Furthermore, the eye movements of one-year-old observers anticipate the actions of other people but not the actions of objects (Falck-Yttr, Gredeback, & von Hofsten, 2006).

Additional evidence in support of the hypothesis that the human visual system becomes increasingly tuned for analyses of human motion comes from tasks comparing infants' visual sensitivity to point-light displays of human and animal motions. At the age of three months, infants demonstrate equivalent patterns of visual sensitivity to the motions of a point-light person, spider, and cat. By five months of age, patterns of visual sensitivity shift significantly, and infants demonstrate greater sensitivity to point-light displays of human motion than to point-light displays of animal motion (Pinto, 2006). This pattern of "perceptual narrowing" suggests that the infant visual system becomes specialized or tuned for the detection of canonical human motion (Pinto, 2006). Specialized perceptual tuning is supported by fMRI data indicating that pSTS activity becomes increasingly tuned to human motion as typical children age (Carter & Pelphrey, 2006).

Interestingly, specialized tuning of the pSTS for the analysis of human actions is not guaranteed. For example, in children with Autism Spectrum Disorder (ASD), pSTS activity does not become selective for human movement (Pelphrey & Carter, 2008). Consistent with this, observers with ASD do not show enhanced visual sensitivity to human motion relative to object motion (Kaiser, Delmolino, Tanaka, & Shiffrar, 2010). It remains to be determined whether this lack of tuning results from the social isolation associated with living with ASD or whether this lack of tuning gives rise to that social isolation, or both.

In typical observers, visual sensitivity to simple, unmasked point-light displays of human movement starts to reach adult levels at around five years of age (Pavlova et al., 2001; Blake, Turner, Smoski, Pozdol, & Stone, 2003). For point-light walkers hidden within point-light masks, walker detection performance improves significantly between the ages of six and nine, and again from year nine to adulthood (Freire, Lewis, Maurer, & Blake, 2006). When visual sensitivity to point-light human and object motion are compared in typical children, enhanced visual sensitivity to human motion in both masked and unmasked displays is clearly evident at seven years of age and likely appears sooner (Kaiser et al., 2010).

An interesting wrinkle in our understanding of the development of visual sensitivity to the human body results when one compares the developmental timelines between preferences for coherent point-light displays of human/animal motion, summarized above, and preferences for coherent, static depictions of the human body. Evidence from a standard preferential looking paradigm suggests that before the age of about eighteen months, infants do not reliably differentiate between coherent and scrambled static drawings of the human body (Slaughter, Heron, & Sim, 2002). The way in which static body stimuli are experimentally scrambled is likely important here. When scrambling is so extensive that limbs cannot be detected, ERP data suggest that young infants do differentiate between coherent and scrambled depictions of static body postures (Gliga & Dehaene-Lambertz, 2005). However, when bodies are scrambled at the level of limbs, such that, for example, a scrambled body consists of a model with legs coming out of its head and arms coming out of its torso, then infants as old as fifteen months of age show no preference for coherent static human bodies. Given that human infants only a few days old preferentially attend to point-light displays of coherent chicken motion (Simion et al., 2008) as well as to coherent static human faces (Johnson, Dziurawiec, Ellis, & Morton, 1991), it is surprising that infants as old as fifteen months still do not preferentially attend to coherent static depictions of the human form (Slaughter et al., 2002). This is rather unexpected when one considers the obvious fact that faces are always attached to bodies. In addition, and as summarized above, the results of several developmental studies with point-light displays of animal motion point to differentiated visual sensitivity to coherent and scrambled biological motion—stimuli that differ in their underlying form (Simion et al., 2008; Vallortigara et al., 2005). One potential explanation is that motion is the key facilitator of visual sensitivity to the human form (Christie & Slaughter, 2010). Another possibility is that, like preferences for faces (e.g., Johnson et al., 1991), preferences for static depictions of coherent human bodies are present early in human development and then decline. A third possibility is that the sophistication of global motion processes underlying human body percepts evolves more slowly than previously assumed.

5 But Why Do Percepts of Human Motion and Object Motion Differ?

The evidence summarized above supports the hypothesis that the visual analysis of human movement and object movement differs in fundamental ways. The next obvious question is *why*. Put another way, what are the origins of our visual sensitivity to other people's actions? There are at least three basic factors that might differentiate, at least for human observers, human motion from object motion. First, human motion represents the only category of visual motion that human observers can both produce and perceive. A human observer can execute patterns of movement that clearly approximate the actions of other people. But the same observer cannot produce actions that duplicate the spatiotemporal characteristics of the motion trajectories of windblown trees, melting ice cream, speeding cars, shifting sands, or crashing waves. If the human visual system taps the motor system for information that facilitates the visual analysis of human action, then this perception-action linkage should differentiate visual analyses of, and visual sensitivity to, other people's actions, relative to analyses of and sensitivity to other categories of visual motion.

Second, as inherently social creatures, typical observers have a lifetime of experience watching other people act and paying close attention to those actions. Of course, we also have a lifetime of experience watching the wind blow shrubbery about and watching waves crash at the shoreline. But we pay relatively little attention to the specifics of those types of movement. Thus, it seems reasonable to propose that differences in visual experience might also differentiate visual percepts of human motion and object motion.

Finally, the ways in which people move their bodies convey extensive social and emotional information that the movements of objects do not. Thus, social and emotional analyses may also differentiate the perception of objects and people. In the following sections, findings are reviewed that support the impact of all three of these factors—motor experience, visual experience, and social-emotional processes—in differentiating the perception of human motion from the perception of object motion. In other chapters in this volume, arguments are also posited for evolutionary pressures that might further differentiate perceptual analyses of human movement from other categories of movement.

5.1 Motor Experience

In the last twenty years, the results of numerous studies have converged to suggest that the visual perception of human movement is dependent on the observer's own motor experience (see Prinz, 1997, and Wilson & Knoblich, 2005, for theoretical overviews). For example, the production of simple hand (and arm) movements by humans is described by the two-thirds power law, which defines the relationship between the hand trajectory's instantaneous velocity and radius of curvature (Viviani, 2002). Human observers demonstrate accurate visual motion percepts when dynamic stimuli conform to this fundamental law of motor production, but not when the same stimuli violate this law (Viviani, 2002). Another law of motor production, known as Fitts's law, defines how quickly a person can move between two targets as being a function of target width and the distance between the targets. Visual percepts of apparent motion between targets are consistent with this motor law (Grosjean et al., 2007). Furthermore, the perception of motor outcomes, such as when throwing a dart, reflects each observer's own motor processes (Knoblich & Flach,

2001). Such evidence indicates that motor processes systematically constrain the visual perception of human movement. Indeed, it has been convincingly argued that the human visual system is optimized for the analysis of human-generated movements (Viviani, 2002).

Additional support for the hypothesis that motor processes impact the visual perception of human movement comes from interference effects during the simultaneous production and perception of human actions. Some studies have shown that the performance of actions by an observer interferes with that observer's perception of other people performing similar actions. For example, visual sensitivity to a point-light walker's gait speed is significantly lower while observers walk than when they stand or ride a bicycle during task performance (Jacobs & Shiffrar, 2005). Action perception also interferes with action production. For example, the variability of an individual's sinusoidal arm movements increases during the observation of another person's sinusoidal arm movements (Kilner, Paulignan, & Blakemore, 2003). Consistent with the importance of velocity profiles, this interference effect depends on the similarity between the velocity profiles of simultaneously observed and produced arm movements (Kilner, Hamilton, & Blakemore, 2007).

Motor learning, per se, has been shown to selectively enhance visual sensitivity to the actions of others. For instance, observers can improve their sensitivity to point-light displays of unusual gaits by learning to execute those gaits while blindfolded (Casile & Giese, 2006). Finally, research with patient populations further supports the importance of motor constraints on the perception of human motion. For example, hemiplegic patients with motor system lesions exhibit degraded visual sensitivity to point-light actions that correspond to actions performed with their compromised limbs relative to visual sensitivity to point-light actions that correspond to their functional limbs (Serino et al., 2010). Case-study evidence has documented an individual born without hands, and lacking internal representations of those absent hands, who consistently perceives paths of apparent hand rotation that are physically impossible (Funk et al., 2005). In other words, this individual perceives hands, which he has never had, in the same way that handed observers perceive objects.

Important developmental research also supports the impact of the observer's motor capabilities on the perception of other people's actions. For example, children with motor impairments resulting from Down syndrome show decrements in their visual sensitivity to point-light displays of human motion relative to age-matched control observers (Virji-Babul, Kerns, Zhou, Kapur, & Shiffrar, 2006). Learning a new motor skill also changes infant observers' percepts of that same motor skill in others (Sommerville & Woodward, 2005; Woodward, 2009; see also chapter 15 in this volume). Importantly, infants also show a preference for attending to actions that they themselves can currently perform. When given the opportunity to look at a point-light walker or a point-light crawler, infants who crawl prefer to look at the point-light crawler, and infants who can walk prefer to look at the point-light walker (Sanefuji, Ohgami, & Hashiya, 2008; see also chapter 14 in this volume). Since crawling infants see other people walking much more than they see other people crawling, this attentional bias cannot be attributed to visual experience. Instead, this combination of results provides compelling support for a linkage between action production and action perception. In sum, motor experience shapes visual sensitivity to other people's actions.

5.2 Visual Experience

In addition to motor experience, visual sensitivity to human motion is also dependent on visual experience. In fact, computational modeling suggests that various aspects of visual sensitivity to human motion can be explained by visual experience alone (Giese & Poggio, 2003). As inherently social animals, humans typically spend more time watching other people act than they spend watching drifting clouds or swaying plants. The results of several psychophysical studies suggest that such biased patterns of visual experience have perceptual consequences. In one such study, observers viewed point-light displays of human walkers and rated the degree to which each figure looked human (Bulthoff, Bulthoff, & Sinha, 1998). As long as the point-light walker displays retained their normal two-dimensional projection, observers rated the walkers as human, even when the points-lights had three-dimensional structural anomalies. Such data suggest that visual experience is sufficient to override substantial depth distortions.

However, substantial levels of visual experience are needed to alter visual percepts of human movement. For example, observers in one study viewed point-light displays of friends and strangers walking with both commonly occurring and unusual gaits (Jacobs, Pinto, & Shiffrar, 2004). Overall, observers most accurately identified walkers who performed their naturally occurring gaits. However, enhanced sensitivity to frequently observed actions required over a dozen hours per week of face-to-face, real-world interaction between the observer and the person depicted as a point-light walker. Visual experience also influences levels of neural activity in area STSp during the observation of masked point-light people in motion (Grossman, Blake, & Kim, 2004).

The results detailed in the previous paragraphs suggest that both visual experience and motor experience contribute to our visual sensitivity to human movement. Studies comparing the relative contributions of these two sources of information indicate that motor experience is the larger contributor. For example, when observers view point-light displays of themselves and others performing various actions, they are best able to identify their own actions and least able to identify the actions of strangers (Loula et al., 2005). Enhanced sensitivity to self-generated actions can be attributed to contributions from the observer's own motor system (Knoblich & Flach, 2001). Visual experience cannot account for enhanced visual sensitivity to one's own actions (Prasad & Shiffrar, 2009).

5.3 Social and Emotional Processes

Anyone who has ever watched a silent Charlie Chaplin film knows that extensive social and emotional information is available from bodily movements alone. Research confirms this intuition. For example, naïve observers are above chance in their ability to detect a walking person's gender (e.g., Barclay, Cutting, & Kozlowski, 1978; Pollick et al., 2005) and identity (e.g., Jacobs et al., 2004; Loula et al., 2005; Jokisch et al., 2006) from point-light displays. Even more impressively, untrained observers can also detect, at least at above-chance levels, a person's potential reproductive fitness (Brown et al., 2005), social dominance (Montepare & Zebrowitz-McArthur, 1988), and vulnerability to attack (Gunns et al., 2002) in point-light displays. Obviously, human movements express extensive social information that the human visual system can readily extract.

As a class, animate entities act with intention. Untrained observers can detect the intentions of point-light defined people in action. For instance, observers can detect when a moving point-light actor intends to deceive the observer about his or her gender (Runeson & Frykholm, 1983). In another study, point-light displays were created of people who lifted an empty box normally and in a manner that erroneously suggested that the box was heavy. Again, observers accurately detected the deceitful lifts (Runeson & Frykholm, 1983). Visual sensitivity to the intention to deceive is experience dependent. For example, the accuracy with which a stationary observer can determine whether a point-light defined basketball player intends to pass an invisible basketball or intends to fake a pass depends upon the observer's motor and/or visual experience with the game of basketball (Sebanz & Shiffrar, 2009).

Affective states can also be reliably detected in point-light displays of human movement (Atkinson et al., 2004; Pollick et al., 2001). Social context influences the perception of affective states in point-light actors. For example, when a point-light defined person expresses some emotional state during an interaction with another person, that emotional state is most accurately detected when displays show both individuals rather than only the individual expressing the emotion (Clarke et al., 2005).

Neurophysiological evidence supports the hypothesis that social and emotional cues influence visual analyses of human motion. For example, the STSp responds more strongly during the perception of emotional, rather than instrumental, actions (Gallagher & Frith, 2004). Psychophysical evidence indicates that typical observers are most sensitive to the presence of a coherent point-light walker in a mask when that walker is in an angry emotional state rather than a neutral, happy, sad, or fearful emotional state (Chouchourelou et al., 2006). This behavioral finding is consistent with the existence of recurrent processing between the STSp and the amygdala, which appears to be primed for the detection of potential threats (Amaral, 2003). An angry person is clearly a threatening stimulus. Thus, these results suggest that emotional processes automatically contribute to, and indeed help to define, visual sensitivity to the actions of other people. Since people express emotional states and behave socially, and physical objects do not, it appears likely that emotional processes help to differentiate visual analyses of human and object motions.

6 Conclusions

The results summarized in this chapter suggest that human observers of all ages typically process, experience, and react to the movements of other people differently from the way they process, experience, and react to the movements of objects. This differentiation starts early in observers' lives, and it both expands and becomes more refined as observers gain experience observing and interacting with the physical and social worlds. Psychophysical studies with adults and infants suggest that motor experience plays a particularly influential role in shaping our perceptions of other people's actions. Observers also demonstrate high levels of visual sensitivity to the movements of animals that are shaped like, and can move like, the observer. This, along with studies of patient populations, suggests that the visual differentiation of animate and inanimate motions may depend on the physical similarity between seen bodies and the observer's own body. In other words, animate entities may simply be entities that move like the observer.

Acknowledgments

Thanks are due to the Simons Foundation (grant #94915) for funding and support.

References

Adelson, E. H., & Movshon, J. A. (1982). Phenomenal coherence of moving visual patterns. *Nature, 300,* 523–525.

Amaral, D. G. (2003). The amygdala, social behavior, and danger detection. *Annals of the New York Academy of Sciences, 1000,* 337–347.

Atkinson, A. P., Dittrich, W. H., Gemmell, A. J., & Young, A. W. (2004). Emotion perception from dynamic and static body expressions in point-light and full-light displays. *Perception, 33,* 717–746.

Barclay, C., Cutting, J., & Kozlowski, L. (1978). Temporal and spatial factors in gait perception that influence gender recognition. *Perception & Psychophysics, 23,* 145–152.

Beauchamp, M. S., Lee, K. E., Haxby, J. V., & Martin, A. (2003). fMRI responses to video and point-light displays of moving humans and manipulable objects. *Journal of Cognitive Neuroscience, 15,* 991–1001.

Bertenthal, B. I., & Pinto, J. (1994). Global processing of biological motions. *Psychological Science, 5,* 221–225.

Blake, R., & Shiffrar, M. (2007). Perception of human motion. *Annual Review of Psychology, 58,* 47–73.

Blake, R., Turner, L., Smoski, M., Pozdol, S., & Stone, W. (2003). Visual recognition of biological motion is impaired in children with autism. *Psychological Science, 14,* 151–157.

Brown, W. M., Cronk, L., Grochow, K., Jacobson, A., Liu, C. K., et al. (2005). Dance reveals symmetry especially in young men. *Nature, 438,* 148–150.

Bulthoff, I., Bulthoff, H., & Sinha, P. (1998). Top-down influences on stereoscopic depth-perception. *Nature Neuroscience, 1,* 254–257.

Carter, E. J., & Pelphrey, K. A. (2006). School-aged children exhibit domain-specific responses to biological motion. *Social Neuroscience, 1,* 396–411.

Casile, A., & Giese, M. A. (2006). Non-visual motor learning influences the recognition of biological motion. *Current Biology, 16,* 69–74.

Chouchourelou, A., Matsuka, T., Harber, K., & Shiffrar, M. (2006). The visual analysis of emotional actions. *Social Neuroscience, 1,* 63–74.

Chouchourelou, A., Golden, A., & Shiffrar, M. (2012). What does "biological motion" really mean? Differentiating visual percepts of human, animal, and non-biological motions. In K. Johnson & M. Shiffrar (Eds.), *Visual perception of the human body*. Oxford: Oxford University Press.

Christie, T., & Slaughter, V. (2010). Movement contributes to infants' recognition of the human form. *Cognition, 114*, 329–337.

Clarke, T. J., Bradshaw, M. F., Field, D. T., Hampson, S. E., & Rose, D. (2005). The perception of emotion from body movement in point-light displays of interpersonal dialogue. *Perception, 34*, 1171–1180.

Cohen, L. R. (2002). The role of experience in the perception of biological motion. *Dissertation Abstracts International, B: The Sciences and Engineering, 63*, 3049.

Falck-Yttr, T., Gredeback, G., & von Hofsten, C. (2006). Infants predict other people's action goals. *Nature Neuroscience, 9*, 878–879.

Fox, R., & McDaniel, C. (1982). The perception of biological motion by human infants. *Science, 218*, 486–487.

Freire, A., Lewis, T. L., Maurer, D., & Blake, R. (2006). The development of sensitivity to biological motion in noise. *Perception, 35*, 647–657.

Funk, M., Shiffrar, M., & Brugger, P. (2005). Hand movement observation by individuals born without hands: Phantom limb experience constrains visual limb perception. *Experimental Brain Research, 164*, 341–346.

Gallagher, H. L., & Frith, C. D. (2004). Dissociable neural pathways for the perception and recognition of expressive and instrumental gestures. *Neuropsychologia, 42*, 1725–1736.

Giese, M. A., & Poggio, T. (2003). Neural mechanisms for the recognition of biological movements. *Nature Reviews: Neuroscience, 4*, 179–192.

Gliga, T., & Dehaene-Lambertz, G. (2005). Structural encoding of body and face in human infants and adults. *Journal of Neuroscience, 17*, 1328–1340.

Grosjean, M., Shiffrar, M., & Knoblich, G. (2007). Fitts' law holds for action perception. *Psychological Science, 18*, 95–99.

Grossman, E. D., Battelli, L., & Pascual-Leone, A. (2005). Repetitive TMS over STSp disrupts perception of biological motion. *Vision Research, 45*, 2847–2853.

Grossman, E. D., & Blake, R. (2001). Brain activity evoked by inverted and imagined biological motion. *Vision Research, 41*, 1475–1482.

Grossman, E. D., Blake, R., & Kim, C.-Y. (2004). Learning to see biological motion: Brain activity parallels behavior. *Journal of Cognitive Neuroscience, 16*, 1669–1679.

Gunns, R. E., Johnston, L., & Hudson, S. (2002). Victim selection and kinematics: A point-light investigation of vulnerability to attack. *Journal of Nonverbal Behavior, 26*, 129–158.

Heptulla-Chatterjee, S., Freyd, J., & Shiffrar, M. (1996). Configural processing in the perception of apparent biological motion. *Journal of Experimental Psychology: Human Perception and Performance, 22*, 916–929.

Hildreth, E. (1984). *The measurement of visual motion*. Cambridge, MA: MIT Press.

Hirai, M., & Hiraki, K. (2005). An event-related potentials study of biological motion perception in human infants. *Brain Research: Cognitive Brain Research, 22*, 301–304.

Hiris, E., Krebeck, A., Edmonds, J., & Stout, A. (2005). What learning to see arbitrary motion tells us about biological motion perception. *Journal of Experimental Psychology: Human Perception and Performance, 31*, 1096–1106.

Jacobs, A., Pinto, J., & Shiffrar, M. (2004). Experience, context, and the visual perception of human movement. *Journal of Experimental Psychology: Human Perception and Performance, 30*, 822–835.

Jacobs, A., & Shiffrar, M. (2005). Walking perception by walking observers. *Journal of Experimental Psychology: Human Perception and Performance, 31*, 157–169.

Johansson, G. (1964). Perception of motion and changing form. *Scandinavian Journal of Psychology, 5*, 181–208.

Johansson, G. (1973). Visual perception of biological motion and a model for its analysis. *Perception & Psychophysics, 14*, 201–211.

Johansson, G. (1976). Spatio-temporal differentiation and integration in visual motion perception: An experimental and theoretical analysis of calculus-like functions in visual data processing. *Psychological Research, 38*, 379–393.

Johnson, M., Dziurawiec, S., Ellis, H., & Morton, J. (1991). Newborns' preferential tracking of face-like stimuli and its subsequent decline. *Cognition, 40*, 1–19.

Jokisch, D., Daum, I., & Troje, N. (2006). Self-recognition versus recognition of others by biological motion: Viewpoint-dependent effects. *Perception, 35*, 911–920.

Kaiser, M., Delmolino, L., Tanaka, J., & Shiffrar, M. (2010). Comparison of visual sensitivity to human and object motion in autism spectrum disorder. *Autism Research, 34*, 191–195.

Kilner, J., Hamilton, A. F., & Blakemore, S.-J. (2007). Interference effect of observed human motion on action is due to velocity profile of biological motion. *Social Neuroscience, 2*, 158–166.

Kilner, J., Paulignan, Y., & Blakemore, S.-J. (2003). An interference effect of observed biological motion on action. *Current Biology, 13*, 522–525.

Knoblich, G., & Flach, R. (2001). Predicting the effects of actions: Interactions of perception and action. *Psychological Science, 12*, 467–472.

Loula, F., Prasad, S., Harber, K., & Shiffrar, M. (2005). Recognizing people from their movement. *Journal of Experimental Psychology: Human Perception and Performance, 31*, 210–220.

Mahajan, N., & Woodward, A. (2009). Seven month old infants selectively reproduce the goals of animate but not inanimate agents. *Infancy, 14*, 667–679.

Marr, D. (1982). *Vision: A computational investigation into the human representation and processing of visual information*. San Francisco: W. H. Freeman.

Mather, G., & West, S. (1993). Recognition of animal locomotion from dynamic point-light displays. *Perception, 22*, 759–766.

Montepare, J. M., & Zebrowitz-McArthur, L. A. (1988). Impressions of people created by age-related qualities of their gaits. *Journal of Personality and Social Psychology, 55*, 547–556.

Pavlova, M., Krägeloh-Mann, I., Sokolov, A., & Birbaumer, N. (2001). Recognition of point-light biological motion displays by young children. *Perception, 30*, 925–933.

Pelphrey, K. A., & Carter, E. J. (2008). Brain mechanisms for social perception: Lessons from autism and typical development. *Annals of the New York Academy of Sciences, 1145*, 283–299.

Pinto, J. (2006). Developing body representations: A review of infants' responses to biological-motion displays. In G. Knoblich, M. Grosjean, I. Thornton, and M. Shiffrar (Eds.), *Perception of the human body from the inside out* (pp. 305–322). Oxford: Oxford University Press.

Pinto, J., & Shiffrar, M. (1999). Subconfigurations of the human form in the perception of biological motion displays. *Acta Psychologica, 102*, 293–318.

Pinto, J., & Shiffrar, M. (2009). The visual perception of human and animal motion in point-light displays. *Social Neuroscience, 4*, 332–346.

Pollick, F. E., Kay, J. W., Heim, K., & Stringer, R. (2005). Gender recognition from point-light walkers. *Journal of Experimental Psychology: Human Perception and Performance, 31*, 1247–1265.

Pollick, F. E., Paterson, H. M., Bruderlin, A., & Sanford, A. J. (2001). Perceiving affect from arm movement. *Cognition, 82*, 51–61.

Prasad, S., & Shiffrar, M. (2009). Viewpoint and the recognition of people from their movements. *Journal of Experimental Psychology: Human Perception and Performance, 35*, 39–49.

Price, K. J., Shiffrar, M., & Kerns, K. (2012). Movement perception and movement production in Asperger's syndrome. *Research in Autism Spectrum Disorders, 6,* 391–398.

Prinz, W. (1997). Perception and action planning. *European Journal of Cognitive Psychology, 9,* 129–154.

Puce, A., & Perrett, D. (2003). Electrophysiology and brain imaging of biological motion. *Philosophical Transactions of the Royal Society, Series B: Biological Sciences, 358,* 435–445.

Pyles, J. A., Garcia, J. O., Hoffman, D. D., & Grossman, E. D. (2007). Visual perception and the neural correlates of novel "biological motion." *Vision Research, 47,* 2786–2797.

Reid, V. M., Hoehl, S., & Striano, T. (2006). The perception of biological motion by infants: An event related potential study. *Neuroscience Letters, 395,* 211–214.

Runeson, S., & Frykholm, G. (1983). Kinematic specification of dynamics as an informational bias for person-and-action perception: Expectation, gender recognition, and deceptive intent. *Journal of Experimental Psychology: General, 112,* 585–615.

Rutherford, M. D., Pennington, B. F., & Rogers, S. J. (2006). The perception of animacy in young children with autism. *Journal of Autism and Developmental Disorders, 36,* 983–992.

Sanefuji, W., Ohgami, H., & Hashiya, K. (2008). Detection of the relevant type of locomotion in infancy: Crawlers versus walkers. *Infant Behavior and Development, 31,* 624–628.

Saxe, J. G. (1855). *Poems by John Godfrey Saxe.* Boston: Ticknor & Fields.

Saygin, A. P. (2007). Superior temporal and premotor brain areas necessary for biological motion perception. *Brain, 130,* 2452–2461.

Sebanz, N., & Shiffrar, M. (2009). Detecting deception in a bluffing body: The role of expertise. *Psychonomic Bulletin & Review, 16,* 170–175.

Serino, A., Casavecchia, C., DeFilippo, L., Coccia, M., Shiffrar, M., & Ladavas, E. (2009). Lesions to the motor system affect action perception. *Journal of Cognitive Neuroscience, 223,* 413–426.

Shepard, R. N. (1984). Ecological constraints on internal representation: Resonant kinematics of perceiving, imagining, thinking, and dreaming. *Psychological Review, 91,* 417–447.

Shiffrar, M., & Freyd, J. (1990). Apparent motion of the human body. *Psychological Science, 1,* 257–264.

Shiffrar, M., & Freyd, J. (1993). Timing and apparent motion path choice with human body photographs. *Psychological Science, 4,* 379–384.

Shiffrar, M., Lichtey, L., & Heptulla-Chatterjee, S. (1997). The perception of biological motion across apertures. *Perception & Psychophysics, 59,* 51–59.

Shiffrar, M., & Lorenceau, J. (1996). Increased motion linking across edges with decreased luminance contrast, edge width, and duration. *Vision Research, 36,* 2061–2067.

Shiffrar, M., & Pavel, M. (1991). Percepts of rigid motion within and across apertures. *Journal of Experimental Psychology: Human Perception and Performance, 17,* 749–761.

Simion, F., Regolin, L., & Bulf, H. (2008). A predisposition for biological motion in the newborn baby. *Proceedings of the National Academy of Sciences of the United States of America, 105,* 809–813.

Slaughter, V., Heron, M., & Sim, S. (2002). Development of preferences for the human body shape in infancy. *Cognition, 85,* B71–B81.

Sommerville, J. A., & Woodward, A. L. (2005). Pulling out the intentional structure of action: The relation between action processing and action production in infancy. *Cognition, 95,* 1–30.

Stevens, J. A., Fonlupt, P., Shiffrar, M., & Decety, J. (2000). New aspects of motion perception: Selective neural encoding of apparent human movements. *Neuroreport, 11,* 109–115.

Thelen, E., Fisher, D. M., & Ridley-Johnson, R. (1984). The relationship between physical growth and a newborn reflex. *Infant Behavior and Development, 7,* 479–493.

Thornton, I., Pinto, J., & Shiffrar, M. (1998). The visual perception of human locomotion. *Cognitive Neuropsychology, 15,* 535–552.

Tremoulet, P., & Feldman, J. (2000). Perception of animacy from the motion of a single object. *Perception, 29,* 943–951.

Ulloa, E. R., & Pineda, J. A. (2007). Recognition of point-light biological motion: Mu rhythms and mirror neuron activity. *Behavioural Brain Research, 183,* 188–194.

Vallortigara, G., Regolin, L., & Marconato, F. (2005). Visually inexperienced chicks exhibit spontaneous preference for biological motion patterns. *PLoS Biology, 3,* e208.

Virji-Babul, N., Cheung, T., Weeks, D., Kerns, K., & Shiffrar, M. (2007). Neural activity involved in the perception of human and meaningful object motion. *Neuroreport, 18,* 1125–1128.

Virji-Babul, N., Kerns, K., Zhou, E., Kapur, A., & Shiffrar, M. (2006). Perceptual-motor deficits in children with Down syndrome: Implications for intervention. *Down Syndrome Research and Practice, 10*: 74–82.

Viviani, P. (2002). Motor competence in the perception of dynamic events: a tutorial. In W. Prinz & B. Hommel (Eds.), *Common mechanisms in perception and action: Attention and performance* (Vol. XIX, pp. 406–442). Oxford: Oxford University Press.

Wilson, M., & Knoblich, G. (2005). The case for motor involvement in perceiving conspecifics. *Psychological Bulletin, 131*, 460–473.

Woodward, A. (2009). Infants' grasp of other's intentions. *Current Directions in Psychological Science, 18*, 53–57.

II THE PERCEPTION OF ANIMACY AND INTENTIONAL BEHAVIOR

Section Introduction: The Perception of Animacy and Intentional Behavior

M. D. Rutherford and Valerie A. Kuhlmeier

Animacy perception is the perception and categorization of an entity as a living, and by some accounts intentional, being. In the natural world, an animate entity could be a person, or it could be as phylogenetically distant from us as a mouse, insect, or amoeba. In the following seven chapters, stimuli are highly stylized but are seen as animate and even intentional by human observers. Indeed, animacy can be portrayed by stimuli as visually simple as a dot, arrow, or triangle—or even a small cardboard box. Of theoretical import to this work is the examination of the characteristics of motion of these stimuli that result in the percept of animacy and the detection of goal-directed actions and underlying dispositions.

The scientific division between the study of animate motion and the study of the kind of biological motion portrayed in the point-light walkers described in section 1 is largely artificial and a result of historic happenstance. In his chapter in section 1, Troje describes the historic twists and turns in the use of point-light walkers, which currently are used predominantly by those studying the visual system's ability to quickly detect, monitor, and use socially relevant information. The extent to which the perception of both whole-body motion (typically examined by those studying animacy perception) and within-body motion information (portrayed by the point-light walkers of those studying biological motion perception) rely on the same psychological mechanisms is an empirical question.

The ability of the human visual system to detect animacy and then use this information to inform inference and predict behavior is a growing area of inquiry. Such detection and discrimination is fundamental to all social cognition—both developmentally and in adult processing. In real time, before any kind of person perception, theory-of-mind processing, emotion understanding, or even predator detection can take place, one must determine what things in the environment are alive. In development, infants presumably must first have some ability to discriminate animate beings from inanimate ones in order to attend specifically to those objects to which social cognition should apply.

The foundations of this line of inquiry stem from work by researchers such as Heider and Simmel (1944) and Bassili (1976), in which figures as simple as moving circles and triangles were seen, by most observers, as alive and engaging in goal-directed activities such as approaching, chasing, and even harassing, based only on the motion paths. Decades later, Premack and Premack (1997) showed that infants see certain motion paths and interactions of two moving circles as positively or negatively valenced, and Gergely, Nadasdy, Csibra, and Bíró (1995) developed a model to explain infant attribution of goal-directed action that could apply to the simple approach behavior of a circle.

The research presented in the seven chapters of this section builds from these foundations. Exciting and fundamental questions in this field are still wide open. For example, little is known about the different stages involved in the processing of animate information. Compared to what is known about the visual processing of motion and the apparent distinction between "what" and "where" processing in the visual field, little is known about animacy detection in the brain. For example, work presented here examines the specific cues that are involved in animacy detection and looks for experiments that are capable of isolating cues and probing their effects, as well as determining the influence of bottom-up and top-down perceptual and cognitive mechanisms (e.g., chapters by Rutherford, McAleer and Love, Frankenhuis and Barrett, and Scholl). Additionally, researchers want to know more about how these experimental findings generalize to real-time social perception and how animacy perception happens in a complex, naturalistic environment (e.g., chapters by McAleer and Love, Frankenhuis and Barrett).

Research approaching these questions is supported and complemented by developmental and neuroscience frameworks. The former has focused attention on the role of cognitive and motoric development as well as the early competencies involved in infants' detection and interpretation of animate motion and social interaction (e.g., chapters by Luo and Choi, and Kuhlmeier; this topic is continued in section 4 of this volume). Neuroscience frameworks help to enhance these proposals by elucidating the mechanisms of goal attribution; this elucidation includes a critical consideration of the role of proposed systems that directly match observed action to self-produced action (e.g., chapter by Hamilton and Ramsey). Together, these research approaches expand our understanding of what may be the foundation of social psychology: the perception of animacy and goals.

References

Bassili, J. N. (1976). Temporal and spatial contingencies in the perception of social events. *Journal of Personality and Social Psychology, 33,* 680–685.

Gergely, G., Nadasdy, Z., Csibra, G., & Bíró, S. (1995). Taking the intentional stance at 12 months of age. *Cognition, 56,* 165–193.

Heider, F., & Simmel, M. (1944). An experimental study of apparent behavior. *American Journal of Psychology, 57,* 243–259.

Premack, D., & Premack, A. J. (1997). Infants attribute value to the goal-directed actions of self-propelled objects. *Journal of Cognitive Neuroscience, 9,* 848–856.

6 Evidence for Specialized Perception of Animate Motion

M. D. Rutherford

Abstract

Adults are able to easily extract motion information from even very simple stimuli in order to perceive animate motion and infer agency. This ability develops early in infancy, is a human universal, and is irresistible when certain motion cues are presented. Theorists have been intrigued by the idea that underlying this perception is specialized cognitive machinery designed for just this purpose, though there is not broad consensus that this is the case. Here I argue in favor of the idea that only the existence of specialized social perceptual psychology explains the performance of typical and autistic observers. Accounts based on processing by general-learning mechanisms cannot account for findings in this area.

The Nature of Developmentally Privileged Psychological Processes

In this chapter, I will propose that humans have and use specialized perceptual processes designed to perceive animate motion and detect animate beings in the environment based on motion cues. Before reviewing empirical evidence that speaks to this question, I will first discuss the notion of specialized psychological processes, a discussion that is necessarily understood in the context of evolution by natural selection and best appreciated in the context of development. It may be useful here to discuss what is meant or implied when we talk about specialized perceptual processes. To be clear, what I mean, specifically, is that these psychological processes were designed via natural selection because the ability to solve the specific problem in question conferred a selection advantage to individuals who held that psychological attribute. In other words, an individual who was able to discriminate animate from inanimate objects had a competitive selection advantage over those who did not, so mutations that supported these psychological processes increased in frequency over generations. This idea is meant to contrast with the possibility that social perceptual skills—such as animacy perception, face perception, and intentionality perception—are achieved by more "domain general" psychological processes. That is, it refutes the idea that the same general visual processes that solve all other visual problems, including nonsocial problems, give rise to these social percepts as well.

The notion of specialization, or specialized cognitive processes, necessarily refers to a function; the process is specialized in service of a particular function. Let me clarify: As is the case regarding the function of any bodily organ, the function of any psychological process is the purpose for which it was shaped and selected for by natural selection. The function of the heart is to pump blood, the function of the retina is to interpret differences in light entering the eyes into information available to the brain, and the purpose of social cognition is to allow us to successfully interact with other people—all for biological advantage. For the purposes of this chapter, I will take this evolutionary history to define an adaptation's function, even if the adaptation is currently used for a different purpose.

I will briefly review the ideas of core knowledge and domain specificity by way of laying the groundwork for our understanding of the specialized perception of animate motion. The concept of core knowledge is an important and relatively recent idea in developmental science that captures the notion of cognitive function designed in our evolutionarily relevant circumstances (Carey, 1995; Carey & Spelke, 1992, pp. 169–200; Siegler & Crowley, 1994, pp. 194–226; Spelke & Kinzler, 2007, pp. 89–96). Core knowledge is the set of privileged domains of knowledge that children learn easily, without explicit tutoring, and at predictable ages, by virtue of a cognitive preparedness that is specific to those domains. Importantly, these domains of core knowledge are the domains in which cognitive aptitude would have conferred a fitness advantage in our evolutionary history.

Broadly speaking, the core knowledge concept has been recruited to address the basic epistemological problem of how the torrent of sensations and stimuli that a developing child has access to can be processed so as to be useful information. Approached from a software engineering point of view, how is it possible that a baby could have knowledge? One possible solution to this fundamental knowledge-acquisition problem is to create certain core domains with sufficient, reliably developing structures (including concepts and attentional priorities) that result in useful conceptual development.

Developmental psychologists who favor the core domain perspective propose that there are certain domains of knowledge—those that were advantageous in the environment in which humans evolved—that children learn easily by virtue of cognitive preparedness that is specific to those domains. Because children throughout the world develop competencies in these areas at roughly the same age and without specific tutoring, these core domains of knowledge are thought to be human universals.

One implication of this developmental and evolutionary perspective on knowledge acquisition, and an idea embraced by core-knowledge theorists, is that certain areas of specialized cognitive competency are domain specific. A domain is a body of knowledge about a given topic that includes information about what entities are included in the domain as well as rules that describe how the entities in the domain behave. Each domain of core knowledge covers a different area of expertise. For example, an intuitive understanding about biology includes information about what entities are to be considered (plants and animals) and the processes that apply to entities in that domain (e.g., growth, death, inheritance). The domain also has limits: knowledge of what entities are excluded such that rules that apply within the domain are not used to make inferences outside of the domain (Spelke & Kinzler, 2007, pp. 89–96). These specific information-processing strategies have been shaped by natural selection and focus on areas of knowledge that were fitness-relevant in the environment in which our ancestors evolved.

The specific mechanisms that are employed in service of development may differ from domain to domain. For example, one area in which there is clearly a rich endowment of specialized learning mechanisms is language acquisition. The learning mechanisms employed in language acquisition (e.g., the auditory learning of consonants [Jusczyk, 1995, pp. 263–301], rules of word learning [Markman, 1989], and attention to infant-directed speech [Cooper & Aslin, 1990, pp. 1584–1595]) are unique to that domain. The psychological processes that allow one to acquire one's native language are different from the psychological processes that allow one to acquire, for example, knowledge about faces. Furthermore, each of these specialized kinds of learning is different from more domain-general learning processes, such as classical and operant conditioning. Indeed, Chomsky's (1965) proposal that children have a specialized "language acquisition device" that allows language learning supported his demonstration that conditioning and imitation were unable to account for language learning.

In addition to language, there are other areas of learning that are thought to be domains of core knowledge in developing humans: intuitive physics (Baillargeon, 1998, pp. 503–529; Kellman & Spelke, 1983, pp. 483–524; Spelke, 1985, pp. 89–114); an intuitive understanding of biology (Springer & Keil, 1989, pp. 637–648; Wellman & Gelman, 1998); an intuitive understanding of number, addition, and subtraction (Wynn, 1995, pp. 172–177); and perhaps several specific types of core knowledge in the social domain. The focus of this chapter is the functionally specialized social perception of animacy as an area of core knowledge.

The Importance of Social Perception

One area in which adults and developing children have a tremendous amount of intuitive knowledge is social understanding. Humans have an evolutionary history of group living and intense social interaction; as a result, we have psychological processes designed to help us understand and navigate the social world. Indeed, humans are an obligate social species: in the environment in which our ancestors lived, or the environment of evolutionary adaptedness (EEA), people had to live with others in order to avoid the risk of malnutrition, predation, or even death; even in modern environments, people generally find extended periods of isolation extremely aversive. In the environment in which our ancestors evolved, living in a community was not just fun and interesting (though it was that); it was a matter of life or death. Our social cognitive adaptations reflect this.

Lone individuals eventually come to crave social contact. The need for other people is so compelling that a person may hallucinate social contact if real companions are not available for extended periods (Adams, 1997). Furthermore, Harlow's work with rhesus monkeys showed that in that species total social isolation can, over an extended period of time, result in severe maldevelopment and even death, even when physical and nutritional requirements are met (Harlow, Dodsworth, & Harlow, 1965).

Humans have specialized psychological processes designed to help them identify, remember, and interact with other people. In addition, the human mind has coevolved adaptations that are appropriate for solving adaptive problems on multiple sides of a relationship. For example, we read facial expressions and produce facial expressions; we seek maternal care as infants and provide parental security as adults.

Given that living with others was essentially mandatory for our ancestors, social skills were crucial for successful negotiating, getting one's needs met, and avoiding exploitation. Any adult living in a community has to solicit friendships, monitor allegiances, be aware of insults, avoid insulting others, and monitor and fulfill obligations. Infants and young children also need to manage their social world, since humans, born immature compared to most other species, are completely dependent on care from others. There is evidence of developmentally early social competence: Newborns show a preference for social stimuli within hours of birth and exchange smiles with their caregiver at six weeks.

The need for specialized social cognitive processing is so great in humans that some researchers credit this evolutionary drive for humankind's cognitive sophistication. The "social brain hypothesis" is the hypothesis that the comparatively large size of the human brain, as well as the exceptional innovation and general intelligence of humans, is an evolutionary result of the social challenges inherent to living in large groups (Dunbar, 1998a, pp. 178–190; Jolly, 1966, pp. 501–506). The fact that human intelligence is demonstrably greater than that of other animals, including other mammals and primates, is a result of the complex challenges of social life (Jolly, 1966, pp. 501–506).

Compared to other social animals, humans live in large groups. Our human ancestors evolved in groups of about 150 (Dunbar, 1998b; Dunbar & Spoors, 1995, pp. 273–290). Baboons groups, by way of comparison, range from a few individuals (i.e., five to ten) to as many as 100. Even the great apes rarely live in groups larger than 50. Living with 150 or so other people is cognitively taxing when it comes to individuating, remembering, having mental impressions of, tracking favors and insults from, and following what is important to each individual. Thus, humans have many complex and specialized cognitive processes that support social cognition.

Note that the claim that there are specialized psychological processes for animacy perception is not a commitment to any particular developmental course. In particular, it is not a claim that these processes mature without environmental input that is specific to the domain in question. Face perception development may rely, in part, on visual access to faces, and the development of animacy perception may rely, in part, on access to visual cues of animacy, for example. There is no reason to expect that natural selection selects for a developmental course that does not take advantage of environmental cues that are reliably present in a species-typical environment. Indeed, one should expect that statistically reliable information can be taken advantage of by natural selection.

Seeing Animacy

Fundamental to any social information processing is the ability to identify those entities that are animate. Past research on the detection of animate entities can be broadly divided into two categories of focus: (1) the features of the object in question, such as the presence of a face, eyes, or particular features of the limbs (Guajardo & Woodward, 2004, pp. 361–384); and (2) motion cues (Heider & Simmel, 1944, pp. 243–259; Scholl & Tremoulet, 2000, pp. 299–309). Adults easily and spontaneously perceive animacy based on motion cues. An early and compelling demonstration of the effects of motion cues was a dynamic scenario showing simple geometric figures moving about in a way that created the perception of goals, emotions, intentions, and even personality traits (Heider & Simmel, 1944, pp. 243–259). Michotte also provided early and simple demonstrations (Michotte, 1950, pp. 114–125) and suggested that simple motion cues provide the foundation of animacy perception (Michotte, 1963). Since then, many others have likewise shown that people have complex social attributions in response to the movement of simple geometric figures (Hashimoto, 1966, pp. 1–26; Morris & Peng, 1994, pp. 949–971; Rime, 1985, pp. 241–260).

Early in infancy, the human mind can make social inferences from motion cues (Gergely et al., 1995, pp. 165–193; Hamlin, Wynn, & Bloom, 2007, pp. 557–560; Luo & Baillargeon, 2005, pp. 601–608; Rochat, Morgan, & Carpenter, 1997, pp. 441–465; Southgate & Csibra, 2009, pp. 1794–1798). Converging lines of evidence strongly suggest that the motion of simple geometric shapes is perceived as animate by the human visual system and can elicit percepts as complex as intentionality (Gergely, Nadasdy, Csibra, & Bíró, 1995, pp. 165–193), chasing (Rochat, Morgan, & Carpenter, 1997, pp. 441–465), and helping and hindering (Kuhlmeier, Wynn, & Bloom, 2003, pp. 402–408). Furthermore, there is evidence that these percepts are found cross-culturally (Barrett et al., 2005, pp. 313–331).

Although much of the past research in this area has relied on fairly complex displays, including apparently dramatic narratives that are acted out by simple geometric figures, the motion cues necessary to create the perception of animacy are surprisingly simple. To date, only a few studies have focused on these very simple motion displays and attempted to identify the most basic and rudimentary motion cues (e.g., speed and acceleration) that are perceived by the human visual system as animate. The use of simple motion cues has clear advantages: It allows researchers to manipulate exact motion parameters, both individual and relational, and examine the effects of these manipulations on the perception of animacy. Recent evidence suggests that the adult visual system is designed to perceive and extract social information based on quite simple motion cues (Scholl & Tremoulet, 2000, pp. 299–309); for example, the mere acceleration of an object is enough to elicit a report of the perception of animacy (Tremoulet

& Feldman, 2000, pp. 943–951). Indeed, it has been claimed that the relevant motion cue with respect to animacy perception is apparent self-propulsion (Dasser, Ulbaek, & Premack, 1989, pp. 365–367) that can manifest as acceleration or, in some cases, deceleration (Dittrich & Lea, 1994, pp. 253–268).

Specialized Perception

At this point, I will review some recent work that provides empirical support for the idea that there are, indeed, specialized psychological processes designed to perceive animate motion. I will show that there is an association between perceived speed and perceived animacy that persists even when the perceived speed difference is illusory. I will show that this association can be dissociated in ways that can be explained in the context of a visual system that evolved in a terrestrial environment. Finally, I will describe some deficits and anomalies in the perception of animacy in groups of people who have autism spectrum disorders (ASD)—a condition in which nonsocial cognitive function can be selectively preserved.

The Association and Dissociation of Speed and Animacy

We know from previous work that simple motions are cues that the human visual system uses to detect animacy. Recent empirical work has taken steps toward identifying the most rudimentary motion cues that are used by the human visual system to perceive animacy. To this end, Tremoulet and Feldman examined the effects of acceleration and changes in direction on the perception of animacy. They found that greater acceleration and sharper changes in direction increased the likelihood that a small moving dot would be perceived as animate (Tremoulet & Feldman, 2000, pp. 943–951).

Paul Szego and I (Szego & Rutherford, 2007, 2008) followed up on this work. We wondered whether speed alone, without acceleration, would be a cue to animacy. We discovered (1) an association between perceived speed (with no acceleration) and perceived animacy, (2) that this applied even in cases where the perception of speed differences was illusory, and (3) that there is a functional dissociation between perceived speed and perceived animacy in the context of a gravitational field.

In one early experiment, designed to test whether differences in constant speed could be used as an animacy cue, observers watched two objects travel across two circles of the same size at different speeds. We oriented the computer screen horizontally so as to minimize the perception of

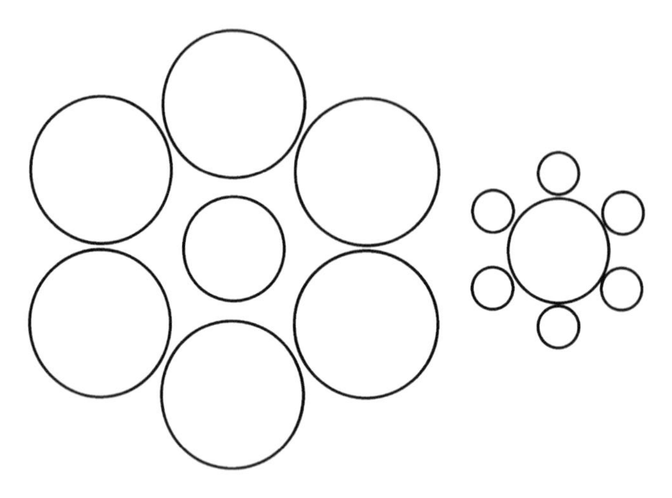

Figure 6.1
The Ebbinghaus illusion.

motion in a gravitational field. We used a two-interval forced-choice design, in which observers had to respond to the question, "Which dot looks alive?" More participants responded that the faster dot looked alive. Indeed, the faster the dot, the greater the proportion of observers agreeing that the dot was alive; this suggests that there is, indeed, an association between perceived speed and perceived animacy in the absence of any acceleration.

Next, we created a display in which there was an apparent speed difference between two dots, in the absence of any actual speed difference, in order to test the association between apparent speed differences and the perception of animacy. We used the Ebbinghaus illusion (figure 6.1) in order to create the illusion that dots moving across the center circles were traveling different distances. In the Ebbinghaus illusion, the two center circles are equal in size, but are immediately surrounded by larger or smaller circles making them appear smaller and larger, respectively. The dot traversing the apparently larger circle will thus appear to have traveled a greater distance in the same amount of time, creating the illusion of greater speed.

In a two-interval forced-choice task, observers watched a display, such as the one shown in figure 6.1, as a dot moved first across one of the two inner circles and then across the other. Observers were told to think of the two sets of circles as different types of flowers. In the animacy condition, they were told that one of the two dots was a bug and one was a piece of dirt; they were to report which of the two was the bug. In the speed condition, observers were asked, "Which dot is faster?" In this experiment, there were no actual speed differences (Szego & Rutherford, 2007, pp. 1–7).

We found that in the speed condition, significantly more people saw the dot moving across the apparently larger circle as faster (68% vs. 32%), and in the animacy condition, significantly more people saw the dot moving across the apparently larger circle as alive (64% vs. 36%). Remember that in all cases, people were comparing two dots that did not differ in speed, size, or duration of display. We concluded that the differences in the perception of animacy were driven by the differences in the perception of speed. People saw the dot moving across the apparently larger area as alive because they saw it as faster.

Note that this use of an illusory speed difference has some clear advantages over some other methods. Specifically, if we had asked which of two dots was alive when one of the two dots actually was moving faster (or had accelerated, or had turned a sharper corner), it would be fairly easy for participants to infer our intent. The demand characteristics of such an

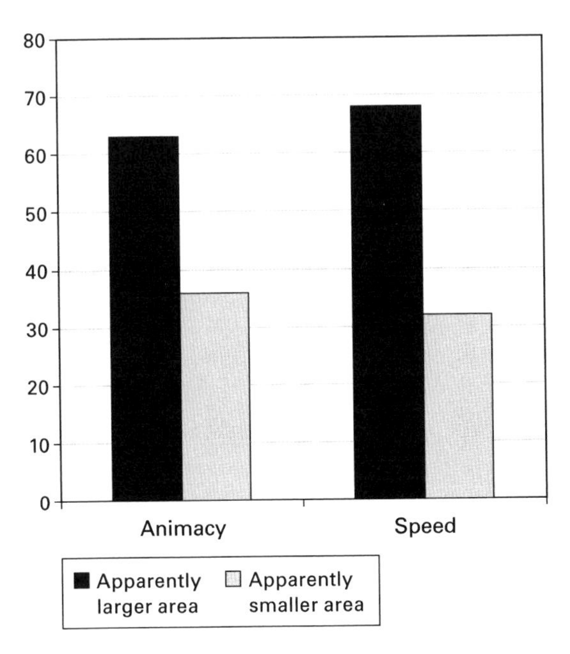

Figure 6.2
The percent of participants who identified the dot moving across the apparently larger, or apparently smaller, circle, when asked which dot was alive or faster.

explicitly discernible difference in stimuli would lead the compliant participant to reasonably respond that the visibly faster dot (or the dot that turned at the sharper angle) was the one that was supposed to be alive. Here, since there was, in fact, no difference in speed between the two dots displayed, such demand characteristics cannot explain our results.

Having shown that there is an association between the perception of speed and the perception of animacy even in cases where speed differences are illusory, we wanted to show an adaptive dissociation between speed and animacy. Evolutionary solutions to adaptive problems can be expected to take into account statistical regularities found in the environment of evolutionary adaptedness. We reasoned that since the human visual system evolved in the terrestrial gravitational field, it should be designed to take into account predictable regularities that result from the fact that objects are affected by the gravitational field. By creating the perception of a gravitational field, we created a context in which objects with the fastest perceived speed and objects that were more likely to be perceived as animate were not the same (Szego & Rutherford, 2008, pp. 299–374).

Since human visual strategies for perceiving animacy evolved in a terrestrial environment, the perception of speed and the perception of animacy may dissociate in the sense that animacy is attributed to fast objects if and when the objects' speed is attributed to self-propulsion, but not when the speed can be attributed to some external force or to gravity (e.g., speedily falling objects may not be seen as animate). Conversely, objects that rise in a gravitational field, whether they do so speedily or not, may be seen as animate since their movement in opposition to gravity needs some explanation.

Again, we employed a task that involved the discrimination of objects that were perceived as alive from those that were not. This experiment again employed a two-interval forced-choice task, and it included a within-subject comparison between animacy perception and speed perception. In each trial, the participant saw a pair of dots (in sequence). In the animacy trials, participants were asked which of the two was alive; in the speed trials, they were asked which of the two was faster. In all cases, the two dots that they compared were traveling at the same speed, but one was traveling up a vertically oriented computer screen, and the other was traveling down.

As predicted, we found a dissociation between perceived speed and perceived animacy. A majority of participants reported that the dots moving up the screen were animate compared to dots moving down the screen. However, a majority of participants saw the dots moving down the

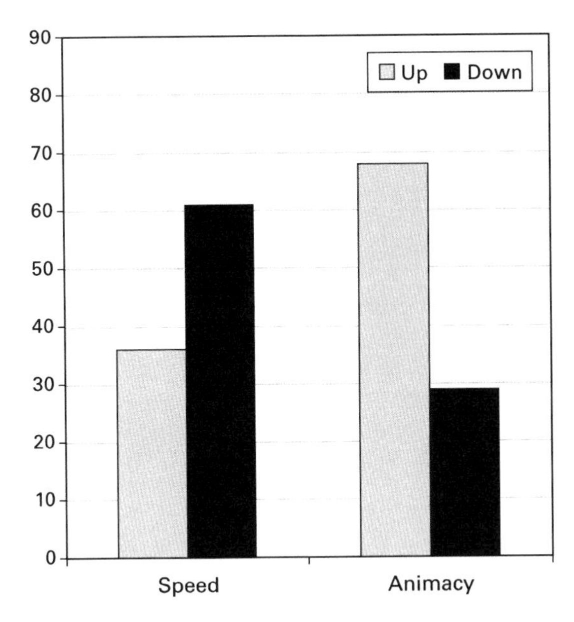

Figure 6.3
Percent of participants who identified the dot moving up, or down, the vertical computer screen, when asked which dot was alive or faster.

computer screen as faster compared to dots moving up the computer screen. There was a significant interaction between condition (speed or animacy) and direction (rising or falling) such that dots that were traveling upward were more likely to be seen as both alive and slower.

In the terrestrial environment, objects moving upward must have a power source; when there is a lack of any external propulsion, such objects are perceived as alive. Objects falling to the ground need no internal power source to explain their motion; such objects are seen as inert regardless of perceived speed (Szego & Rutherford, 2008, pp. 299–374).

Why Consider Autism?

There is evidence that (1) specific motion cues are perceived by adults as animate, (2) these percepts are generated quickly and automatically without any deliberation, (3) infants from early on attribute animacy to objects that are seen in motion, and (4) these phenomena are cross-cultural. Underlying the discussion of these observations is the question of whether the perception of animacy and agency is mediated by specialized psychological processes. One body of evidence that speaks to this question derives from studies looking at perceptions of animacy in children and adults with autism. If individuals with autism spectrum disorders do not experience the same perception of animacy when viewing displays that reliably evoke such a percept in typical people, but at the same time perform normally on visual tasks matched for complexity and executive demands, then this can be taken as evidence that a specialized psychological process is absent or maldeveloped in those with ASD. Similarly, evidence that viewing an animate motion display results in typical viewers experiencing the perception of animacy, while that perception is characteristically different in a group with ASD (while comparable nonsocial processing is typical), could be taken as evidence that the psychological processes underlying animacy perception in typical people is specialized; the percept is not a result of the same general psychological processes that underlie nonsocial processing.

There is evidence that those with ASD perform atypically on tasks that measure the perception of social motion. For example, children with ASD have deficits with respect to the perception of emotions, but not the perception of actions portrayed by point-light walkers (Moore, Hobson, & Lee, 1997, pp. 401–423). Others have also found that children with ASD show a deficit relative to matched controls, though they still perform above chance on tasks measuring the perception of biological motion itself (Blake et al., 2003, pp. 151–157). And when asked to describe Heider & Simmel's (1944) classic scene starring a moving triangle and a circle, those with ASD are less likely than controls to describe the scene in social terms (Abell, Happe, & Frith, 2000, p. 1–20; Klin, 2000, p. 831–846).

The Spontaneous and Nonspontaneous Use of Specialized Social Perception

If neurotypical children use specialized psychological processes to perceive animate motion, and such perceptual processes are engaged automatically, then one might expect to see group differences with respect to the automaticity and spontaneity of the perception of animate motion in a group of children with autism. Bruce Pennington, Sally Rogers, and I found evidence that although children with ASD could be trained to discriminate animate from inanimate motion, they apparently did not do so spontaneously (Rutherford, Pennington, & Rogers, 2006, pp. 893–992).

We tested twenty-three children with autism (average age five years, eight months), eighteen mental-age-matched children with other developmental disabilities (average age five years, seven months), and eighteen mental-age-matched typically developing children (average age three years, nine months) on an animacy-perception task and a control task in which they had to judge the relative weight of two balls.

In the animacy-perception task, children watched two black circles moving about on a computer touch screen. One circle was moving as if self-propelled; the other only moved in response to gravity or being touched. When the circles stopped moving at the end of the scenario, the child was asked to pick one by touching it on the computer screen. The child was rewarded for touching the one that was animate: the circle turned red and the child got a treat.

Similarly, in the control condition, children watched two black circles moving about in such a way that it was apparent which was heavier. For instance, in one scenario, the circles were on a teeter-totter; one side was shown lowering while the other was rising. Again, when the circles came to rest at the end of the display, the child was asked to pick one and was rewarded for touching the one that was heavier.

There were two sets of scenarios: eight training and twelve test scenarios. During the training phase, the child was rewarded for touching the one that was animate or heavier, depending on the condition. Children were presented with such scenarios until they got six in a row correct.

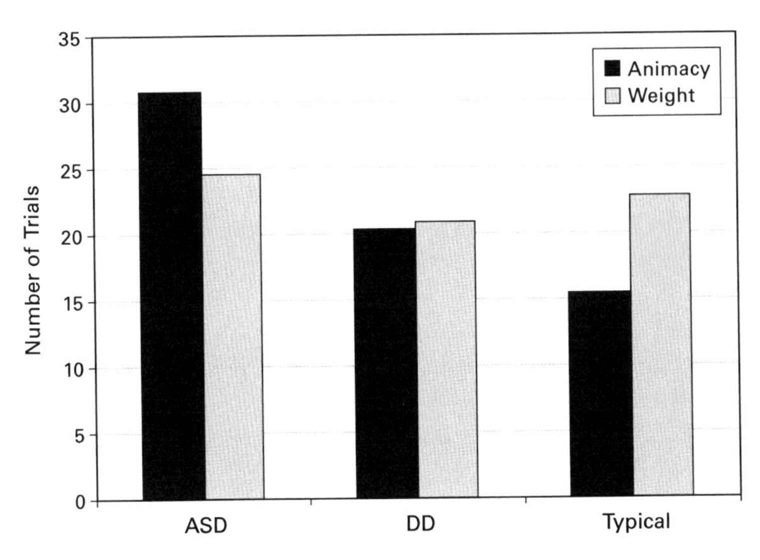

Figure 6.4
The number of trials it took for children with autism, children with other developmental disorders, and matched control children to reach criteria on the animacy detection task and the control task.

In the training session, the dependent measure was the number of trials it took a child to reach criterion, which was six trials in a row correct. Children with ASD initially showed a deficit in categorizing animate and inanimate objects based on motion cues, taking more trials than control groups to reach criterion.

Following the training set of trials, children proceeded to the test trials. The twelve animacy test scenarios were six paired temporal mirrors of each other. That is to say, if played backward, the opposite circle would become the animate circle. This was meant to control for a number of low-level factors such as the amount of time in motion and position on the screen. In the control test trials, the child was again shown a display in which the circles could be distinguished based on weight, and the task was to touch the heavier of the two. There were twelve control test scenarios.

In the test trials, there were no group differences. The children with ASD were as accurate as the control groups at selecting which of two moving circles was animate.

These results are consistent with the idea that social perceptual processes are preserved in autism, since the group results in the test phase are not significantly different, but that these processes are not engaged spontaneously. Likewise, these results are also consistent with the idea that those

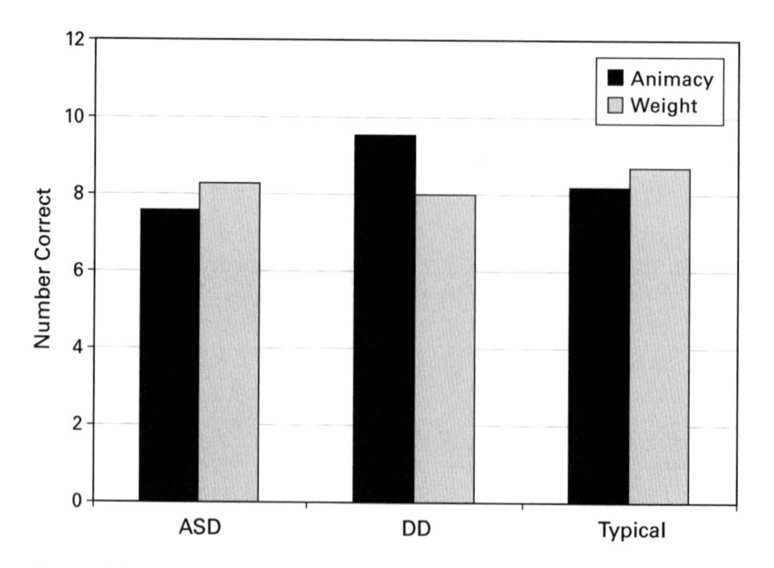

Figure 6.5
The number of correct trials children in each group got on the animacy detection and control test trials.

with ASD are able to learn an effective strategy for solving the animacy detection problem. In either case, there are significant group differences during the training phase of the animacy perception task, which suggests that the spontaneous processing of animate motion is different in the two groups. This evidence supports the idea that animacy perception in the typical group relies on dedicated perceptual processes specialized for this purpose.

The Use of Deliberate Heuristics to Perceive Biological Motion

In another recent study, Niko Troje and I tested for evidence of specialized social perceptual processing using a comparison between a group of adults with autism spectrum disorders and a matched control group (Rutherford & Troje, 2011). There is some reason to believe that the social perceptual processes used during the perception of point-light walker displays are specialized. First, accurate perception of point-light walkers is subject to an inversion effect (Troje & Westhoff, 2006, pp. 821–824). Second, there is evidence that dedicated brain areas are used during the perception of point-light walker displays (Grossman et al., 2000, pp. 711–720; Wheaton et al., 2001, pp. 401–406). A comparison of typical performance on a point-light walker task to that of observers with ASD might reveal evidence of specialized social perception among typical participants.

Our participants completed two types of tasks. In the first task, they had to choose which of two displays showed a coherent point-light walker in a mask of dots. In the second task, they had to indicate the direction of the masked walker when, in half of the displays, the walker was actually scrambled. In the first task, which was a detection task, correct completion of the task required the observer to perceptually organize the moving dots into a coherent walking figure; in the second task, which was a direction task, the observer would have to infer directional information from local dots rather than from global information—especially in the scrambled trials. Walkers were depicted as humans, cats, and pigeons.

We found three lines of evidence supporting the idea that the perception of biological motion is mediated by specialized psychological processes. These three lines of evidence (group differences in the relationship between IQ and biological motion perception, group differences in the inversion effect, and group differences in the effect of species) are described in the following paragraphs. We interpret our findings as being consistent with the idea that those with autism are not using these specialized processes, but instead may be using more general and perhaps more deliberate cognitive processing to solve these biological motion tasks.

The relationship between IQ and social perception in ASD. In our exploration of the perception of biological motion in those with ASD, we found that in the direction task, there was a statistically significant positive correlation between IQ scores and performance for the ASD group alone. Recall that our direction task requires observers to perceive direction based, at least for some trials, on information that is only available at the local, not the global, level. Even performance on just the scrambled trials showed a positive correlation between performance and IQ. For the ASD group alone, performance on the direction task was predicted by verbal IQ ($r = 0.57$, $p < 0.05$) and full scale IQ ($r = 0.70$, $p < 0.01$). There was no significant correlation for the control group.

We take this group contrast as evidence that the control group was using specialized psychological processes to solve the social perceptual problem, whereas the ASD group was not relying entirely on these specialized processes. It is possible that the ASD group makes use, in part, of specialized social perceptual processes and, in part, on more general cognitive pro-

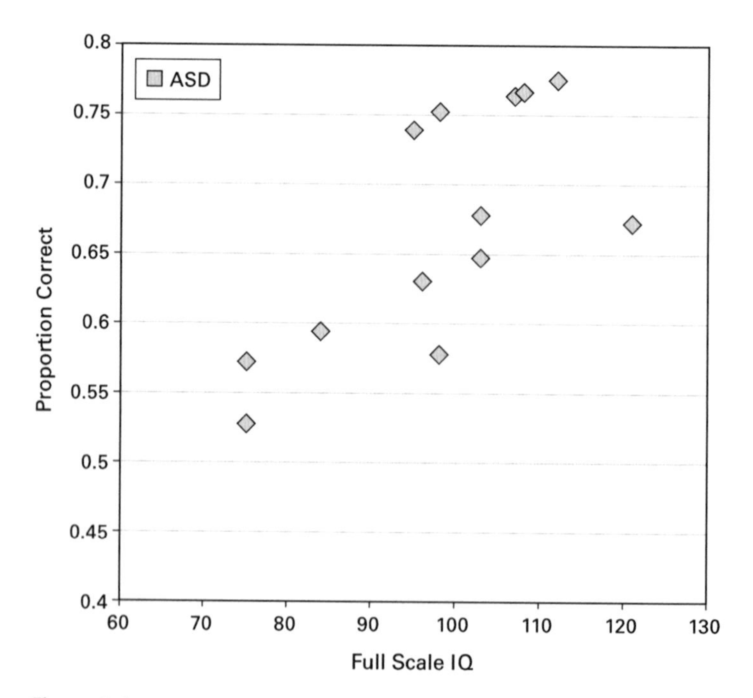

Figure 6.6
Proportion of trials each participant with ASD got correct, and their IQ, illustrating the correlation between these two measures.

cesses. It is also possible that the ASD group relies entirely on general cognitive processes to solve this task; either is consistent with these data. These findings are consistent with the idea that those in the ASD group with higher IQs were able to create deliberate compensatory strategies that were used to solve the biological motion perception task. For those in the control group, who by hypothesis were using specialized social perceptual processes to solve the problem, general intelligence was not related to performance.

That said, there is another possible reason for this difference between the groups. It is possible that a correlation between IQ and the perception of direction in biological motion was measurable for the ASD group alone because the control group uniformly understood the directions and thus performed well on the task, whereas there was IQ-related variance in task understanding in the ASD group. The following contrast favors the former, rather than the latter, possibility.

Group differences in the cost of inversion. Some visual tasks are performed much more easily when the stimuli are presented upright than when they are presented upside down, a phenomenon known as an inversion effect. In general, inversion effects have been taken as evidence of specialized psychological processing. Reliable inversion effects have been found in biological motion perception; this is true both with respect to the extraction of global coherence necessary in our detection task and in the perception of direction based on local information needed in our direction task (Troje & Westhoff, 2006, pp. 821–824). And there is ample evidence of inversion effects in face processing (Valentine, 1988, pp. 471–491; Yin, 1969, pp. 141–145), which is also taken as evidence of specialized processing in that domain.

In our work on the perception of biological motion across groups with and without ASD, we found an interesting interaction between IQ and the inversion effect. The inversion effects were estimated by subtracting the average proportion correct on the inverted trials from the average proportion correct on the upright trials. When considering the scrambled trials in the direction tasks—the trials in which task-relevant information was only available in local elements—we found a significant negative correlation between verbal IQ and the inversion effect in the ASD group ($r = -0.58$, $p = 0.37$) but not in the control group ($r = 0.46$, *n.s.*).

In other words, there was a higher cost of inversion for lower IQ people with ASD than there was for higher IQ people with ASD. This is consistent with the idea that those with ASD solve the perceptual problem using relatively general cognitive processes, and those in the control group solve the problem using specialized social perceptual strategies that are unrelated to general intelligence. Those in the control group were uniformly affected by inversion, which is a sign that they were using specialized cognitive processes. In the ASD group, individuals with higher verbal IQs were less affected by inversion; this perhaps indicates that the strategy they were employing to solve the problem was more general. The difference in this interaction between groups may reflect the difference between a psychological process designed to solve a specific problem in our evolutionary past and intelligence more generally.

Group differences in the effect of species. A third interesting contrast between the ASD group and the control group was the difference in the effect of the animal modeled by the point-light walker. In the direction task, there was a significant group-by-species interaction, meaning that which species was represented by the point-light walker affected performance in the typical group, but not in the ASD group. Those in the control group performed slightly better at trials showing humans than they did at those showing cats, and they did better with trials showing cats than they did with those showing pigeons. Participants in the ASD group did not show this advantage for human walkers.

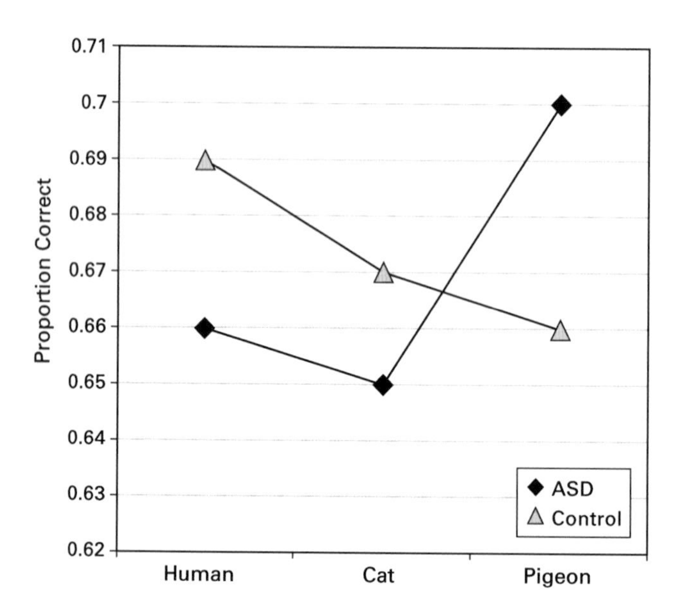

Figure 6.7
Proportion of trials participants in each group got correct for trials displaying humans, cats, and pigeons.

As a group, then, those with ASD were able to apply whatever perceptual strategy they were using to various animal species without cost of moving further, taxonomically, from the human species. The control group did pay a cost when they viewed nonhuman and nonmammalian images; this result is consistent with the view that the control observers alone were using social perceptual mechanisms specialized for viewing human action.

Conclusions and Discussion

The experiments described in the previous section, considered as a group, collectively provide evidence for the idea that the human visual system has specialized perceptual machinery dedicated to perceiving animacy based on motion cues. These studies, consistent with previous work, show an association between perceived speed and perceived animacy. Unlike previous work, we show that by manipulating the context, we can create a dissociation between perceived speed and perceived animacy. We showed that although children with ASD can discriminate animate from inanimate motion after training, they do not do so spontaneously—unlike the control group. And we show evidence suggesting that adults with ASD may use deliberate rather than intuitive cognitive strategies to perform tasks involving biological motion perception. The existence of specialized social cognitive psychology can account for these findings.

Importantly, there is no apparently straightforward way for general learning processes or associations to account for these results. The relationship between IQ and social perceptual performance in a biological motion task for a group of people with ASD suggests that general cognitive processes are involved in successful task completion for that group; a lack of such an association for the control group strongly suggests that the task is completed by a specialized perceptual process not measured by IQ testing. Similarly, results from studying children with ASD seem to suggest that they are learning some task-relevant skill in the laboratory that allows them to succeed in the animacy perception task, but the control groups are not: The children in the control group, presumably relying on animacy-perception psychology, successfully complete the task without training. Finally, results showing an association between illusory speed and perceived animacy, and those showing that the direction of gravity predicts animacy perception, are hard to explain by appealing to general cognitive or perceptual processing. This result points to animacy perception as a result of some low-level processes of the visual system.

Earlier I commented that a claim that specialized psychological processes underlie performance on a social perceptual task is not a commitment to any particular view of its development. One might think that this chapter is advocating a nativist stance, or claiming that animacy perception is "innate," but that extension of the argument is not warranted. Let's consider this claim in the context of the finding that children with ASD are less likely to spontaneously discriminate animate from inanimate objects. Is it the case that children with ASD have disabled social perceptual processes that are unable to develop typically even with adequate experience? Or is it the case that children developing with ASD are less likely to see animate motion and therefore are unable to develop animacy perception as typical children do? Either is possible.

To elaborate on the latter, there is evidence that children with autism show a lack of "social orienting" early in development (Osterling & Dawson, 1994, pp. 247–257). As young children, those developing with autism are less likely to look at faces and orient toward social information than those who are developing typically. Proponents of the social-orienting view of autistic development believe that the resulting lack of social input contributes to the relative maldevelopment of social cognitive processes (Mundy & Neal, 2001, pp. 139–168). Specialized social perceptual processes may or may not be developmentally dependent on exposure to a specific type of visual stimuli during development.

It might be pointed out that there has been some discussion about what criteria a researcher might use to identify a psychological process that is specialized—and that is specialized by virtue of its evolutionary history. My assertion that there may be a functionally specialized, social perceptual psychology is related to the discussion on modularity in the Fodorian sense. Fodor originally proposed "modules" as mechanisms that process specific kinds of information in highly constrained ways; he intended for the idea to apply to peripheral systems (Fodor, 1983). Subsequently, others have proposed that in fact the mind might be replete with modularity and include, for example, modules designed for language processing (Pinker, 1994) and social exchange (Cosmides & Tooby, 1992, pp. 163–228). Thus, the ideas proposed in this chapter are more in agreement with these latter descriptions of domain-specific modularity than with modularity in the strictly Fodorian sense.

Social cognitive processes were shaped by strong adaptive pressures in our evolutionary past resulting from our living in large, complex social groups. Animacy perception is fundamental to social cognition; it develops early and is seen cross-culturally. Evidence presented in this chapter shows that animacy perception is designed to operate in a gravitational context in ways that cannot be explained by more general-purpose perceptual processes. Individuals with ASD, perhaps recruiting more general perceptual and cognitive processes in the perception of animacy, may perceive animacy differently from the way that neurotypical people do. Animacy perception seems to be accomplished using perceptual processes designed for this purpose.

References

Abell, F., Happe, F., & Frith, U. (2000). Do triangles play tricks? Attribution of mental states to animated shapes in normal and abnormal development. *Journal of Cognition and Development, 15*, 1–20.

Adams, D. (1997). *Chasing liquid mountains*. Sydney: Macmillan.

Baillargeon, R. (1998). Infants' understanding of the physical world. In M. Sabourin, F. Craik, & M. Robert (Eds.), *Advances in psychological science* (Vol. 2). London: Psychology Press.

Barrett, H. C., Todd, P. M., Miller, G. F., & Blythe, P. W. (2005). Accurate judgments of intention from motion cues alone: A cross-cultural study. *Evolution and Human Behavior, 26*, 313–331.

Blake, R., Turner, L. M., Smoski, M. J., Pozdol, S. L., & Stone, W. L. (2003). Visual recognition of biological motion is impaired in children with autism. *Psychological Science, 14*(2), 151–157.

Carey, S. (1995). On the origin of causal understanding. In D. Sperber, D. Premack, & A. J. Premack (Eds.), *Causal cognition: A multi-disciplinary Debate*. New York: Oxford University Press.

Carey, S., & Spelke, E. (1994). Domain-specific knowledge and conceptual change. In L. A. Hirschfeld & S. A. Gelman (Eds.), *Mapping the mind: Domain specificity in cognition and culture* (pp. 169–200). Cambridge: Cambridge University Press.

Chomsky, N. (1965). *Aspects of the theory of syntax*. Cambridge, MA: MIT Press.

Cooper, R. P., & Aslin, R. N. (1990). Preference for infant-directed speech in the first month after birth. *Child Development, 61*, 1584–1595.

Cosmides, L., & Tooby, J. (1992). Cognitive adaptations for social exchange. In J. H. Barkow, L. Cosmides, & J. Tooby (Eds.), *The adapted mind: Evolutionary psychology and the generation of culture*. New York: Oxford University Press.

Dasser, V., Ulbaek, I., & Premack, D. (1989). The perception of intention. *Science, 243*(4889), 365–367.

Dittrich, W. H., & Lea, S. E. G. (1994). Visual perception of intentional motion. *Perception, 23*, 253–268.

Dunbar, R. I. M. (1998a). *Grooming, gossip, and the evolution of language*. Cambridge, MA: Harvard University Press.

Dunbar, R. I. M. (1998b). The social brain hypothesis. *Evolutionary Anthropology, 6*(5), 178–190.

Dunbar, R. I. M., & Spoors, M. (1995). Social networks, support cliques, and kinship. *Human Nature, 6*, 273–290.

Fodor, J. A. (1983). *The modularity of mind*. Cambridge, MA: MIT Press.

Gergely, G., Nadasdy, Z., Csibra, G., & Bíró, S. (1995). Taking the intentional stance at 12 months of age. *Cognition, 56*(2), 165–193.

Grossman, E., Donnelly, M., Price, R., Pickens, D., Morgan, V., Neighbor, G., et al. (2000). Brain areas involved in perception of biological motion. *Journal of Cognitive Neuroscience, 12*(5), 711–720.

Guajardo, J. J., & Woodward, A. L. (2004). Is agency skin deep? Surface attributes influence infants' sensitivity to goal-directed action. *Infancy, 6*, 361–384.

Hamlin, J. K., Wynn, K., & Bloom, P. (2007). Social evaluation by preverbal infants. *Nature, 450*, 557–560.

Harlow, H. F., Dodsworth, R. O., & Harlow, M. K. (1965). Total social isolation in monkeys. *Proceedings of the National Academy of Science of the United States of America, 54*(1), 90–97.

Hashimoto, H. (1966). A phenomenal analysis of social perception. *Journal of Child Development, 2*, 1–26.

Heider, F., & Simmel, M. (1944). An experimental study of apparent behavior. *American Journal of Psychology, 57*, 243–259.

Jolly, A. (1966). Lemur social behavior and primate intelligence. *Science, 153*(3735), 501–506.

Jusczyk, P. (1995). Language acquisition: Speech sounds and phonological development. In J. L. Miller & P. D. Eimas (Eds.), *Handbook of perception and cognition* (Vol. 2): *Speech, language, and communication*. Orlando, FL: Academic Press.

Kellman, P. J., & Spelke, E. S. (1983). Perception of partly occluded objects in infancy. *Cognitive Development, 15*, 483–524.

Klin, A. (2000). Attributing social meaning to ambiguous visual stimuli in higher-functioning autism and Asperger syndrome: The Social Attribution Task. *Journal of Child Psychology and Psychiatry, and Allied Disciplines, 41*(7), 831–846.

Kuhlmeier, V. A., Wynn, K., & Bloom, P. (2003). Attribution of dispositional states by 12-month-olds. *Psychological Science, 14*, 402–408.

Luo, Y., & Baillargeon, R. (2005). Can a self-propelled box have a goal? Psychological reasoning in 5-month-old infants. *Psychological Science, 16*, 601–608.

Markman, E. (1989). *Categories and naming in children: Problems of induction*. Cambridge, MA: MIT Press.

Michotte, A. (1950). The emotions regarded as functional connections. In M. Reymert (Ed.), *Feelings and emotions: The Mooseheart symposium*. New York: McGraw-Hill.

Michotte, A. (1963). *The perception of causality*. Oxford: Basic Books.

Moore, D. G., Hobson, R. P., & Lee, P. W. (1997). Components of person perception: An investigation with autistic, non-autistic retarded, and typically developing children and adolescents. *British Journal of Developmental Psychology, 15*(4), 401–423.

Morris, M. W., & Peng, K. (1994). Culture and cause: American and Chinese attributions for social and physical events. *Journal of Personality and Social Psychology, 67*(6), 949–971.

Mundy, P., & Neal, R. A. (2001). Neural plasticity, joint attention, and a transactional social-orienting model of autism. In L. M. Glidden (Ed.), *International review of research in mental retardation: Autism*. San Diego, CA, US: Academic Press.

Osterling, J., & Dawson, G. (1994). Early recognition of children with autism: A study of first birthday home videotapes. *Journal of Autism and Developmental Disorders, 24*(3), 247–257.

Pinker, S. (1994). *The language instinct*. New York: William Morrow.

Rime, B. (1985). The perception of interpersonal emotions originated by patterns of movement. *Motivation and Emotion, 9*(3), 241–260.

Rochat, P., Morgan, R., & Carpenter, M. (1997). Young infants' sensitivity to movement information specifying social causality. *Cognitive Development, 12*(4), 441–465.

Rutherford, M. D., Pennington, B. F., & Rogers, S. J. (2006). The perception of animacy in young children with autism. *Journal of Autism and Developmental Disorders, 36*, 893–992.

Rutherford, M. D., & Troje, N. F. (2011). IQ predicts biological motion perception in autism spectrum disorders. *Journal of Autism and Developmental Disorders, 42*(4), 557–565.

Scholl, B. J., & Tremoulet, P. (2000). Perceptual causality and animacy. *Trends in Cognitive Sciences, 4*(8), 299–309.

Siegler, R. S., & Crowley, K. (1994). Constraints on learning in nonprivileged domains. *Cognitive Psychology, 27,* 194–226.

Southgate, V., & Csibra, G. (2009). Inferring the outcome of an ongoing novel action at 13 months. *Developmental Psychology, 45,* 1794–1798.

Spelke, E., & Kinzler, K. D. (2007). Core knowledge. *Developmental Science, 10*(1), 89–96.

Spelke, E. S. (1985). Perception of unity, persistence, and identity: Thoughts on infants' conceptions of objects. In J. Mehler & R. Fox (Eds.), *Neonate cognition: Beyond the blooming, buzzing confusion*. Hillsdale, NJ: Erlbaum.

Springer, K., & Keil, F. C. (1989). On the development of biologically specific beliefs: The case of inheritance. *Child Development, 60,* 637–648.

Szego, P. A., & Rutherford, M. D. (2007). Actual and illusory differences in constant speed influence the perception of animacy similarly. *Journal of Vision, 7*(12), 1–7.

Szego, P. A., & Rutherford, M. D. (2008). Dissociating the perception of speed and the perception of animacy: A functional approach. *Evolution and Human Behavior, 29*(5), 299–374.

Tremoulet, P., & Feldman, J. (2000). Perception of animacy from the motion of a single object. *Perception, 29*(8), 943–951.

Troje, N. F., & Westhoff, C. (2006). The inversion effect in biological motion perception: Evidence for a "life detector"? *Current Biology, 16,* 821–824.

Valentine, T. (1988). Upside-down faces: a review of the effect of inversion upon face recognition. *British Journal of Psychology, 79,* 471–491.

Wellman, H. M., & Gelman, S. A. (1998). Knowledge acquisition in foundational domains. In D. Kuhn & R. S. Siegler (Eds.), *Handbook of child psychology* (Vol. 2). *Cognition, perception, and language*. New York: Wiley.

Wheaton, K. J., Pipingas, A., Silberstein, R. B., & Puce, A. (2001). Human neural responses elicited to observing the actions of others. *Visual Neuroscience, 18*(3), 401–406.

Wynn, K. (1995). Infants possess a system of numerical knowledge. *Current Directions in Psychological Science, 4,* 172–177.

Yin, R. (1969). Looking at upside-down faces. *Journal of Experimental Psychology, 81,* 141–145.

7 Perceiving Intention in Animacy Displays Created from Human Motion

Phil McAleer and Scott A. Love

Abstract

Typically, the actions of agents in classical animacy displays are synthetically created, thus forming artificial displays of biological movement. Therefore, the link between the motion in animacy displays and that of actual biological motion is unclear. In this chapter we will look at work being done to clarify this relationship. We will first discuss a modern approach to the creation of animacy displays whereby full-video displays of human interactions are reduced into simple animacy displays; this results in animate shapes whose motions are directly derived from human actions. Second, we will review what is known about the ability of typically developed adults and people with autism spectrum disorders to perceive the intentionality within these displays. Finally, we will explore the effects that motion parameters such as speed and acceleration, measured directly from original human actions, have on the perception of intent; fMRI studies that connect neural networks to motion parameters, and the resultant perception of animacy and intention, will also be examined.

Ever since Heider and Simmel (1944) maneuvered three cut-out shapes across a light box almost seventy years ago, the premise for the creation of animacy displays has remained fairly stable; only the technology has evolved over time. In contrast to the study of biological motion, where the predominant practice has remained true to Johansson's (1973) technique of using markers on a human body, animacy displays are commonly synthesized via complex computer algorithms or clever animation techniques (e.g., Bloom & Veres, 1999; Gao & Scholl, 2011; Morito, Tanabe, Kochiyama, & Sadato, 2009; Tremoulet & Feldman, 2000; Zacks, 2004). It is perhaps the wonder of animacy perception that from these synthetically created displays are such strong and vivid sensations of intent and desires evoked. Yet it is proposed that without the connection to actual human motion, it is difficult to fully relate the findings from animacy studies to the reality of how people interact or understand each other's intentions and actions. It is this connection between human motion and the perception of intention in animacy displays that is the focal point of this chapter.

Prior to McAleer and Pollick (2008), the closest that animacy displays had come to being created from human movement was arguably in the work of Blythe, Todd, and Miller (1999) and Barrett, Todd, Miller, and Blythe (2005). Instead of utilizing computer algorithms to generate the movement of the agents in the displays, these two studies relied on displays created via participants controlling animated "ants" onscreen; these displays were then used by the participants to act out a specific set of intentions derived from an extensive meta-analysis of evolutionary literature: Chasing, Courting, Fighting, Following, Guarding, and Playing.[1] These intentions were shown to be universal across both culture (by comparing results from Western adults and Shuar adults of Amazonian Ecuador) and age (by comparing results from various ages of children and adults) (Barrett et al., 2005). These six intentions form the basis of the majority of our work presented in this chapter.

It is perhaps prudent to propose here that animacy perception studies tend to fall into two categories: (1) Those that aim to determine the changes in movement that lead to the percept of animacy and (2) those that ask how we can relate the perception of intent in these animated displays to real-world examples. By no means are these two categories exclusive of one another, though it can be argued that to achieve their goals, each should require different displays. The first category would require careful manipulation of kinematics and aesthetics to compare and contrast the independent effect of manipulating an individual variable; for example, the work of both Tremoulet and Feldman (2000, 2006) and Szego and Rutherford (2007, 2008) highlight how the perception of an "alive" agent can be produced or adjusted via minute manipulations of a solitary agent. The second category, on the other hand, requires comparison of actual intentions; the closer the displays appear and act in comparison to what actually occurs, the better. It is in this second category, of establishing intentions from animacy displays, that we grounded our work.

Creating the Displays

Before highlighting our work on intention perception, it is imperative to describe the method that we use to create our animacy displays and how it changed the standards in the field. Via a somewhat fortuitous viewing of the movement of a dancer reduced to a bounding red box, we quickly realized that EyesWeb software (www.eyesweb.org) (Camurri, De Poli, Leman, & Volpe, 2001; Camurri, Mazzarino, & Volpe, 2004; Camurri, Trocca, & Volpe, 2002), established by the InfoMus group at the University of Genoa, Italy, could be adapted to create animacy displays that would be directly derived from human motion. In our initial experiments, we sought to establish whether these displays evoke the perception of animacy by exploring whether participants would freely describe the resultant displays in terms of human motion. Using a video display of two dancers performing a sequence of modern dance, the technique worked as follows: (1) Using EyesWeb, establish a contrast between the human actors and the background in original video; (2) perform a background subtraction based on this contrast to leave a silhouette image of the dancers on the screen; (3) track the center of mass of each silhouette figure to obtain central positional coordinates of each actor; and (4) replot the coordinates as the center of a geometric shape, such as a rectangle, circle, or triangle (see figure 7.1). In this process, steps 1 through 3 were performed via processing tools in EyesWeb, and step 4 was performed via MATLAB (MathWorks, Natick, MA).

The important contributions of this subtraction/reduction[2] method were twofold. In the first instance, it was a method that could bridge the gap from animated movement to real-world scenarios. Second, it opened up a question that had received little coverage in prior animacy perception

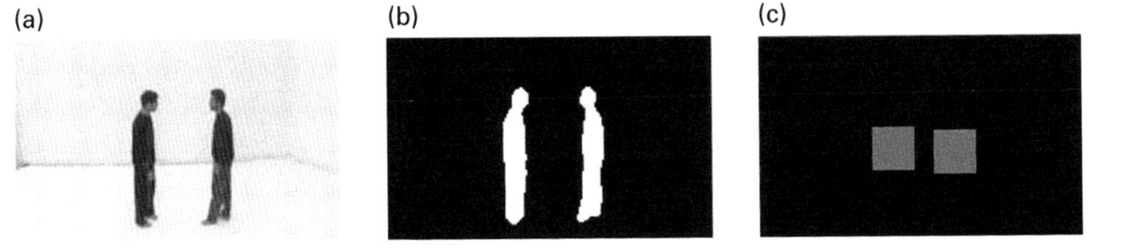

Figure 7.1
For the systematic reduction method, the original video enters the process at Position (a), where it is converted to the silhouette image via background subtraction (b). From there, the number of silhouettes is counted and tracked at the barycenter, and a resultant image is created in which the actors are represented by squares (c). The points in the patch at which the coordinates for each actor are extracted, as in the present research, are circled. The section between the circles and Point C is not used in the present experiments but does show the output capabilities of EyesWeb. Reprinted from McAleer and Pollick (2008) with permission.

research—that of viewpoint. Previously, research had followed the example of Heider and Simmel (1944) and utilized animacy displays depicting an overhead perspective (Bassili, 1976; Blakemore et al., 2003; Bloom & Veres, 1999; Blythe et al., 1999; Castelli, Frith, Happé, & Frith, 2002; Gelman, Durgin, & Kaufman, 1995; Tremoulet & Feldman, 2000). A few studies on the other hand had made use of side view displays or ambiguous displays (Csibra, Gergely, Bíró, Koós, & Brockbank, 1999; Gergely, Nádasdy, Csibra, & Bíró, 1995; Kuhlmeier, Wynn, & Bloom, 2003; Szego & Rutherford, 2007; 2008), yet none directly questioned the importance of viewpoint for the perception of animacy or intent. The subtraction method we proposed could begin to address this question of viewpoint. In our work, discussed here, we compared the perception of animacy in displays where actors were filmed from both overhead and side-view perspectives.

Prior to addressing viewpoint, our first goal was to establish that people would perceive the displays created using our method as animate. To do so, we compared free response data to (1) the video display of two dancers performing modern dance and (2) the derived animacy display of this scene, which was depicted by two red rectangles on a black screen (McAleer, Mazzarino, Volpe, Camurri, Paterson, Smith, & Pollick, 2004). In our analysis, we looked for terms and statements indicating that participants had attributed human movements and characteristics (e.g., chased, followed, touched) and emotions (e.g., happy, sad) to the displays. Surprisingly, no animate terms were used to describe the movement of the two rectangles. In fact, no terms indicative of the shapes being perceived as alive were found; all descriptions appeared to relate to random motion. We found this surprising because the movements were in fact human, and we had hypothesized that they would be perceived so. This was again the case in our second experiment when we compared a video of a solitary dance to its derived animacy display (one white rectangle on a black background). We hypothesized from these null results that perhaps the original movements of the dancers, in both the solo and dyadic examples, were movements that would not be familiar to the general population; thus, when reduced to an animacy display, the movements would simply appear somewhat random or Brownian (Brown, 1828). We concluded that perhaps, in the absence of an available schema (Rumelhart, 1980) to explain the movement in the animacy displays, participants viewed the movements as incoherent and random rather than as human; this suggests that only actions that we have a clear understanding of can be perceived as human/animate when depicted in animacy displays.

The Universal Intents and Viewpoint

Our second series of experiments turned to the universal intentions of Blythe et al. (1999) and Barrett et al. (2005) mentioned in the previous section. If it was the case that participants failed to perceive animacy in the dance scenarios due to a lack of knowledge of the movements, then by contrast they should have no difficulty perceiving animacy in displays derived from movements and intentions proposed to be universal and known by all. Furthermore, we addressed the question of viewpoint (see previous section).

Two actors were filmed simultaneously on a custom-built square stage, from both an overhead and side-view perspective, while they performed loosely scripted actions relating to the six intentions of Chasing, Flirting, Fighting, Following, Guarding, and Playing (McAleer & Pollick, 2008); for our studies, we changed the term "Courting" to "Flirting" as we felt the latter would be more easily understood by both the actors used for creating the displays and the eventual participants of the experiments. The directions for each intention can be seen in table 7.1.

One video example was filmed for each of the six intentions, and the animacy displays derived from these six were displayed as white circles on a gray background. As mentioned, the reduction method tracks the center of mass of the silhouette image from the original display. Thus, to create the overhead displays, actors were tracked in the center of their heads. To create the side-view displays, actors were tracked approximately in the center of their stomachs (figure 7.2.).

Both the animacy displays and video displays of the six intentions, from both perspectives, were shown to participants. The participants were to select the intention that best described the movement in the displays in a six-alternative forced-choice (6AFC) task. Results revealed that participants could perceive the intended intention in both the video and animacy displays from both perspectives (side-view and overhead)—but with varying degrees of accuracy. In the video displays, accuracy levels were all significantly above chance (16.667%) for both perspectives: For example, displays of Following were recognized at levels greater than 90% irrespective of viewpoint, yet the side-view display of Playing was recognized at only a 40% accuracy level. In comparison, the degree of accuracy in terms of recognition of the intent in the animacy displays was weaker across both perspectives: Only the Following and Fighting displays from the overhead perspective were recognized at levels greater than 80%; in

Table 7.1

Instructions for how the six intentions were to be loosely performed by the actors. Reprinted from McAleer, Kay, Pollick, and Rutherford (2011a) with permission.

Intention	Description
CHASING	*The two actors started a foot apart against the back wall of the stage, both facing in the same direction. Actors ran at a ¾ pace in a clockwise direction doing laps of the stage but never caught each other. Actors would run in a figure of eight, every couple of laps. They continued for approximately 30 seconds. Actors never came into contact with each other, as the chase was stopped prior to this occurring.*
FIGHTING	*Actors started at opposite sides of the stage, along the midline, facing inwards toward each other, perpendicular to the side camera. They walked toward each other and stopped about a foot apart where they then proceeded in a clockwise circle facing each other as though measuring each other up for a battle/fight. After a couple of complete circles one actor made an aggressive movement towards the other actor who jumped backward to get away from the aggressor. The roles were then reversed and the interaction was repeated for 30 seconds upon which the actor on the right fell back and down to the ground. No harmful, physical contact occurred.*
FLIRTING	*Actors stood with bodies facing toward the side camera about five feet apart. Actor on the right moved to the actor on left and circled them twice then moved back to their original starting point. Actor on left then repeated the action of previous actor. On completion of second circle by actor two around actor one, both actors moved off together to top left of stage. Though the actors were in very close proximity to one another, no direct contact was made. However, no explicit instruction was given to prevent direct contact.*
FOLLOWING	*Same directions as* **Chasing** *except performed at a walking pace.*
GUARDING	*One actor started on far left and one started on far right, along midline of stage, facing perpendicular to the side camera. Actors were informed that the actor on the right was to pretend that they had a ball behind them that the actor on the left wanted. Actor on left was informed to try and get the ball, whilst actor on right was told to prevent this. This was performed for five minutes prior to recording, using a real ball, in order to make the actors familiar with the situation. Actors were allowed to come into contact if they felt it was warranted, but never actually touched each other.*
PLAYING	*Actors start in middle of the stage facing towards each other, two feet apart, perpendicular to the side camera. Actors were instructed to play a game of "tag," moving freely around the stage, switching roles when the game required them to do so. To make a "tag," actors were instructed that they were required to make contact with at least one hand on the other person. This continued for approximately 30 seconds.*

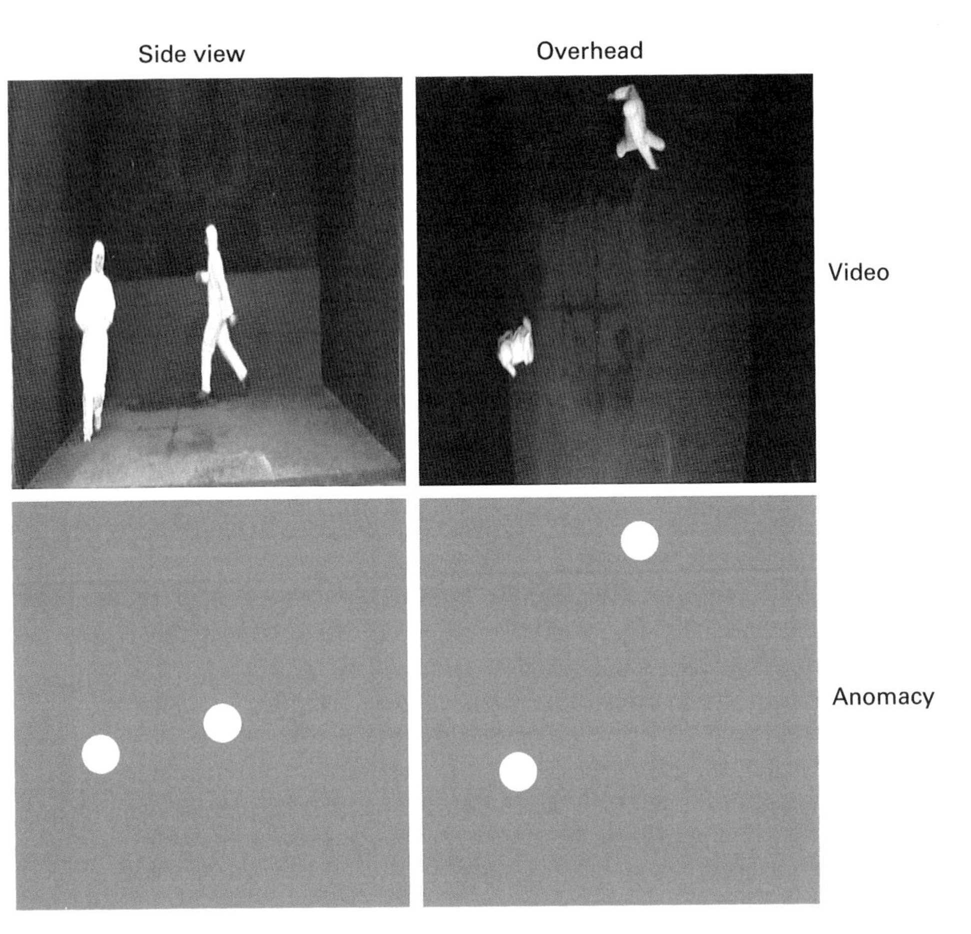

Figure 7.2
Examples of the original footage and animacy displays after being processed through EyesWeb and MatLab. Images have been squared for presentation purposes. Reprinted from McAleer, Kay, Pollick, and Rutherford (2011a) with permission.

contrast, Guarding and Following displays from the side-view perspective fell below chance performance. In terms of a potential cause for the poor recognition levels, we argued that this perhaps was largely owing to confusion (e.g., the labeling of a Playing display as a Fighting display) in the perception of the original video displays resulting in larger confusions in participants' perception of the animacy displays. If each video example of a particular intent was consistently recognized 100 percent (i.e., a chase was always seen as such, etc.), then perhaps the confusion during animacy displays would have been reduced. Furthermore, Blythe et al. (1999) found that participants had a bias toward responding that representations in animacy displays depict play; the researchers related this bias to the involvement of playing in the learning of other actions and intentions as a child. In our study (McAleer & Pollick, 2008), however, it was difficult to establish any systematic bias toward one intention or another. Again, a potential cause may have been poor levels of recognition of the desired intent in some of the original video displays.

In the global sense, our experiment showed that animacy displays could be derived from videos of actual human movement and that certain intentions could be perceived in such displays. The experiment made use of an alternative forced choice (AFC) task as opposed to the free-response method that we had used previously. One characteristic of the AFC task that is worth noting is that it does, by its nature, force participants into reporting a perceived intent whether the display looks intentional or not. In other words, it is possible that participants never viewed the movements of the agents as animate but were not given the chance to respond as such. However, if this were the case, then response accuracy would be approaching chance level for all displays. Furthermore, data reported at the 2005 Cognitive Neuroscience Society conference (McAleer, Mazzarino, Volpe, Camurri, Paterson, Smith, & Pollick, 2005) clearly showed that participants would use animate terms to describe our animacy displays in a free-response task. That said, it was found that participants would offer only brief, direct descriptions (e.g., "it is a chase") and did not offer elaborate responses akin to those in Heider and Simmel (1944). From this we suggested that our stimuli did indeed evoke a perception of animacy and intent, but that future research should look into the effect of having altering intentions in a single display; whereas Heider and Simmel (1944) showed long displays of evolving story lines, we showed only simple displays of singular intents. We proposed that it was perhaps this evolution of intent within displays that results in elaborate and spontaneous free responses.

Furthermore, and returning to the question of viewpoint, the results showed that participants were much more accurate in perceiving the appropriate intent in displays viewed from an overhead perspective as opposed to a side view, with close to 70 percent accuracy for the former and approximately 50 percent accuracy for the latter. Regarding the counterintuitive nature of this finding, we argue that although the overhead viewpoint is a perspective that we are not generally accustomed to in either our ordinary lives or our evolutionary history, it perhaps gives more information as to the relative distance between agents. For example, the agents in the overhead Chasing display could clearly be seen to be moving fully around the display in a circular fashion, but the agents in the side-view animacy displays appear to be moving linearly back-and-forth across the screen. Although we initially regarded this bias toward overhead displays as a curious preference, there are indeed examples from the real world—particularly the sports world—where an elevated viewpoint would be advantageous: for example, it is often reported by football (soccer) commentators that the manager of a team is sitting high in the stadium for the first half of the game to view the formation of the opposing team and gain insight as to how they intend to play. Establishing the true effect of viewpoint can offer insight into how we, as humans, determine intent and make use of differing viewpoints.

In a second experiment (McAleer & Pollick, 2008), we questioned whether there truly was a bias or preference for the overhead perspective. One possible alternative explanation was that the increased levels of accuracy in the viewing of overhead displays was due to a lack of understanding of which perspective to take in order to appropriately perceive the side-view displays: that is, participants may have watched the side-view displays as though from an overhead perspective, which would have hampered their ability to understand the intent presented by the displays. To test this possibility, we created additional variables for the animacy displays: (1) Identity/occlusion (we gave indication of occlusion, or which agent was closer to the front of the screen, by coloring one circle black and one white) and (2) context (we outlined the display to suggest a viewpoint) (figure 7.3). For the overhead displays, context was added by surrounding the agents with four white lines to suggest a floor viewed from above; in the side-view displays, we added three white lines to suggest a floor and two sidewalls viewed from eye level.

A new set of participants viewed the animacy displays using four conditions that the addition of the two new variables allowed: (1) No identity, no context; (2) identity, no context; (3) no identity, context; and

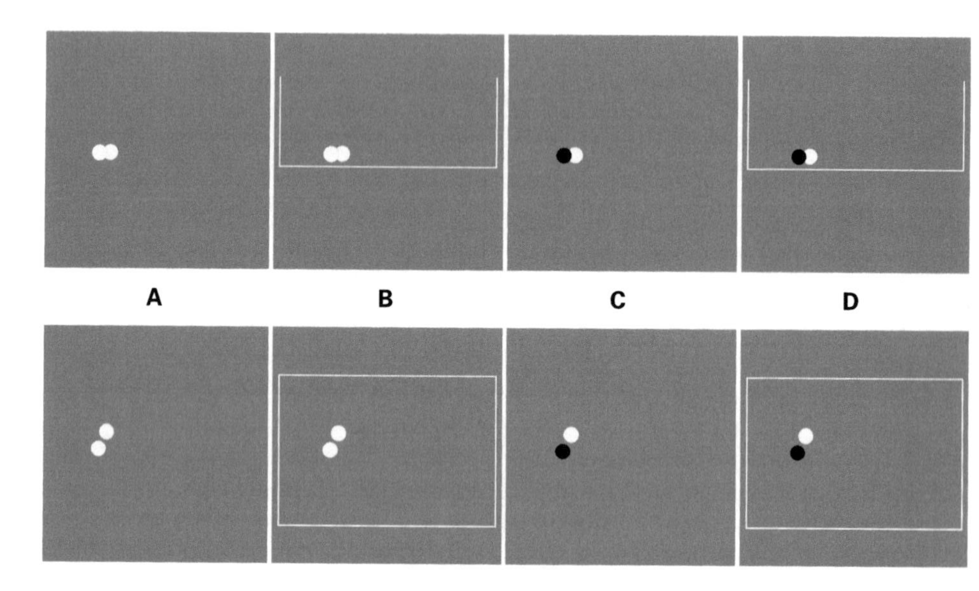

Figure 7.3
The additional visual cues with the top panel showing side view displays and the bottom panel showing the overhead displays. The four conditions are (A) no occlusion, no context (NONC); (B) no occlusion, context present (NOC); (C) occlusion present, no context (ONC); and (D) occlusion and context present (OC). Reprinted from McAleer and Pollick (2008) with permission.

(4) identity, context. The task was again a 6AFC. Results showed that neither identity nor context had any influence on the participants' ability to perceive the displayed intents and again revealed an overall increased ability to perceive the appropriate intent in overhead, as opposed to side-view, displays. A final experiment that directly asked participants to indicate which viewpoint a display was perceived from (2AFC—overhead or side view) indicated that participants had no difficulties in perceiving the appropriate perspective with accuracy above 80 percent for both views. Thus, our comparison of overhead and side-on viewpoints robustly indicates what we termed a "preference" for the overhead. However, there are two caveats to this statement. The first is that, though we called it a preference, no direct measure of this was gathered; it would possibly be better termed in respect to improved performance when overhead displays were viewed, because it is this improved performance that we have clear evidence of. The second, and related, caveat is that the direct overhead view may not be the optimal viewpoint but is simply significantly better than a side-on view. Attempts to derive the canonical viewpoint for animacy

displays should involve experimentation with numerous angles or perhaps allow for manipulation of the angle by the participant to arrive at a view that he or she is most comfortable with.

As a side issue, it would be interesting to study our internal representations of viewpoint. Perhaps we internally imagine animate displays, or indeed real-world movements, with an overhead perspective, thus converting, in our minds, our natural view into a viewpoint that allows for increased information that we can use to interpret intents and actions. In children, we could test this theory using a looking-time paradigm: for example, a child could view an intention from a side-view perspective and then be shown either the same intent or a different intent from the overhead view, with looking time being used to judge the child's expectations. The question of how to get at this same concept in adults is perhaps more taxing, yet nevertheless deeply interesting.

A First Impression

Having established that people can successfully derive the intention of two agents within the displays described in the previous section, an important follow-up is to establish the amount of time it takes for a person to arrive at this judgment of intention. In the real world, it would be somewhat ineffective if, as humans, we could only recognize a fight after the first punch had been thrown. The ability to predict intent, or to perceive intent rapidly, would serve us better. Tom Troscianko and colleagues investigated the ability to predict intent by having participants view video displays of people taken from closed-circuit television cameras and asking them to predict if a particular person's intent was violent or nonviolent. Their work indicated that the viewers were in fact able to predict when violence was about to occur (Troscianko et al., 2004). Furthermore, participants were able to verbalize the specific body movements and actions of the persons in the displays that gave the impression that a violent event was imminent. But participants had viewed fifteen seconds of the display prior to making a judgment, so that study cannot speak to the rapidness of such judgments. However, we are aware that we are capable of making rapid judgments about people; for instance, face perception can be used to attribute a social characteristic such as trustworthiness to a person within less than 100 ms of exposure (Willis & Todorov, 2006; Todorov, Pakrashi, & Oosterhof, 2009). Therefore, it is reasonable to expect that people can make rapid judgments about intent.

We asked a group of thirteen participants (McAleer, 2006) to judge intent in the animacy displays described above over exposure durations of one, five, and ten seconds. Based on the bias in viewpoint (see previous section), only the overhead displays were shown. Results showed that even after one second of exposure, participants could successfully identify all intentions except Guarding. After a five-second exposure, all intentions were successfully recognized at levels greater than chance; there was no significant improvement with an increase in exposure duration to ten seconds. Indeed, comparing across experiments, accuracy levels after five seconds were comparable to, if not better than, the accuracy levels of participants who had seen the full-duration displays (approximately thirty seconds) in previous tests (McAleer et al., 2005).

Our results are a clear indication that people can make rapid judgments about intent. As these intents can be distinguished at short durations, we would suggest that the visual system is attuned to recognizing them quickly. It is possible that this process was adapted for survival reasons; this validates theories that the intentions we studied (see second paragraph of this chapter) are generic in human nature and have in the past served the purpose of maintaining survival and increasing the chances of reproduction (Blythe et al., 1999; Barrett et al., 2005). However, our question as to the speed with which we make these judgments should not end here. In terms of visual processing, one second is a long time, and as in face perception (Todorov et al., 2009), it is possible that the judgment of intent may occur within milliseconds instead of seconds. That said, it must be remembered that face studies generally use static images that allow for very rapid judgments, whereas in the case of animacy displays, a sufficient duration of time is required for the motion information to be viewed. Intent surely can't be judged from static displays of circles?

More accurate studies of the dynamics of the perception of intent could be achieved using highly time-sensitive methodologies such as electroencephalography (EEG) and magnetoencephalography (MEG). Moreover, such techniques could prove extremely fruitful in understanding not only the timing of intent perception but also its functional location via modern source localization techniques. To our knowledge, few if any studies have exploited these techniques to investigate animacy or intent perception.

The Kinematics

The movements that underlie intentions, and how those movements change over time, must be discovered if we are to fully comprehend intent perception. Using the three-dimensional coordinates (positional x and y, across time) of the original human interactions used to create the animacy displays, we attempted to identify the fluctuations in the kinematics that define each intention (McAleer, 2006). Kinematic information, beyond offering high-level descriptions of how the displays appear, has been shown to give insight into intent perception (Blythe et al., 1999; Zacks, 2004).

Using a classification by elimination (CBE) technique, in which kinematics are entered into an algorithm to establish the minimum number required for accurate recognition of an intent, Blythe et al. (1999) showed that absolute velocity (i.e., the forward velocity of an agent with respect to its background) was the best classifier for intentions, followed by, in decreasing order, relative angle, relative velocity, relative heading, relative vorticity, absolute vorticity, and relative distance. Zacks (2004) investigated the motion properties in animacy displays to (1) establish how people segment sequences of actions into events and (2) determine how motion properties relate to action segments. He explored a set of motion cues that included position, speed, acceleration, distance, relative speed, and relative distance. Zacks used a stepwise regression to establish that the relative distance of the agents (i.e., how far apart the agents were, as well as their acceleration) accounted for the highest proportion of variance. In accordance with previous authors, Zacks concluded that observers use the motion properties of displays to determine the intentions portrayed by the agents, but made a case for the inclusion of prior knowledge regarding context, experience, and so forth. From this, it appears that, in understanding the intentions of agents, the motion properties that seem most pertinent relate to (1) the absolute velocity and acceleration of an agent and (2) the relative distance, velocity, and acceleration between agents.

Following Zacks' method (2004), we focused our analysis on the speed, acceleration, and relative distance of the actors in our displays (full duration). The motion properties examined were adapted from Zacks (2004) and described as follows:

Speed the speed of each agent or the magnitude of the agent's instantaneous velocity, where the velocity of an agent, in both the x and y directions, is calculated by numerical differentiation of the position of an agent (i.e., its positional x and y location)
Acceleration the magnitude of each agent's acceleration
Relative Distance (Distance) the distance between two agents
Relative Speed (RelSpeed) the speed at which agents were moving toward or away from each other, as calculated by numerical differentiation of the Relative Distance

Relative Acceleration (RelAcc) the acceleration at which agents were moving toward or away from each other, as calculated by numerical differentiation of the Relative Speed

For all parameters, the mean and standard deviation (SD) were calculated and normalized to obtain scale-free parameters for all six intentions (full duration) and both viewpoints. As a result of this, we obtained a total of twenty-six predictors; these can be broken down into four categories:

(1) *Position Parameters* X1Mean; Y1Mean; X2Mean; Y2Mean; X1SD; Y1SD; X2SD; Y2SD; RelXMean; RelYMean; RelXSD; RelYSD
(2) *(Relative) Distance Parameters* DistanceMean; DistanceSD
(3) *Speed Parameters* Speed1Mean; Speed1SD; Speed2Mean; Speed2SD
(4) *Relative Speed & Acceleration Parameters* RelSpeedMean; RelSpeedSD; RelAccMean; RelAccSD

These predictors were repeatedly entered into a forward stepwise regression paradigm until changes in the Akaike information criterion (AIC) (Akaike, 1974) were no longer significant. The dependant variable for the regression models was the percentage of participants who identified a display as a specific intention in the second experiment of McAleer and Pollick (2008) (e.g., the number of times any display was called "chase"). Data were collapsed across presentation conditions (i.e., whether occlusion or contextual cues were present) as these made only small differences to the ability to recognize intentions. As a general rule of thumb, predictors added into the regression model higher than the fourth position explain very little variance (Howell, 1997). The first five predictors in the regression models for each intention, for both viewpoints, can be seen in tables 7.2 (overhead) and 7.3 (side view).

Considering overhead displays, Chasing tended to be successfully recognized if (1) the person chasing was moving fast and (2) the two people were not accelerating relative to each other. Accurate classification of Fighting was reliant on (1) the speed of one person showing little variation and (2) large variations in the speed of each agent increasing relatively to the other. The recognition of Flirting was reliant on positional predictors—particularly the stationary lateral position of one agent for the first moments of the displays. Following was categorized by (1) the mean distance between the two agents remaining low and (2) the relative speed between the two agents not fluctuating. Similar to Flirting, positional predictors determined Guarding, but the low speed of the person doing the guarding was a factor as well. Finally, Playing was signaled by (1) large variations in the relative speed between the two agents and (2) large relative accelerations. Many of these descriptions appear to make intuitive sense.

Table 7.2
The first five predictors entered into a regression equation to explain the variance of the intention recognition data of overhead displays. Column on left indicates the intention shown; column on right indicates parameters. Symbols + and – indicate whether the correlation between the parameter and behavioural data was positive or negative. Values in parenthesis indicate Beta values.

Full Duration	Overhead				
Chasing	+Speed1Mean –(+22.46)	–RelAccMean (–11.56)	–X1Mean (–4.33)	–RelAccSD (–1.94)	–Y1Mean (–0.393)
Fighting	+RelSpeedSD (+31.79)	–Speed1SD (–11.05)	+Y2Mean (+6.98)	–RelSpeedMean (–2.63)	+X1Mean (+1.09)
Flirting	–X1Mean (–22.44)	–RelYMean (–14.44)	–Y2SD (–10.42)	–RelSpeedMean (–3.34)	+Y1Mean (+2.95)
Following	–DistanceMean (–27.05)	–RelSpeedSD (–15.64)	+RelYSD (+3.62)	+X1SD (+1.19)	+X1Mean (+0.13)
Guarding	–RelXMean (–19.22)	–Speed2Mean (–8.49)	–Y1Mean (–6.1)	–RelSpeedMean (–1.39)	–X1Mean (–0.6)
Playing	+RelSpeedSD (+32.41)	+Y2Mean (+19.45)	+RelAccMean (+8.71)	+Y1SD (+2.81)	–X1Mean (–0.7)

Table 7.3

The first five predictors entered into a regression equation to explain the variance of the intention recognition data of side view displays. Column on left indicates the intention shown; column on right indicates parameters. Symbols + and – indicate whether the correlation between the parameter and behavioral data was positive or negative. Values in parenthesis indicate Beta values.

Full Duration	Side View				
Chasing	+Speed2Mean (+8.56)	–X1Mean (–5.68)	+Speed1Mean (+4.59)	–Y2Mean (–1.28)	+Y1Mean (+0.31)
Fighting	–X1SD (–33.03)	+RelXMean (+16.98)	+RelSpeedMean (+3.9)	–RelYMean (–1.97)	+X1Mean (+0.05)
Flirting	–Speed1Mean (–20.9)	–X1Mean (–12.05)	+Speed1SD (+4.46)	+Y2Mean (+4.4)	+Y1Mean (+0.18)
Following	+Y2SD (+46.69)	+Y1Mean (+11.68)	+Speed2SD (+64.19)	–Speed1SD (–52.02)	+X1Mean (+0.02)
Guarding	–RelXMean (–10.9)	+X2SD (+4.99)	+Acc1SD (+56.92)	–Acc1Mean (–55.34)	–X1Mean (–0.05)
Playing	+RelAccMean (+5.97)	–RelYSD (–4.83)	–RelYMean (–2.27)	–Speed1Mean (–0.52)	+X1Mean (+0.04)

The results of the regression modeling for the side-view displays are based on the relative positioning of the agents, as well as changes in speed and acceleration. Similar to Chasing in the overhead, a side-view display was likely to be called Chasing if the speed of both the chaser and the person being chased was high. Fighting was categorized by (1) small changes in the lateral position of one agent and (2) the agents moving at fast speeds relative to each other. Flirting was predicted in the side view if (1) the speed of one person was almost stationary at the start and (2) there were large changes in the speed of that person. A successful categorization of Following was based on (1) large changes in the horizontal position, (2) large changes in the speed of the person being followed, and (3) changes in the horizontal position of the follower. Guarding was characterized by (1) changes in the lateral position of the person doing the guarding and (2) changes in the acceleration of the person doing the attacking. Finally, Playing was characterized by (1) large accelerations by each agent relative to the other and (2) minimal changes in the horizontal dimension of the agents.

Overall, the stepwise regression modeling shows that, in general, the main predictors of the intentions within the animacy displays—for overhead displays at least—are speed and acceleration, both absolute and relative, with distance and positional predictors also contributing. These results are broadly consistent with previous findings (Barrett et al., 2005; Blythe et al., 1999; Zacks, 2004). However, in this instance we find that the speed of an agent is more predictive than changes in acceleration—though this finding may be unique to our collection of stimuli. One cautionary note would be that this analysis was based on the single example of each intention, which we presented in the experiments described at the beginning of this section. Even though our database of recorded movements was not extensive enough to draw firm conclusions, we did obtain initial impressions of the key movements that participants might use to recognize intent. That said, the use of multiple examples of each intention, from numerous actors, may produce differing results and is an extremely worthwhile line of further study.

Plotting the Intentions

To obtain an understanding of what the kinematics that define intention look like, and to help us understand some of the findings of previous studies, we explored subjective impressions of the three-dimensional plots of the agents' positional coordinates (McAleer, 2006). A key finding of the experiments outlined in the previous section was that participants were better at categorizing intentions when the displays were shown from an overhead perspective. We proposed that the reason for this was that the overhead displays revealed more information as to the relationship between the agents (McAleer & Pollick, 2008). In contrast, we proposed that because the side-view displays only revealed motion in a horizontal plane—with depth and distal information being unavailable—the participants found it harder to make accurate judgments when viewing these displays. Three-dimensional plots for each intention in both viewpoints can be seen in figure 7.4.

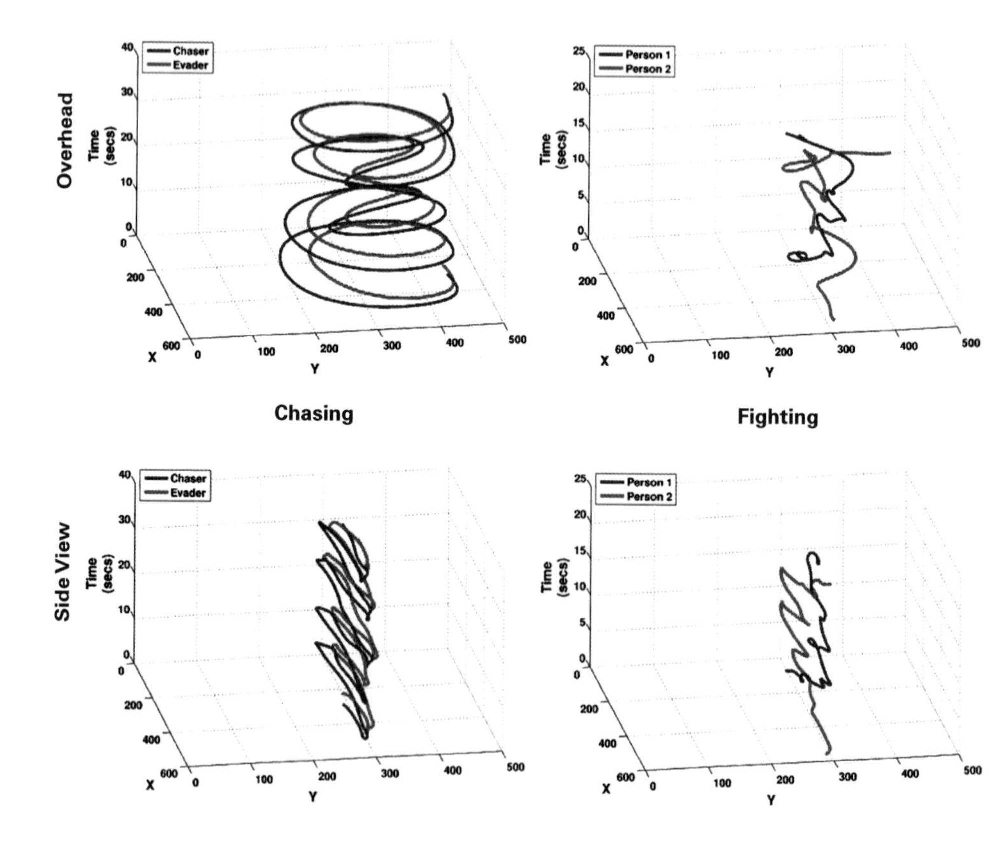

Figure 7.4
3D plots representing the positional co-ordinates of actors performing all six intentions for both viewpoints: overhead (top plots) and side view (bottom plots). Vertical axis shows time (seconds). Actor/Agent being chased is shown in red. In order, from left to right, the intentions are Chasing, Fighting, Flirting, Following, Guarding, and Playing.

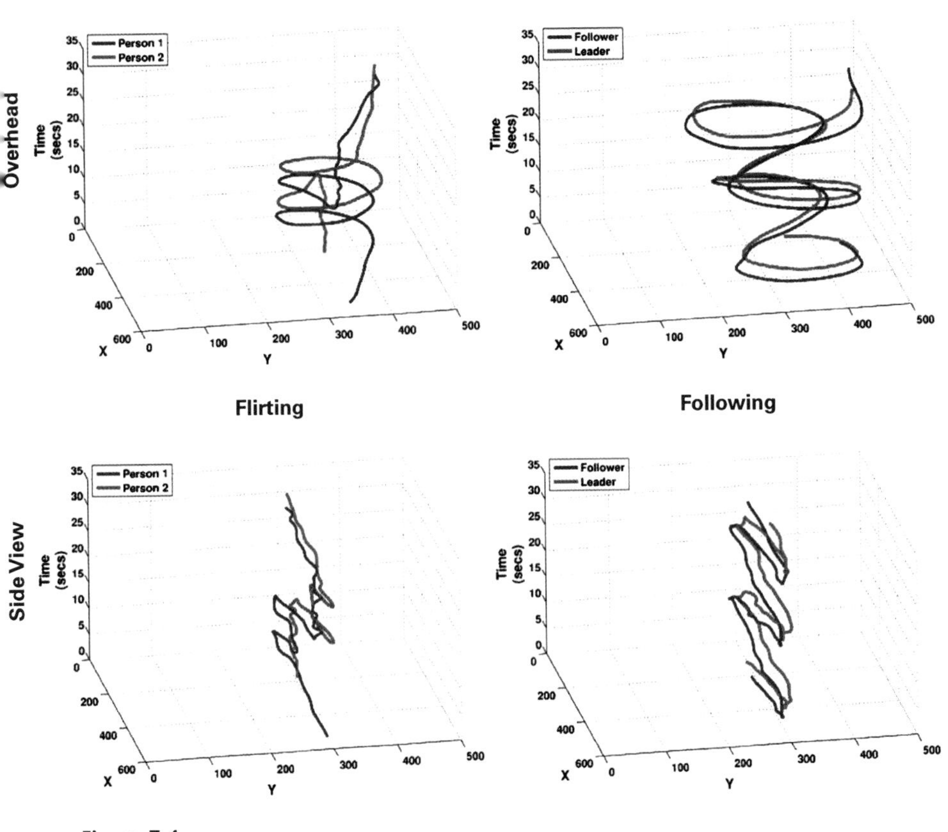

Figure 7.4
(continued)

The subjective analysis of the three-dimensional plots confirms our hypotheses. The overhead plots show both agents traveling in varying directions around both the x and y space. These plots are made up of (1) loops and curves that indicate agents moving in a circular motion and (2) straight lines that indicate agents moving in a directed manner. The side-view plots, on the other hand, only show movement in the x dimension. This would explain why agents in the side-view displays appeared to move only left and right in straight lines. As a result of these differences in appearance, we would have expected participants to be better at judging intentions from the overhead viewpoint because more differentiating information is available from that viewpoint. It remains unclear whether the preference for overhead displays is a true preference or whether additional distal information in the side-view displays (e.g., agents increasing

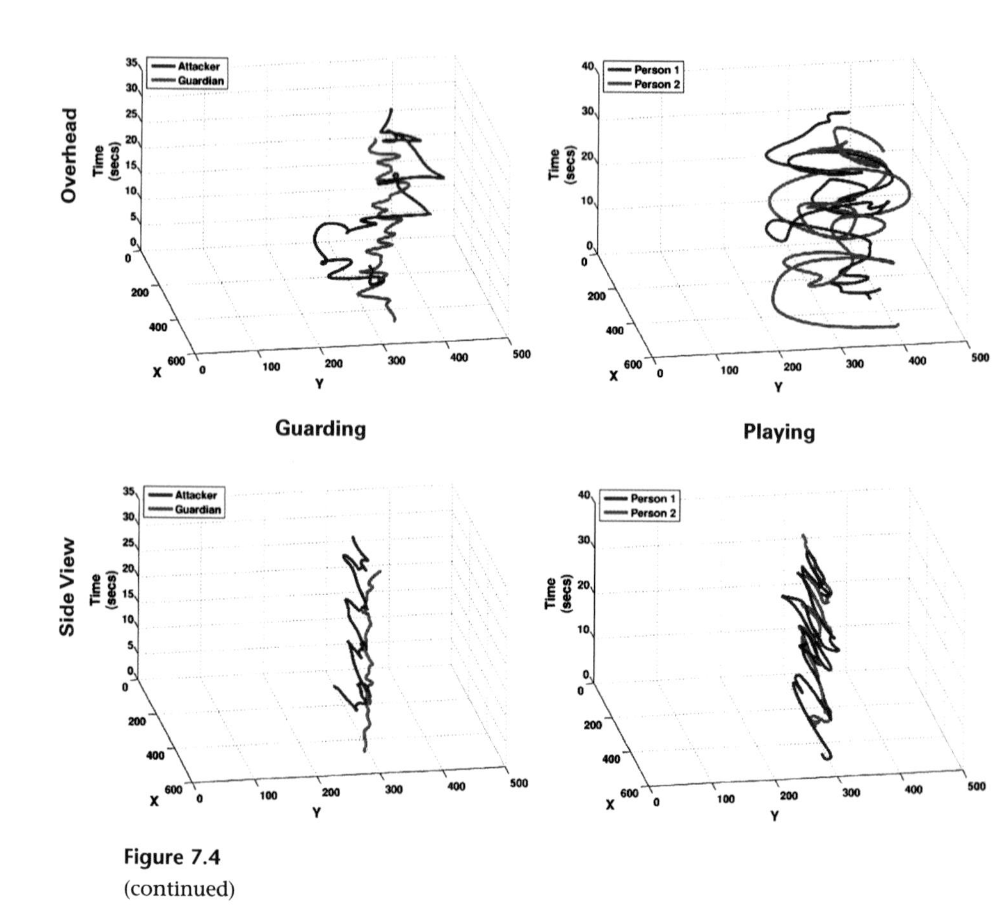

Figure 7.4
(continued)

and decreasing in size with depth) would have enabled better intention perception.

In looking at individual intentions, we focused solely on the overhead plots; this is because the side-view plots' similarities resulted in relatively little motion information that could be used for distinguishing specific intentions. In the overhead displays, the plots of Chasing and Following look similar; both have smooth circular motion, with one agent's motion pattern being almost an exact copy of the other. The person who is following or chasing has a tighter graph; this indicates that the displays' followers and chasers move in smaller circles than the persons being followed or chased. This makes ecological sense: the person following/chasing sees where the other is heading and alters his or her direction accordingly (showing predictive ability). The overhead plot of Flirting shows (1) an agent standing still while a second agent circles him or her and (2) the reversal of roles. Participants may use this reciprocal action to distinguish the intention being displayed. Finally, the overhead plot of Playing shows various types of movement, such as loops, curves, and quick straight-line bursts. The movements that make up Playing consist of a mixture of elements from other intentions: the looped and curved motion seen in Following and Chasing as well as the quick movements in straight lines seen in Fighting and Guarding. Interestingly, perhaps this mixture of movements in the Playing displays lead to the observer bias toward Playing witnessed by Blythe, Todd, and Miller (1999). They argued that participants are biased toward thinking that they see Playing in the displays because play is developmentally antecedent: via play we learn all other intentions. In addition to that, our plots would suggest that participants might incorrectly categorize a display as Playing because the movements of the agents involved overlap with the movements used in other intentions. Again, though, it must be stated that despite these kinematic findings and subjective plot analyses being consistent with previous studies, they were based on a limited sample of displays. The findings and analyses just described should thus be viewed as a springboard for advancing our understanding of how we perceive and interpret individual intentions in dyadic interactions.

Autism Spectrum Disorders

As discussed in chapter 6 of the present volume, studies into autism spectrum disorders (ASDs) that examine the proposed deficits in social understanding and intention perception that appear to be part of ASD have continually made use of animacy displays. As in animacy studies in general, many of these studies have made common use of free-response tasks and displays synthetically created without the use of natural human movement (Abell et al., 2000; Bowler & Thommen, 2000; Castelli et al., 2002; Castelli, 2006; Klin, 2000).

In collaboration with Mel Rutherford, we sought to address the use of free-response tasks and synthetic-movement displays in ASD studies (McAleer et al., 2011a, 2011b). We hypothesized that an aspect of the synthetic movement in previous studies' displays could potentially have had an impact on the ability of people with ASD to perceive the intention in them. Furthermore, an eye-tracking study by Klein and colleagues (Klein et al., 2009) suggested that the part of the agent that the participant looked at had a clear impact on the anthropomorphic terms used to describe the displays: participants that tracked a moving triangle toward its leading edge were more inclined to use anthropomorphic terms that those that tracked the triangle at its center. Thus we proposed that our highly diminished displays, which showed moving shapes without leading edges, would create more of a level playing field across an ASD group and controls. Regarding free-response methods, it is not clear that they are appropriate for use in ASD studies (Klein et al., 2009) where language is known to be a barrier. Gao et al. (2009) point out that, in general, participants often write descriptions that they believe the experimenter wants; this tendency would clearly lead to group differences in performance even in the absence of animacy perception differences. Finally, we looked to ask again the question of the effect of viewpoint and whether the preference for overhead displays was held in an ASD population as it was in our previous experiments.

We asked fourteen ASD adult participants and fourteen age- and IQ-matched, typically developing controls to judge the intent in our displays (as before): we showed all six Blythe et al. (1999) intentions, from both viewpoints, in both the animacy and original video format. Each participant saw each display once. We hypothesized two possible outcomes of this study: (1) If the manner of displays or task had no impact, then the results would be comparable to the findings of previous ASD animacy studies; or (2) if the creation of animacy displays from human motion and the use of alternative forced choice did have an impact, then results across groups should be largely comparable. We found no significant differences across groups in any condition or interaction of conditions (McAleer et al., 2011a). Furthermore, we confirmed this lack of group differences via computational permutation and randomization analyses based on Monte Carlo simulations (Fisher, 1922; Westfall & Young, 1989). In addition, we found that the actual response patterns, in terms of confusion between one intention and another, were strikingly similar across experimental population groups. Finally, none of these results were explainable by differences in IQ scores.

In terms of the question of viewpoint, this study failed to show significantly improved performance by participants in the identification of intentions shown in the overhead displays. This is somewhat surprising based on our previous findings. It was our assumption that showing each display only once may have had an impact on the outcome. We suggested that when participants view a display once, their identification of the intention is mainly based on perceptual information; those who view a display a second time utilize cognitive strategies when making their identification. On removing this second viewing from our design, we perhaps stopped the development of any strategy or cognitive reasoning; we propose that it is this reasoning that results in the advantage, or preference, for overhead displays. Thus, the influence of the viewpoint perspective of animacy displays in ASD studies remains an open question.

Regarding ASDs, we concluded that the methodology in our study would indicate that the deficit in social understanding witnessed in ASD studies was magnified by previous techniques such as synthetic displays or free response. Yet before taking that as fact, we grounded our displays in previous literature, the most relevant of which being the Happé-Frith displays (Abell et al., 2000; Castelli et al., 2000). Of the three categories of displays suggested by Happé and Frith (random motion, goal-directed motion, and Theory of Mind motion), we proposed that the majority of the Blythe et al. (1999) intentions (Chasing, Fighting, Following, and—to some extent—Guarding and Playing) would fall into the category of goal-directed motion, based on the criteria set out in Abell et al. (2000). However, the intention of Flirting, by the same criteria, would be classified as a Theory of Mind action, particularly given the similarity in concept to the "seducing" display of Abell et al. (2000). Taken together, this suggests that the Blythe et al. (1999) intentions fall on a continuum between the goal-directed and Theory of Mind intentions of Abell et al. (2000). Moreover, if this is true, then we suggest our study highlights that people with ASD do in fact recognize the intentions depicted within the displays and understand them, but it is perhaps the ability to apply reasoning to the actions that is diminished. We propose this based on the ability of people with ASD in this study to accurately determine the intent within a display when given appropriate options. From there, we feel the question must fall to discovering at what point the system begins to break down.

Future Questions 1: fMRI

The key point of our autism study was that there were no significant differences in the ability to determine the intentions in the displays across groups. Interestingly, this is not evidence of a lack of a deficit, but perhaps evidence that the ASD population uses a compensatory strategy—possibly via a neurological adaptation—to make up for a deficit. At least in terms of biological motion perception research involving point-light walker tasks (Johansson, 1973), recent brain-imaging studies using functional Magnetic Resonance Imaging (fMRI) have shown that even though ASD and control populations have comparable behavioral results, the two groups utilize distinctly differing neural mechanisms (Freitag et al., 2008; Herrington et al., 2007; McKay et al., 2010; 2011; cf. Simmons et al., 2009). To distinguish if this is what is occurring in our animacy study with ASD populations requires extensive research using neuroimaging techniques such as fMRI and EEG—a key future direction for this area of research.

Previous neuroimaging studies using a range of animacy displays have suggested the existence of a socializing or mentalizing network consisting of regions including the medial prefrontal cortex, the temporoparietal junction, superior temporal sulcus, anterior cingulate cortex and the fusiform gyrus (Blakemore et al., 2003; Frith & Frith, 2003; Iacoboni et al., 2004; Saxe & Kanwisher, 2003; Schultz et al., 2004; Schultz et al., 2005). All of these regions have been shown to increase levels of activation when participants are required to either understand the intention within a display or apply Theory of Mind to a situation. Furthermore, Zacks (2004) showed that similar brain regions—namely the posterior superior temporal sulcus (pSTS) and the human motion complex (hMT+)—are used, as in biological motion tasks, to comprehend the actions of Heider and Simmelesque displays. Our hypothesis for studies into the brain regions used in the attribution of mental states by people with ASD would be that the low-level visual processing areas function in ASD participants as in controls, but at this point the system becomes disrupted by not feeding forward to the mentalizing network. This is in fact in line with early findings from Kuzmanovic and colleagues (Kuzmanovic, Schillbach, Georgescu, Santos, Shah, Bente, Fink, & Vogeley, 2010).

Stepping away from ASD studies, our current neuroimaging work looks to bridge the findings from previous fMRI experiments using synthetic animacy displays that depict real-world situations (McAleer, Becirspahic, & Love, 2011; McAleer, Becirspahic, Pollick, Paterson, Latinus, Belin, & Love, 2011). Conceptually, with the use of our ability to transcend from actual video footage to a derived animacy display via our reduction technique, we can explore the brain regions that are involved in perceiving intent in both video and animacy display conditions and, in turn, compare those brain regions to regions previously defined as being used for the perception of intent in synthetic displays. We feel that the first step in achieving this should be to establish a standardized animacy localizer for fMRI studies—one that would locate all involved regions, akin to localizers already established that locate the brain regions used in face processing (Kanwisher, McDermott & Chun, 1997), voice processing (Belin et al., 2000), and biological motion perception (Grossman & Blake, 2001). We propose that displays based on those by Tremoulet and Feldman (2000) would be the most efficient for this purpose because they, compared to other synthetic displays, allow for a high level of control over low-level visual properties of the stimuli (e.g., motion levels across conditions); indeed, this problem of lack of control of visual properties in animacy displays is of current concern (Schultz et al., 2005; Gao, Scholl, & McCarthy, 2010). From the areas localized through the use of these controlled displays, we could then establish which are involved in the attribution of intent in both real-world scenarios and the animacy displays derived from them via our reductive technique (McAleer & Pollick, 2008). Moreover, when topped with an understanding of the kinematics of these displays, we will greatly improve our current understanding of intent perception via animacy displays.

Future Questions 2: Updating the Method

As with all methods of stimuli creation, advancements in technology allow for updates and improvements that facilitate the technique. We no longer place lights on joints such as Johansson (1973) did, or move cutout shapes on a light box such as Heider and Simmel (1944) had done; the eight years that have passed since we first created our method for animacy stimuli have allowed for vast improvements. The most obvious of these improvements has been in the ability to track the motion of an actor.

Our system relies on the ability to (1) record the actor's movement on a video camera, (2) strip the movement and background from the video using EyesWeb, and (3) recreate the displays using Matlab. Nowadays this process may seem rather laborious. Fortunately, with the advent of motion capture suites such as Vicon (www.vicon.com) and their use in biological motion stimuli production (e.g., Piwek, Petrini, & Pollick, 2010), our method of animacy stimuli production can be greatly sped up and improved. In theory, the positioning of a motion marker directly in the center of an actor's head, who is in turn within a motion capture suite, should effectively result in a display that is equivalent to the overhead animacy displays that we have created using our system. Likewise, a marker around the actor's stomach or hips would result in an image similar to those in our side-view displays. Furthermore, given the number of possible cameras in a suite—usually somewhere between six and eighteen—and the possible number of marker positions on the human body, we open up a whole new level in the ability to create animacy displays derived from human motion. Using this advancement to our system, we could easily establish displays to help us address questions such as whether to change the tracked location of the side-view displays; perhaps tracking the head is more effective than tracking the middle. Furthermore, the number of agents used in a display becomes limited only by the amount of space available; this would allow us to move beyond the simple interactions or solitary movements we have thus far explored.

The final key advantage to this design would be in the reduction of time taken to create displays. Our method of using video cameras, though effective, is slow due to the numerous stages involved, including time spent on the video processing to enhance the resultant positional coordinates. Using a motion capture suite would allow a database of movements, at least at the coordinate level, to be created in a matter of hours rather than days. Having such a database will be fundamental to completing our understanding of how we perceive intent and the kinematics we make use of to do so.

The Importance of Real-World Displays

Throughout this chapter we have proposed the hypothesis that without a direct connection between animacy and actual human movement, it is impossible to fully relate the phenomenon being studied to real-world situations. To that end, the work offered in this chapter has attempted to show what can be achieved and learned via a starting point of human actors. However, the question of whether our hypothesis is correct or not is still open to debate, which we shall now briefly comment on.

We humans are social animals; this is evident in our desire and willingness to project goals and intentions onto the movements of geometric shapes. So far, we have witnessed and discussed evidence that our judgment and perception of intent is (1) rapid, (2) available in populations where other social deficits are found to occur and thus is fundamental to us, and (3) universal across countries and populations. Combining these issues, it is probable that our projection of animacy is a leftover from early on in our evolutionary development, when we needed to the ability to determine which things in our environment were living, and what their intentions were, in order to survive. For the most part, the who, what, where, when, and how of animacy can be derived from synthetic displays and movements. Yet eventually a dead end must be overcome when the movement of a shape, synthesized in the lab, can no longer tell us anything important about the intention of the man or woman running toward us in the street or approaching us in a bar. Thus, it is our belief that establishing a bridge between real-world movements and animacy perception will readily help us past this dead end.

Often we hear that it is the little things that make us human. It is perhaps these "little things" that the subtraction method of animacy production helps us to find.

Conclusion

There are indeed exciting times ahead for research into animacy and intention perception. The method we have laid down and summarized in this chapter should be a useful addition to the current literature. Yet at no point do we envision a time in which all animacy stimuli are created in the fashion described herein. Indeed, synthetic methods are by far the better technique for research where the goal is to explore minute variations in a condition such as speed or direction. However, when the goal is to establish how humans understand intent or fail to do so, or when we bridge from animated intent to real-world intent, the creation of animacy displays from human motion becomes paramount.

Notes

1. Throughout this chapter, intentions and displays are referred to in the capitalized format (e.g., Chasing); when we are making reference to an actual action, then the lower-case format will be used (e.g., chasing). For example: "the displays of Chasing were well recognized"; "the second actor was chasing the first actor at an increasing velocity."

2. In terms of labeling the method we use to create animacy displays, we often refer to it as either a subtraction method (as the background is subtracted) or a reduction method (as we reduce video displays to animacy displays). For the purpose of the method, the terms are interchangeable.

References

Abell, F., Happe, F., & Frith, U. (2000). Do triangles play tricks? Attribution of mental states to animated shapes in normal and abnormal development. *Cognitive Development, 15*(1), 1–16.

Akaike, H. (1974). New look at statistical-model identification. *IEEE Transactions on Automatic Control, Ac19*(6), 716–723.

Barrett, H. C., Todd, P. M., Miller, G. F., & Blythe, P. W. (2005). Accurate judgments of intention from motion cues alone: A cross-cultural study. *Evolution and Human Behavior, 26*(4), 313–331.

Bassili, J. N. (1976). Temporal and spatial contingencies in perception of social events. *Journal of Personality and Social Psychology, 33*(6), 680–685.

Belin, P., Zatorre, R. J., Lafaille, P., Ahad, P., & Pike, B. (2000). Voice-selective areas in human auditory cortex. *Nature, 403*(6767), 309–312.

Blakemore, S. J., Boyer, P., Pachot-Clouard, M., Meltzoff, A., Segebarth, C., & Decety, J. (2003). The detection of contingency and animacy from simple animations in the human brain. *Cerebral Cortex, 13*(8), 837–844.

Bloom, P., & Veres, C. (1999). The perceived intentionality of groups. *Cognition, 71*(1), B1–B9.

Blythe, P. W., Todd, P. M., & Miller, G. F. (1999). How motion reveals intention: Categorizing social interactions. In G. Gigerenzer et al. (Eds.), *Simple heuristics that make us smart* (pp. 257–285). Oxford: Oxford University Press.

Bowler, D. M., & Thommen, E. (2000). Attribution of mechanical and social causality to animated displays by children with autism. *Autism, 4*, 147–171.

Brown, R. (1828). A brief account of microscopical observations made in the months of June, July, and August, 1827, on the particles contained in the pollen of plants; and on the general existence of active molecules in organic and inorganic bodies. *Philosophical Magazine, 4*, 161–173.

Camurri, A., De Poli, G., Leman, M., & Volpe, G. (2001). A multilayered conceptual framework for expressive gesture applications. Paper presented at the International MOSART Workshop on Current Directions in Computer Music, Barcelona: Pompeu Fabra University, Audiovisual Institute.

Camurri, A., Mazzarino, B., & Volpe, G. (2004). Analysis of expressive gestures: The EyesWeb expressive gesture processing library. In A. Camurri & G. Volpe (Eds.), *Gesture-based communication in human–computer interaction* (pp. 460–467). Berlin: Springer.

Camurri, A., Trocca, R., & Volpe, G. (2002). Interactive systems design: A KANSEI-based approach. Paper presented at the 2002 Conference on New Interfaces for Musical Expression, Limerick, Ireland: University of Limerick, Department of Computer Science and Information Systems.

Castelli, F. (2006). The Valley task: Understanding intention from goal-directed motion in typical development and autism. *British Journal of Developmental Psychology, 24*, 655–668.

Castelli, F., Frith, C., Happé, F., & Frith, U. (2002). Autism, Asperger syndrome, and brain mechanisms for the attribution of mental states to animated shapes. *Brain, 125*, 1839–1849.

Castelli, F., Happé, F., Frith, U., & Frith, C. (2000). Movement and mind: A functional imaging study of perception and interpretation of complex intentional movement patterns. *NeuroImage, 12*(3), 314–325.

Csibra, G., Gergely, G., Bíró, S., Koós, O., & Brockbank, M. (1999). Goal attribution without agency cues: The perception of "pure reason" in infancy. *Cognition, 72*(3), 237–267.

Fisher, R. A. (1922). On the interpretation of x(2) from contingency tables, and the calculation of P. *Journal of the Royal Statistical Society, 85*, 87–94.

Freitag, C. M., Konrad, C., Haeberlen, M., Kleser, C., von Gontard, A., Reith, W., et al. (2008). Perception of biological motion in autism spectrum disorders. *Neuropsychologia, 46*(5), 1480–1494.

Frith, U., & Frith, C. D. (2003). Development and neurophysiology of mentalizing. *Philosophical Transactions of the Royal Society, Series B: Biological Sciences, 358*(1431), 459–473.

Gao, T., Newman, G. E., & Scholl, B. J. (2009). The psychophysics of chasing: A case study in the perception of animacy. *Cognitive Psychology, 59*(2), 154–179.

Gao, T., & Scholl, B. J. (2011). Chasing vs. stalking: Interrupting the perception of animacy. *Journal of Experimental Psychology: Human Perception and Performance, 37*(3), 669–684.

Gao, T., Scholl, B. J., & McCarthy, G. (2010). Distinguishing intentionality from animacy in the posterior superior temporal sulcus. Paper presented at the Annual Meeting of the Society for Neuroscience 2010, San Diego, Nov. 13–17.

Gelman, R., Durgin, F., & Kaufman, L. (1995). Distinguishing between animates and inanimates: Not by motion alone. *Causal Cognition*, 150–184.

Gergely, G., Nadasdy, Z., Csibra, G., & Bíró, S. (1995). Taking the intentional stance at 12 months of age. *Cognition, 56*(2), 165–193.

Grossman, E. D., & Blake, R. (2001). Brain activity evoked by inverted and imagined biological motion. *Vision Research, 41*(10–11), 1475–1482.

Heider, F., & Simmel, M. (1944). An experimental study of apparent behavior. *American Journal of Psychology, 57*, 243–259.

Herrington, J. D., Baron-Cohen, S., Wheelwright, S. J., Singh, K. D., Bullmore, E. I., Brammer, M., et al. (2007). The role of MT+/V5 during biological motion perception in Asperger syndrome: An fMRI study. *Research in Autism Spectrum Disorders, 1*(1), 14–27.

Howell, D. C. (1997). *Statistical methods for psychologists* (5th Ed.). Pacific Grove, CA: Wadsworth.

Iacoboni, M., Lieberman, M. D., Knowlton, B. J., Molnar-Szakacs, I., Moritz, M., Throop, C. J., et al. (2004). Watching social interactions produces dorsomedial prefrontal and medial parietal BOLD fMRI signal increases compared to a resting baseline. *NeuroImage, 21*(3), 1167–1173.

Johansson, G. (1973). Visual-perception of biological motion and a model for its analysis. *Perception & Psychophysics, 14*(2), 201–211.

Kanwisher, N., McDermott, J., & Chun, M. M. (1997). The fusiform face area: A module in human extrastriate cortex specialized for face perception. *Journal of Neuroscience, 17*(11), 4302–4311.

Klein, A. M., Zwickel, J., Prinz, W., & Frith, U. (2009). Animated triangles: An eye tracking investigation. *Quarterly Journal of Experimental Psychology, 62*(6), 1189–1197.

Klin, A. (2000). Attributing social meaning to ambiguous visual stimuli in higher-functioning autism and Asperger syndrome: The social attribution task. *Journal of Child Psychology and Psychiatry, and Allied Disciplines, 41*(7), 831–846.

Kuhlmeier, V., Wynn, K., & Bloom, P. (2003). Attribution of dispositional states by 12-month-olds. *Psychological Science, 14*(5), 402–408.

Kuzmanovic, B., Schilbach, L., Georgescu, A. L., Santos, N., Shah, N. J., Bente, G., et al. (2010). Animacy perception in adults with high functioning autism: An fMRI study. The 40th Society for Neuroscience 2010 Annual Meeting, Nov. 13–17, 2010, San Diego, California.

McAleer, P. (2006). *Understanding intentions in animacy displays derived from human motion.* PhD thesis, University of Glasgow, Glasgow.

McAleer, P., Becirspahic, M., & Love, S. (2011). How does your brain see "living" circles: A study of animacy and intention using fMRI. *Perception, 2*(3), 200.

McAleer, P., Becirspahic, M., Pollick, F.E., Paterson, H.M., Latinus, M., Belin, P. & Love, S. (2011). Giving life to circles and rectangles: Animacy, intention and fMRI. *Perception, 40*, ECVP Abstract Supplement, 16.

McAleer, P., Kay, J. W., Pollick, F. E., & Rutherford, M. D. (2011a). Erratum to: Intention perception in high functioning people with autism spectrum disorders using animacy displays derived from human actions. *Journal of Autism and Developmental Disorders, 41*(8), 1064.

McAleer, P., Kay, J. W., Pollick, F. E., & Rutherford, M. D. (2011b). Intention perception in high functioning people with autism spectrum disorders using animacy displays derived from human actions. *Journal of Autism and Developmental Disorders, 41*(8), 1053–1063.

McAleer, P., Mazzarino, B., Volpe, G., Camurri, A., Smith, K., Patterson, H., et al. (2004). Perceiving animacy and arousal in transformed displays of human interaction. Paper presented at the 2nd International Symposium on Measurement, Analysis, and Modeling of Human Functions, Genoa, Italy.

McAleer, P., Mazzarino, B., Volpe, G., Camurri, A., Smith, K., Patterson, H., et al. (2005). Creating animacy displays from scenes of human action. Paper presented at the 2005 Cognitive Neuroscience Society Annual Meeting, New York.

McAleer, P., & Pollick, F. E. (2006). Extracting cues to intention with the use of animacy displays derived from human activity. *Perception, 35*(3), 425.

McAleer, P., & Pollick, F. E. (2008). Understanding intention from minimal displays of human activity. *Behavior Research Methods, 40*(3), 830–839.

McKay, L., McAleer, P., Simmons, D. R., Marjoram, D., Piggot, J., & Pollick, F. E. (2010). Distinct configural processing networks reveal differences in biological motion processing in ASD. Paper presented at the Human Brain Mapping—Annual Meeting, Barcelona, June 6–10.

McKay, L., Simmons, D. R., McAleer, P., Marjoram, D., Piggot, J., & Pollick, F. E. (2012). Do distinct atypical cortical networks process biological motion information in adults with Autism Spectrum Disorders? *NeuroImage, 59*(2), 1524–1533.

Morito, Y., Tanabe, H. C., Kochiyama, T., & Sadato, N. (2009). Neural representation of animacy in the early visual areas: A functional MRI study. *Brain Research Bulletin, 79*(5), 271–280.

Piwek, L., Petrini, K., & Pollick, F. E. (2010). Mutlimodal integration of the auditory and visual signals in dyadic point-light interactions. *Journal of Vision, 10*(7), 788.

Rumelhart, D. E. (1980). On Evaluating Story Grammars. *Cognitive Science, 4*(3), 313–316.

Saxe, R., & Kanwisher, N. (2003). People thinking about thinking people—the role of the temporo-parietal junction in "theory of mind." *NeuroImage, 19*(4), 1835–1842.

Schultz, J., Friston, K. J., O'Doherty, J., Wolpert, D. M., & Frith, C. D. (2005). Activation in posterior superior temporal sulcus parallels parameter inducing the percept of animacy. *Neuron, 45*(4), 625–635.

Schultz, J., Imamizu, H., Kawato, M., & Frith, C. D. (2004). Activation of the human superior temporal gyrus during observation of goal attribution by intentional objects. *Journal of Cognitive Neuroscience, 16*(10), 1695–1705.

Simmons, D. R., Robertson, A. E., McKay, L. S., Toal, E., McAleer, P., & Pollick, F. E. (2009). Vision in autism spectrum disorders. *Vision Research, 49*(22), 2705–2739.

Szego, P. A., & Rutherford, M. D. (2007). Actual and illusory differences in constant speed influence the perception of animacy similarly. *Journal of Vision, 7*(12), 1–7.

Szego, P. A., & Rutherford, M. D. (2008). Dissociating the perception of speed and the perception of animacy: A functional approach. *Evolution and Human Behavior, 29*(5), 335–342.

Todorov, A., Pakrashi, M., & Oosterhof, N. N. (2009). Evaluating faces on trustworthiness after minimal time exposure. *Social Cognition, 27*, 813–833.

Tremoulet, P. D., & Feldman, J. (2000). Perception of animacy from the motion of a single object. *Perception, 29*(8), 943–951.

Tremoulet, P. D., & Feldman, J. (2006). The influence of spatial context and the role of intentionality in the interpretation of animacy from motion. *Perception & Psychophysics, 68*(6), 1047–1058.

Troscianko, T., Holmes, A., Stillman, J., Mirmehdi, M., Wright, D. B., & Wilson, A. (2004). What happens next? The predictability of natural behaviour viewed through CCTV cameras. *Perception, 33*(1), 87–101.

Westfall, P. H., & Young, S. S. (1989). P-value adjustments for multiple tests in multivariate binomial models. *Journal of the American Statistical Association, 84*(407), 780–786.

Willis, J., & Todorov, A. (2006). First impressions: Making up your mind after a 100-ms exposure to a face. *Psychological Science, 17*(7), 592–598.

Zacks, J. M. (2004). Using movement and intentions to understand simple events. *Cognitive Science, 28*(6), 979–1008.

8 Design for Learning: The Case of Chasing

Willem E. Frankenhuis and H. Clark Barrett

Abstract

Often, mental development is viewed as resulting either from domain-general learning mechanisms or from highly specialized modules containing substantial innate knowledge. However, an evolutionary developmental perspective suggests that learning and specialization are not necessarily in opposition. Instead, natural selection can favor learning mechanisms that rely on information from the environment to construct adaptive phenotypes, exploiting recurrent properties of fitness-relevant domains. Here we consider the possibility that early action understanding is centered on domain-specific action schemas that guide attention towards domain-relevant events and motivate learning about those domains. We examine chasing as a case study. We report studies (1) exploring the mechanisms that guide infants' attention to chasing events and (2) examining the inferences and judgments that children and adults make. We argue that these findings are consistent with the possibility that natural selection has built "islands of competence" in early action understanding that serve as kernels for future learning and development.

Introduction

The nature-nurture debate has mostly died in biology, but it is still actively discussed in the social sciences. The reason it has largely disappeared from discussions of biology is that progress in our understanding of human development has revealed that the whole of ontogeny is an adaptive process in which "nature" cannot be teased apart from "nurture." Instead, phenotypes are actively constructed through interactions between the genotype (including genes and gene regulatory systems) and internal and external environments. This view renders distinctions such as innate versus acquired, and nature versus nurture, of limited use. Instead, a more useful concept is *design for development*, which holds that developmental systems are shaped by natural selection to produce adaptive outcomes via interactions with recurring features of developmental environments. Developmental systems are sensitive to specific aspects of the environment and structure their interactions with these elements in adaptive ways (Frankenhuis & Panchanathan, 2011); therefore, the developmentally relevant environment and its interactive effects, rather than being factors separate from the developmental system, are a function of the design features of the developmental system itself (Barrett, 2007; Oyama, Griffiths, & Gray, 2001). In this chapter, we suggest that the same view will be useful for understanding the development of psychological phenotypes as well.

Adaptations are misunderstood in psychology to the extent that they are viewed as inflexible, isolated from other mechanisms, and incompatible with learning (Buller, 2005). Adaptations include systems for learning (Gould & Marler, 1987; Greenough, Black, & Wallace, 1987), such as inferential systems for estimating environmental parameters (e.g., the location, value, and abundance of resources; Dukas, 2008), mechanisms for assessing one's own attributes relative to other individuals (e.g., relative body size, social status; Maynard Smith & Harper, 2003), and networks that selectively form associations between particular objects and events (e.g., foods and nausea; for a review, see Seligman, 1970). Adaptations also include perceptual and motivational biases for attending to fitness-relevant objects, such as animate events (New, Cosmides, & Tooby, 2007). These biases ensure that organisms process environmental information relevant to (1) current decision making (e.g., an approaching predator) and (2) long-term developmental trajectories (e.g., in environments estimated to be dense with predators, organisms may develop specialized morphology and behavior for avoiding predators; Benard, 2004; Harvell, 1990).

Many animals use psychological adaptations to predict future events, including the locations, trajectories, and actions of other agents, who may have conflicting interests (Barrett, 2005). Consider a gazelle spotting a pride of lions. Depending on the pride's predicted trajectory (e.g., Are they moving closer or farther away?), the gazelle may relax or increase its vigilance levels. If the lions are moving closer, but their gaze is averted, the gazelle may estimate it hasn't been detected and freeze, hoping it remains unseen. However, if the lions are fixating on her, the gazelle may flee. This could trigger a chase, in which the lions and gazelle quickly and continuously predict each others' locations and trajectories. What psychological systems enable such behavior prediction? How do these mechanisms evolve and how do they develop?

Inference Based on General Concepts

Developmental psychologists tend to believe that human infants rely on a small set of general concepts to understand and predict the behavior of others. For instance, according to one influential view—known as the "teleological stance" or infants' intuitive theory of "rational action" (Gergely, 2010; Gergely & Csibra, 2003)—infants expect agents to choose ACTIONS consistent with achieving their GOAL in the most efficient manner possible given environmental CONSTRAINTS. Having such a model of rational action would afford infants inferences about how others will behave. For instance, knowing that an agent has the goal of catching another agent, one might predict the chaser's action to be the shortest path to its victim as allowed by environmental constraints (Gergely, Nádasdy, Csibra, & Bíró, 1995). The principle of rationality can also be used to infer goals (Csibra, Bíró, Koós, & Gergely, 2003): if an agent takes the shortest path to a moving object, one might infer that the agent has the goal of gaining proximity to this object. Finally, if one agent has the goal of catching another and takes the shortest path toward the victim except at a particular point, one could infer a constraint at this location (Csibra et al., 2003). Thus, according to the rationality principle, infants interpret and predict the behavior of other agents using a small set of general concepts—GOAL, ACTION, and CONSTRAINT—interconnected by an assumption of efficiency.

Since its inception, the teleological stance perspective has generated novel and unique predictions that—through systematic testing—have resulted in groundbreaking discoveries (for reviews, see Gergely, 2010; Gergely & Csibra, 2003). The approach has clearly made major contributions to the study of developmental science. Nonetheless, we argue that this theory of infants' inferential capacities, while correct in the domains where it has been tested, might in fact be too broad. Although we agree that "rationality" principles are likely to be engineered into infants' early-developing behavior prediction abilities, we suggest that these abilities might consist of a larger number of narrower inference abilities and not a single, broad, teleological stance (Barrett, 2005). According to this proposal, each of these more specific abilities includes teleological principles for understanding the behavior of others, but covers a narrower scope of behavior than does the assumption that agents pursue the most efficient means to their goals. We call these smaller teleological abilities *islands of competence*, to distinguish them from the more continental view of a "stance" (see figure 8.1).

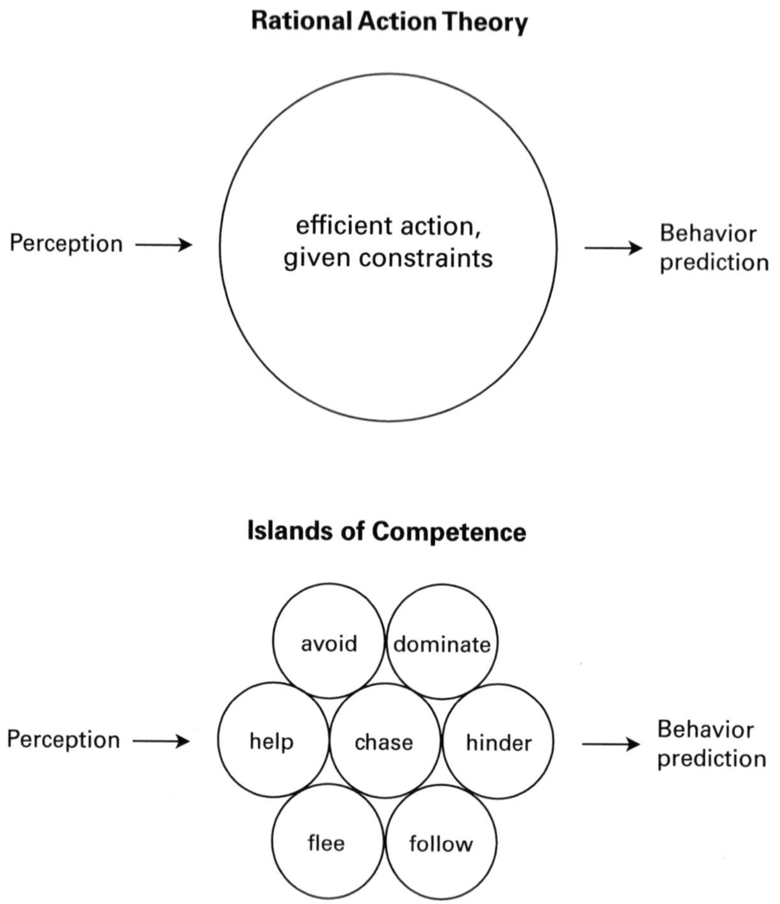

Rational Action Theory

Perception → efficient action, given constraints → Behavior prediction

Islands of Competence

Perception → avoid, dominate, help, chase, hinder, flee, follow → Behavior prediction

Figure 8.1
The theory of rational action versus islands of competence.

The reason for our view is that while concepts like GOAL, ACTION, and CONSTRAINT may be useful for psychologists in individuating *types* of concepts in the mind, they would not by themselves be useful to infants as *tokens* of concepts that could be used to predict actual behavior. General concepts, by themselves, don't sufficiently constrain prediction space. For a behavior prediction system to evolve via natural selection, it must enable infants to make effective predictions about behavior. For example, to make predictions about the behavior of a cat chasing a mouse, an infant must do more than apply a general principle of rationality; he or she must understand the logic of *chasing*, in which the goal of one agent is to catch another agent, and the goal of the second agent is to not be caught. For this reason, we suggest that infants' early-developing action prediction abilities are centered on *action schemas* (e.g., chasing, following, grasping, dominating, searching, etc.), which, while obeying a teleological logic, use a larger set of more specific concepts to do so (Barrett, 2005). The relevant concepts are expected to be different for different domains, but might include agent concepts such as PREDATOR, PREY, MOTHER, and HELPER, goal concepts such as CHASE, ESCAPE, APPROACH, SIGNAL, PROTECT, and constraint concepts such as BARRIER, PATHWAY, CONTAINER, LINE OF SIGHT, and so on. From an evolutionary point of view, the virtue of such specific concepts is that, when properly combined, they can yield more precise predictions about behavior by reducing degrees of freedom in prediction space. Their drawback, of course, is that they yield behavior prediction skills that are patchy, endowing infants with islands of behavior-prediction ability within a much larger sea of *possible* inferences about rational action that, because they don't have the relevant action schemas, infants can't make. However, if a function of early-developing action schemas is learning, we expect action schemas to become more elaborate over the course of an infant's development through (1) refinement of input conditions, (2) learning of new concepts, (3) fine-tuning of inferential processes, and (4) interaction with other knowledge structures, which we will discuss in greater detail below.

Our view predicts that infants will develop action schemas at those ages at which those schemas become fitness-relevant (or became fitness-relevant across evolutionary time). Other schemas that are equally rational may await later development. For example, the logic of chasing might be useful for young infants both because of its role in enabling children to understand chase play—a common form of play that may train adult abilities of pursuit and evasion (Pellegrini, Dupuis, & Smith, 2007; Steen & Owens, 2001)—and because it might enable infants to perceive when something is stalking them or predict the trajectory of animate objects the infant is trying to grasp. Abundant evidence exists for the logic of such action schemas as affiliative approach, grasping, and searching for inanimate objects (such as food or a toy) (Csibra, Gergely, Bíró, Koós, & Brockman, 1999; Gergely et al., 1995; Hamlin et al., 2007; Kuhlmeier et al., 2003; Thomsen et al., 2011; Woodward, 2009; Woodward & Sommerville, 2000). However, there are other action logics, such as those associated with mating and reproduction (e.g., courtship, jealousy) as well as other goals pursued in adulthood (e.g., political maneuvering), which do not develop until later.

The teleological stance is not the only case in which infants might possess islands of competence that only later grow to span a larger conceptual space. The literature on Theory of Mind, too, holds that children reason using a small stock of extremely abstract and general concepts, including BELIEF and DESIRE. However, like GOAL and ACTION, these concepts by themselves would not be sufficient to predict actual behavior without plugging in specific contents—specific beliefs and desires—only some of which children might understand at a given age. GOALS can include things that infants understand, such as chasing and approach, but can also include things that infants don't understand, like becoming wealthy, achieving checkmate in chess, and committing adultery. Similarly, BELIEFS can include things that infants understand, such as beliefs about where food is hidden, and things that infants don't understand, like belief in God or beliefs about the power of free markets to heal ailing economies. When combined with principles of rational action, these beliefs could be used to predict behaviors (like going to church or voting Republican). However, most research on early-developing abilities to predict an agent's behavior is limited to such cases as predicting where an agent will look for food or some other desired object (e.g., a caterpillar looking for an apple or a piece of cheese; Surian, Caldi, & Sperber, 2007). Without looking at the specific belief contents that infants are able to track and the contexts in which they can do so, we risk concluding that a more domain-general ability has developed than actually exists. In other species, research has shown that abilities thought to be general, such as knowledge tracking, are closely tied to certain contexts, such as competition over food (Hare, Call, & Tomasello, 2001). It is possible that children's early-developing abilities in many domains are similarly targeted toward contexts that are useful for the child.

In our view, then, looking at the specific content and contexts that babies are able to understand is just as important as looking at the more abstract features of their conceptual structure. Moreover, we suggest that early-developing action schemas serve as kernels for future learning, including the acquisition of more general concepts. Thus, where many scholars hold that infants start out with general concepts and construct specific understandings later, in this chapter, we propose the opposite.

In the following section, we first provide details about how action schemas might be instantiated in infants' minds in terms of perceptual inputs, cognitive rules, and behavioral outputs, as well as their role in shaping learning. We then discuss chasing as a case study and review empirical work conducted with infants, children, and adults. We will illustrate that existing evidence is consistent with an islands of competence view. In the discussion section, we suggest how our view can be distinguished empirically from other current perspectives. Finally, we also point to several psychological domains, other than chasing, where our model can be tested.

Domain-Specific Action Schemas

A virtue of early-developing islands of competence is that action schemas can help solve frame problems that would be faced by more domain-general systems: that is, action schemas would avoid churning through many possibilities of what might count as "rational" by specifying specific actions that are rational in a given context. In chasing, for example, roles can be assigned to chaser and evader based on their behavior (approach and avoidance, respectively). These in turn lead to predictions about what is rational: for the chaser, systematically reducing the distance to the evader, and for the evader, doing the opposite. Similarly, cues to mutual approach could activate a different schema, an approach schema, in which the goals of agents are to (1) come into contact with each other and (2) find the shortest path in order to do so. In each case, the output of the system is a prediction or expectation that can be measured using, for example, violation of expectation or preferential looking techniques.

Like all evolved specializations, islands of competence evolve under trade-offs. For example, islands of competence may solve frame problems for infants by generating specific expectations in particular cases, but will be less likely to do so effectively outside their domain of application (Barrett & Kurzban, 2006; Cosmides & Tooby, 1994; Frankenhuis & Ploeger, 2007). We would expect such schemas to be in place at the ontogenetic stage at which they resulted, across evolutionary time, in fitness benefits to children. These fitness impacts could come in at least two ways: (1) Benefits of immediate predictions (e.g., predicting where a predator, prey, or conspecific will run, or how a parent might move in response to the child's crying) and (2) learning benefits (note that this is not a strict dichotomy, as predictions about behavior can have learning consequences).

Trade-Offs in the Evolution of Domain-Specific Developmental Designs

How many action schemas should we expect there to be? Here, at least two factors are important. The first is to what extent a particular action schema has a fitness impact on the child at various life stages. For young children, for example, a chasing schema might have more fitness benefits than a mate-competition schema. A second factor has to do with the scope of the schema: how broad or narrow, in spanning possible interaction space, we might expect islands of competence to be. For example, chasing shares much in common with leading and following, both in terms of goals (e.g., staying close to the target) and perceptual properties (e.g., a chaser and evader tend to move in the same direction, as do a leader and follower). However, the goals of the leader (in a following event) and an evader (in a chasing event) are different: The evader is trying to lose the chaser, whereas a leader may want to be followed. Similarly, chasing and playing show perceptual overlap—for instance, play often involves bouts of chasing (Barrett, Todd, Miller, & Blythe, 2005)—yet their goals are different and involve different expectations about behavior—for instance, just how "rational" it is for the evader to allow himself to be caught.

A relevant principle is the principle of functional incompatibility, which was originally developed in thinking about the evolution of multiple memory systems (Sherry & Schachter, 1987). Consider two adaptive functions, X and Y. We might imagine a single mechanism, such as a single action schema, that evolved to handle both functions—for example, making predictions for both chasing and leading/following. Or there could be two mechanisms—one specialized to handle chasing and one for following. In evolutionary terms, the principle of functional incompatibility can be stated as a heuristic: All else being equal, natural selection should favor multiple systems when the net fitness benefit of having multiple systems is greater than the net benefit of having a single system that handles both functions. These benefits will depend on such factors as the costs of building and maintaining multiple systems and the magnitude of the marginal benefit of having multiple systems over a single one (Cosmides & Tooby, 1994). We would expect this marginal benefit to depend on the extent to which different schemas entail different goals. For example, though chase-play and predator-prey interactions often show high degrees of perceptual overlap, their fitness-relevant goals are vastly different; effectively distinguishing them and deriving adequate inferences is crucial. Hence, we might expect infants to be sensitive early on to cues that distinguish cases of chase-play from cases that involve genuine aggression (Smith & Lewis, 1985; Steen & Owens, 2001).

As we mentioned, some of the fitness benefits of early-developing action schemas could come from the generating of immediate expectations about how others will behave. This might be especially true in cases where the child herself is one of the agents in the interaction. For example, cues to hostile approach could prompt the child to cry, thereby avoiding possible harm (e.g., by eliciting parental help). Cues to friendly approach, however, could prompt the child to initiate approach to the other agent. These two responses can have quite different consequences for fitness, and so there are many cases where forming detailed expectations about what another agent will do could benefit even small children.

Islands of Competence May Facilitate Learning

Action schemas can also provide learning benefits; this is true even in the case of interactions in which the child herself is not taking part. For example, by watching predator-prey interactions from a distance, useful information can be gleaned about predator and prey behavior. In general, by watching others, babies can learn the social structure of their world, including who is dominant over whom (Thomsen, Frankenhuis, Ingold-Smith, & Carey, 2011), who is nice and who is nasty (Hamlin, Wynn, & Bloom, 2007; Kuhlmeier, Wynn, & Bloom, 2003), how to interact appropriately with artifacts (Gergely, Bekkering, & Király, 2002), the nuances of local norms of social interaction (Rakoczy, Warneken, & Tomasello, 2008), and more.

Action schemas can facilitate learning in at least two ways. First, they can help guide the infant's attention toward fitness-relevant objects and events. Events can be worth attending to when the behavior of agents satisfies the infant's action predictions, but also in cases where agents' behavior violates the infant's expectations, as this may help tune the underlying action schema by teaching the infant its boundary conditions and exceptions. Second, action schemas can help the child to parse behavior into appropriate categories, which is critical for learning. For example, in order to learn about predation as a separate category of behavior from affiliative interactions, it is important not to blend knowledge gained from watching affiliative approaches between parents and offspring (e.g., a parent approaches a child with goal of helping it) with knowledge gained from watching leopards stalk gazelles (e.g., a leopard approaches a gazelle to kill it).

If our conjecture is correct, and infants' early-developing understanding of behavior is organized around specific islands of competence, then the stock of action schemas that infants initially posses is clearly not the only set of schemas they will ever develop. Surely, adults understand many more contexts of interaction, and make much more nuanced distinctions, than children do. Infants might begin with, for example, relatively simplistic, stereotyped, and separate schemas of chasing and social dominance. At first, they might treat all cases where A approaches B and B flees as the same kind of event—chasing—and all cases where one individual defers to another (e.g., stepping out of the way to let them pass) as dominance (Thomsen et al., 2011). However, by attending to details of interactions, infants can begin to notice fine-grained distinctions between types of actions that might, ultimately, cause them to bifurcate into multiple schemas: for example, one schema for predation-related chasing and another for chasing after a moving inanimate object; one schema for deference out of fear and another for deference based on respect (Henrich & Gil-White, 2001). In some cases, the ability to represent new goal states may develop when behavior is observed that cannot be assimilated into an existing schema, which may result in the spawning of a new schema (Jacobs, 1997). There are many ways in which the development of more sophisticated knowledge can emerge from infants comparing real-world, observed behavior to simpler schemas they already possess (for examples outside the domain of chasing, see Carey, 2009; Gelman, 1990; Mandler, 1992).

Finally, the islands of competence view suggests that rather than observing stage-wise conceptual shifts, such as a shift from a desire-based psychology to a belief-based psychology across all domains, one might see something more akin to Piaget's notion of décalage (Piaget & Inhelder, 1951). At an early age, children might not have a domain-general understanding of belief or a general ability to track beliefs. Rather, they might have narrow competences of belief tracking, such as an ability to track an agent's belief about where a valued object, such as food, is hidden (Onishi & Baillargeon, 2005; Surian et al., 2007). The ability to track other beliefs, such as a person's belief about his mate's fidelity, may appear later. If early-developing skills are organized around islands rather than broad domains, competences that are sometimes viewed as part of the same domain (e.g., interpreting the behavior of other agents teleologically) may actually develop at different points during ontogeny. We now turn to the case of one particular action schema for which we and others have gathered evidence: a chasing schema.

Psychological Design for Chasing

What motion features are characteristic of chasing events, and what mechanisms in humans leverage these properties to interpret and predict behavior? Psychologists have long known that infants are sensitive to motion. For instance, even though the neonate's visual field ranges only 15 to 20 degrees to either side for static stimuli (when the head is still), it can be wider and the distance greater for moving objects (Tronick, 1972). Moreover, infants are especially attuned to biological motion (e.g., Bertenthal, 1993; Fox & McDaniel, 1982). When neonates are presented with two displays side-by-side—one depicting a point-light display of an animal walking and the other depicting nonbiological motion—they tend to navigate their gaze toward the biological motion (Simion, Regolin, & Bulf, 2008). But among the many types of motion infants might attend to, is there anything special about chasing?

Early Development of Perceptions of Chasing

Computational analyses suggest a number of properties that distinguish chasing from other kinds of animate motion. One simple yet effective cue appears to be absolute velocity. Blythe, Todd, and Miller (1999) trained a neural network on three hundred examples of motion trajectories from six intentional categories—pursuing, evading, courting, being courted, fighting, playing—which had been generated by humans controlling a computer mouse (for details, see also Barrett et al., 2005). After training, the network correctly categorized 82 percent of the trajectories in terms of their intention. Out of seven cues, absolute velocity yielded the most accurate categorization, followed by relative angle, relative velocity, relative heading, relative vorticity, absolute vorticity, and relative distance. Given the discriminatory power of absolute velocity, we may expect an early-developing chasing schema to include sensitivity to this cue. (This is not to say that high velocities correlate exclusively with chasing; see below.)

Adults judge faster motion as more animate, whether faster speed results from accelerations (Tremoulet & Feldman, 2000; see also Scholl & Tremoulet, 2000) or from faster constant speed (Szego & Rutherford, 2007), and animacy judgments increase when entities move in a direction that violates gravity (e.g., upward; Szego & Rutherford, 2008). Further, it is easier for adults to detect a chasing event among distracter stimuli when the chaser moves relatively fast; that is, a depiction of a lamb following its mother is more difficult to detect than a depiction of a wolf chasing its prey (Dittrich & Lea, 1994). We recently explored whether infants, too, are sensitive to accelerations. We presented four- and ten-month-old infants with two displays side-by-side, each depicting two moving discs, but in one display, one of the discs would sometimes accelerate. By a substantial margin, nearly all infants looked longer at the accelerating motion (Frankenhuis, House, Barrett, & Johnson, 2013). We note that this finding does not imply that infants perceive accelerations as animate in the way that adults do. Although this is conceivable, our experiment was not designed to test this (for recent reviews on infants' perceptions of animate motion, see Frankenhuis, Barrett, & Johnson, 2013; see also chapter 11 of this volume).

Accelerations are an integral part of chasing events: one agent approaches another, the other accelerates away (possibly triggering the chaser to accelerate as well). Thus, sensitivity to acceleration is consistent with an attentional system designed to orient toward chasing events. However, there are various reasons that accelerations might occur, including accelerations due to gravity (a rock falling) and accelerations due to self-propelled motion (chasing). To the extent that the adaptively relevant (or formal) properties of different stimuli overlap and demand a similar response (e.g., attention-orientation), selection might not produce two distinct mechanisms. On the flip side, to the extent that stimuli *do* require different responses (e.g., continued monitoring may be appropriate for agents, but not for inanimate objects), selection would be expected to favor different responses to these stimuli.

In addition to high velocities and accelerations, at least two relational properties are also characteristic of chasing: attraction (one agent pursuing another, sometimes called "heat-seeking") and fleeing (the evader tries to get away). Rochat, Morgan, and Carpenter (1997) investigated whether three-month-old infants are sensitive to these properties. Infants watched two displays side-by-side, each depicting two discs (blue and red) moving across the screen. The motions of the red and blue discs were identical in both displays, with the exception that in one display the discs would be chasing each other, while in the other they were moving independently. The results showed that three-month-old infants looked longer at the chasing display (although this discrimination was observed only in infants with relatively long attention spans). In the displays used by Rochat and colleagues (1997), two kinds of contingencies in the chase could have generated the infants' preferential looking. First, the chaser took the shortest path to the evader (attraction); second, when the chaser came close, the evader accelerated away (fleeing). Both attraction and fleeing allow for the prediction of one agent's behavior based on the motions of the other agent. It is unclear whether the infants in the study attended to attraction or fleeing, or to both.

We recently gathered data relevant to this question. In our experiment, four- and ten-month old infants watched displays showing attraction without fleeing. Looking times indicated that boys in both age groups preferentially attended to attraction without fleeing over a control display; however, girls did not (Frankenhuis et al., 2013). These results raise the possibility that boys and girls may be differentially sensitive to different properties of chasing (or even of animate motion in general). Another question of interest is: Why did girls preferentially attend to Rochat's contingent display, but not to ours? His displays combined both attraction and fleeing; ours isolated only the former. It is possible that at the ages we tested, girls are sensitive to fleeing without attraction, but unlike boys, not to attraction without fleeing. Future studies might explore infants' perception of fleeing without attraction.

We have thus far discussed the perceptual inputs that orient the visual system toward chasing, focusing on high velocities and accelerations, and two forms of social contingency: attraction and fleeing. In addition, we want to know whether infants, like adults, interpret dynamic motion displays in goal-directed terms (Bassili, 1976, Heider & Simmel, 1944; Michotte, 1963). Rochat, Striano, and Morgan (2004) used a habituation paradigm to investigate whether infants perceive social contingency as goal directed. They showed four- and nine-month-old infants a video of one disc chasing another (e.g., red chasing blue) until looking times decreased. If infants had habituated to the perceptual features of the event, they should have remained uninterested if a role reversal occurred (blue chasing red), since this was perceptually similar to what they had seen before. However, if infants had assigned different goals to the chaser and the evader, they should have regained interest in response to a role reversal, because the agents had changed their goals. When shown the role reversal, infants in the younger age group (four-month-olds) did not regain interest; however, infants in the older age group (nine-month-olds) did increase their looking time, which suggests that they had assigned different goals to the chaser and the evader.

When do infants begin to use inferred intentions to predict a chaser's future trajectory? Csibra and colleagues (2003) presented nine- and twelve-month-old infants with an animation of a large ball moving in the direction of a small ball in a heat-seeking manner, always taking the shortest path toward it. The small ball then moved through a hole in a barrier too small for the large ball to pass through. The large ball would move around the barrier. In the test phase, infants were presented with two different endings: Either the large ball "caught" (contacted) the small ball, or the large ball would slide past the small ball and come to a halt. The results showed that twelve-month-old, but not nine-month-old, infants expected the large ball to contact the small ball (see also Wagner & Carey, 2005). This work shows that by at least twelve months of age, infants can infer the intention to capture another agent based on motion cues alone, and use the inferred intention to predict the trajectory and future locations of a chaser.

Finally, several studies have looked at older children's ability to verbally categorize or describe animate motion displays based on goals. Berry and Springer (1993) showed three-, four-, and five-year-old children four different versions of a chasing display. The first version was identical to the chasing display used by Heider and Simmel (1994). In the second version, form was preserved but motion disrupted, so that children watched static frames from the original tape, sampled at two-second intervals, with each frame lasting two seconds. In the third version form was disrupted but motion preserved, such that the objects were not triangles and a circle anymore, but rather mosaic shapes blending into their background. In the fourth version, both motion and form were disrupted. Children of all age groups tended to interpret the displays in anthropomorphic terms when motion was preserved; however, they did not when motion was disrupted, irrespective of disruptions of form. The same pattern of results was obtained with adults—although adults did provide more anthropomorphic descriptions overall than children (Berry, Misovich, Kean, & Baron, 1992).

Barrett and colleagues (2005) showed three-, four-, and five-year-old German children four motion displays (chasing, following, fighting, and playing) that were identical to those used by Blythe and colleagues (1999). Children watched one display at a time and were asked which of two intentions was depicted on the display (e.g., chasing or following), with one intention always being the correct one. Results showed that four- and five-year-old German children were above chance at categorizing the four types of intentional motion, but three-year-old children were not. Interestingly, children were not above chance for one motion category, chasing—probably, the experimenters surmise, because the experiment used a word for chasing that most German three- to-five-year-olds don't know (*verfolgen*), given prior results showing that children understand chasing from an early age (Csibra et al., 2003; Rochat et al., 2004). These results suggest that language skills used to describe and categorize motion trajectories may develop later than the relevant perceptual and conceptual abilities.

Fine-tuning and Elaboration of the Chasing Schema across the Life Span

How does our understanding of chasing develop into adulthood? As we have mentioned, we believe that an early-developing chasing schema's primary learning function is to orient infants' attention to instances of chasing, thereby allowing them to learn about instances of chasing via input from the world; this is similar to the role that an early-developing face schema or template might play in the development of face perception (Morton & Johnson, 1991). Thus, adults' knowledge of chasing will not be the same as that of infants. It will, in a broad sense, be richer because it is informed by experience. However, there are several ways that adults' inferential structures related to chasing might have narrowed compared to infants' and several ways in which they might have broadened. This is because some kinds of learning entail elaboration of existing knowledge structures and skill sets, which is done by constructing new categories, developing richer interpretations, and so on. Other kinds of learning, however, can entail various kinds of "pruning" (e.g., Kuhl, 2004; Pascalis, de Haan, & Nelson, 2002): Input conditions for activating a schema might become narrower, properties of objects and events might be forgotten, concepts or connections between concepts might erode, and information previously used in decision making might become muted. For example, among many American city-dwellers, an encounter with any spider, whether dangerous or not, might cause an aversive reaction because experience has not yet narrowed the category to just those spiders that are dangerous. Among those who live in spider-rich environments such as the Amazon rainforest, on the other hand, an initially broad category of fearful objects might become narrowed by experience to include just those that merit a fearful reaction. Similarly, in juvenile rhesus macaques, predator alarm calls are initially generated for a larger variety of objects than is appropriate, but this category of objects is gradually narrowed to include only the appropriate set of targets (Cheney & Seyfarth, 1990).

Pruning processes may also play a role in the development of contingency detection. We previously mentioned our finding that boys preferred to attend to "attraction without fleeing" over two discs moving independently (Frankenhuis et al., 2013). We also presented these stimuli to adults, and they did not rate the contingent display as more animate. There could be several reasons for this. Adults might have perceived the motion as contingent, but not as animate. If so, this would not support the idea that adults have narrower input criteria for detecting contingencies in motion; it would just mean that our stimuli do not fit the criteria for being categorized as animate. It is also possible, however, that the input criteria of action schemas' contingency detectors become narrower across development; this narrowing could be the result of real-world experiences (in this case, experiences with actual chasing events). Such perceptual narrowing has been demonstrated in several other areas of perceptual development, in particular phoneme perception (Kuhl, 2004) and face perception (Pascalis et al., 2002).

Prior work has shown that adults tend to judge and describe some kinds of displays of moving objects as animate (Heider & Simmel, 1944; Morris & Peng, 1994; Rimé, Boulanger, Laubin, Richir, & Stroobants, 1985)—even displays showing only a single geometrical shape (Michotte, 1963; Szego & Rutherford, 2007, 2008; Tremoulet & Feldman, 2000)—and researchers have investigated the range of parameters that lead to judgments of animacy and specific categories of animate motion. For instance, Gao, Newman, and Scholl (2009; see also chapter 9 of this volume) explored how objective chasing (the degree to which one shape, the wolf, actually pursues another shape, the sheep) relates to perceived chasing (the degree to which participants detect this behavior). Gao and colleagues found that objective chasing is readily perceived as chasing when the wolf pursues the sheep in a perfectly heat-seeking manner, always moving in the exact direction of the sheep. However, when the wolf's motion deviates even slightly from perfect heat seeking (e.g., it moves in the general direction of the sheep, but not directly toward it), chasing is more difficult to detect—even when the wolf *is* efficient at reducing its distance to the sheep. Percepts of chasing are further impacted by the direction in which the wolf is orienting. When the wolf faces the direction in which it is moving, the detection of chasing is enhanced. In contrast, when the wolf's orientation is random with respect to its trajectory, or when it is surrounded by distracter shapes that also face the sheep, detection of chasing is impaired (Gao, McCarthy, & Scholl, 2010; see also Gao & Scholl, 2011, and chapter 9 of this volume).

Action schemas might also develop as a result of processes such as elaboration of information structures, formation of new connections between knowledge domains, discovery of new properties of objects and events relevant to a particular decision problem, bifurcation of one concept into several more fine-grained ones, and so on. One way in which an island of competence for chasing might grow is through expansion of the scope of knowledge (both contextual and general) that is brought to bear on inferences about intentions, goals, and behavior prediction. For instance, adults have been shown to segment an action stream not only based on bottom-up processing of distinctive sensory characteristics, but also based on top-down effects of knowledge structures, including information about actors' intentions (Zacks, 2004; see also chapter 13 of this volume). So, an adult may know that a chaser might mislead an evader by averting its gaze; the adult might then predict that a chasing event is not over, even though an averted gaze suggests this to the child. Compared to infants, adults in general are likely to use a larger repository of background knowledge (e.g., properties of the predator and prey) and contextual information (e.g., relevant features of the ecology) in making predictions about the outcome of a chasing event (e.g., who will win), often resulting in more accurate behavior prediction.

In their landmark paper from 1944, Heider and Simmel proposed that knowledge might exert top-down influences on intention judgments. Toward the end of their animation, the big triangle (*T*) chases the small triangle (*t*) and the disc (*c*) twice around the house. Participants nearly universally interpreted these movements as chasing. However, when the animation was played in reverse, most participants did not interpret this as *t* and *c* chasing *T*. Heider and Simmel suggest that this difference might result from knowledge of the story line, which shaped the participants' interpretations of intentions. Specifically, they suggest that a history of antagonism between *T* and the pair *t* and *c*, combined with *T* being stronger than *t* and *c*, facilitates interpretations of chasing in the forward animation (p. 254). In contrast, in the reverse animation this combination of features is absent; this absence inhibits the uniform percept of chasing and results in more diverse interpretations. The percept of *T* being stronger may derive from an earlier part in the animation in which *T* pushes *t* backward, as well as from *T* being physically larger than *t* and *c*. Subsequent work has shown that interpretations of agent characteristics (e.g., whether *T* is viewed as aggressive or not) can be manipulated by providing participants with information about the characteristics of *other* agents in the animation (Shor, 1957).

Conclusion

We have reviewed a variety of data consistent with the development of a chasing schema early in childhood that (1) uses perceptual cues of pursuit and evasion to orient infants' attention toward chasing events and (2) enables infants to make predictions about the behavior of agents engaged in chasing. In our view, this evidence is consistent with the idea that infants' abilities to understand and predict intentional action are centered on certain islands of competence, which only later in development are fleshed out into other areas of rational action understanding. Despite the fact that domain-general models of rational action can account for much of the existing data, we have argued on theoretical grounds that, by themselves, general concepts (such as BELIEFS, DESIRES, and CONSTRAINTS) do not allow effective behavior prediction. Although we have not reviewed the evidence here, we think that developmental studies also point to other islands of competence for action understanding, including an approach schema (e.g., Csibra et al., 1999; Gergely et al., 1995), a grasping schema (e.g., Woodward, 2009; Woodward & Sommerville, 2000), a helping schema (Hamlin et al., 2007; Kuhlmeier et al., 2003), and a dominance schema (Thomsen et al., 2011). Of course, there are likely to exist other schemas as well.

We conclude with several caveats to the claims we have made here. First, there exist several pieces of evidence that challenge the view that action understanding is centered on domain-specific action schemas. For example, a recent study by Southgate, Johnson, & Csibra (2008) suggests that infants can make predictive inferences about goals even in a case that appears to be outside the boundaries of interactions that would have occurred over evolutionary time: namely, in a case where a rubbery arm stretches and bends through a maze in order to grasp an object. Consistent with the teleological stance framework, infants expect this bendy arm to bend only when it is forced to do so by the walls of a maze, but not when the barriers are removed: efficient action toward a goal, under constraints. We acknowledge that this is a surprising result, consistent with the teleological stance theory, and a challenge for the islands of competence view since bendy arms are presumably not within the proper evolutionary domain of any such competence. However, we suggest that it is possible that even this display might satisfy the input conditions of either a single evolved schema (e.g., resource search) or some combination of schemas (resource search plus grasping). What is remarkable about the result is that infants are not surprised at seeing an arm bend when they have never seen a real-world arm bend before (the infants in the study were not shown the bendy arm in habituation conditions). Although we would not have predicted that infants should fail to be surprised by this, it is still the case that the *path* the arm takes is the path one would expect using a schema such as chasing or approach. A stronger test of the islands of competence view, perhaps, would be to use some perceptual array where goal satisfaction directly contradicts the predictions of an evolved schema such as chasing or approach. However, we recognize that bendy arms are a challenge for evolved schema theory, and further work is required to address the issue.

Similar concerns might arise from other "unnatural" aspects of laboratory stimuli—such as squares, triangles, and blobs—that generate consistent predictive intuitions in infants, as well as from the fact that overhead views of motion—as in Heider and Simmel (1944), Barrett and colleagues (2005), and others—appear to yield better recognition of intentions than do side-view displays (McAleer & Pollick, 2008). If action understanding is centered on evolved schemas, then why are these schemas activated by evolutionarily novel stimuli?

Again, we can offer only a *post hoc* response, but we suspect that the answer lies in the fact that evolved mechanisms are designed to take certain *relevant* cues as inputs, but not *all* cues. For example, research on early face perception suggests that infants' attention is drawn to facelike stimuli that share only minimal features with real faces, such as the organization of extremely schematic features and, possibly, even just patterns of shading (Morton & Johnson, 1991). Although some would see this as evidence that these early templates are not adaptations for detecting faces (e.g., Nelson, 2001), the use of minimal yet reliable cues is consistent with an evolutionary view. If schemas have a learning function, as we have suggested, then one would not expect them to be endowed with complete knowledge of their target domain. Indeed, such knowledge is what they are designed to acquire. Instead, one would expect them to use a subset of cues that index their target with high cue validity. Consistent with this idea, the use of frugal, high-validity cues appears common in nature, from eyespots to the extremely coarse imprinting rules of baby geese (in essence, imprint on the first large, moving object you see), to frogs' sensitivity to small, dark objects that enter the visual field, stop, and move around intermittently (Lettvin, Maturana, McCulloch, & Pitts, 1959).

What this means is that for chasing, the most predictive cues that a chasing event is occurring do not have to do with the shapes or properties of the chasing objects themselves: it is motion, not the identity of the mover, that makes something a chasing event. If infants are built to *learn* what kinds of things are predators and what kinds of things are prey, for example, it would make sense to equip them with motion detectors that allow them to first recognize the goal of predation and then learn about the features of predators and prey through observation. Similarly, if motion parameters indicative of targeted approach serve as inputs to such a system, and overhead views of motion afford a better view of those parameters, then overhead views might better satisfy the input conditions of the system than, for example, a horizontal view of a predator chasing prey with both moving directly away from the viewer and appearing to be fused in a single unmoving point.

A final caveat, which we wish to stress again, is that the predictions generated by an action schema view are highly overlapping with those of a teleological stance view of action understanding. In essence, everywhere that the action being observed entails the rational, efficient pursuit of goals under constraints—which includes the majority of naturally observed animal behavior—the predictions of both theories overlap. This means that the most informative tests between the theories would involve looking at more studies similar to the bendy arm case, where the action is "rational" under some construals but not expected to fall within the domain of one or multiple evolved action schemas (though this expectation assumes, perhaps unrealistically so, that evolved action schemas would include a full understanding of the anatomical limits of human arms). As it happens, the vast majority of research on infants' understanding of action has looked at cases where the action is not only "rational," but also the kind of action one would expect that an infant might understand—from looking for hidden food, to chasing, to helping someone up a hill, to grasping desired objects. This is not surprising, because developmentalists typically design experimental stimuli to match what they believe infants can understand (as researchers will search for positive findings, not negative ones). However, in order to understand how specific or general cognitive skills are, we need to look not just at prototypical cases where infants are likely to succeed, but also at the boundary conditions where they might fail.

References

Barrett, H. C. (2005). Adaptations to predators and prey. In D. M. Buss (Ed.), *The handbook of evolutionary psychology* (pp. 200–223). New York: Wiley.

Barrett, H. C. (2007). Development as the target of evolution: A computational approach to developmental systems. In S. Gangestad & J. Simpson (Eds.), *The evolution of mind: Fundamental questions and controversies* (pp. 186–192). New York: Guilford.

Barrett, H. C., & Kurzban, R. (2006). Modularity in cognition: Framing the debate. *Psychological Review, 113*, 628–647.

Barrett, H. C., Todd, P. M., Miller, G. F., & Blythe, P. W. (2005). Accurate judgments of intention from motion cues alone: A cross-cultural study. *Evolution and Human Behavior, 26*, 313–331.

Bassili, J. N. (1976). Temporal and spatial contingencies in the perception of social events. *Journal of Personality and Social Psychology, 33*, 680–685.

Benard, M. F. (2004). Predator-induced phenotypic plasticity in organisms with complex life histories. *Annual Review of Ecology Evolution and Systematics, 35*, 651–673.

Berry, D. S., Misovich, S. J., Kean, K. J., & Baron, R. M. (1992). Effects of disruption of structure and motion on perceptions of social causality. *Personality and Social Psychology Bulletin, 18*, 237–244.

Berry, D. S., & Springer, K. (1993). Structure, motion, and preschoolers' perception of social causality. *Ecological Psychology, 5*, 273–283.

Bertenthal, B. I. (1993). Perception of biomechanical motions by infants: Intrinsic image and knowledge-based constraints. In C. Granud (Ed.), *Carnegie symposium on cognition: Visual perception and cognition in infancy* (pp. 175–214). Hillsdale, NJ: Erlbaum.

Blythe, P. W., Todd, P. M., & Miller, G. F. (1999). How motion reveals intention: Categorizing social interactions. In G. Gigerenzer & P. M. Todd, & the ABC Research Group (Eds.), *Simple heuristics that make us smart* (pp. 257–286). New York: Oxford University Press.

Buller, D. J. (2005). *Adapting minds: Evolutionary psychology and the persistent quest for human nature*. Cambridge, MA: MIT Press.

Carey, S. (2009). *The origin of concepts*. New York: Oxford University Press.

Cheney, D., & Seyfarth, R. (1990). *How monkeys see the world: Inside the mind of another species*. Chicago, IL: University of Chicago Press.

Cosmides, L., & Tooby, J. (1994). Origins of domain-specificity: The evolution of functional organization. In L. A. Hirschfeld & S. A. Gelman (Eds.), *Mapping the mind: Domain specificity in cognition and culture* (pp. 85–116). Cambridge: Cambridge University Press.

Csibra, G., Bíró, S., Koós, O., & Gergely, G. (2003). One-year-old infants use teleological representations of actions productively. *Cognitive Science, 27*, 111–133.

Csibra, G., Gergely, G., Bíró, S., Koós, O., & Brockman, O. (1999). Goal attribution without agency cues: The perception of "pure reason" in infancy. *Cognition, 72*, 237–267.

Dittrich, W. H., & Lea, S. E. G. (1994). Visual perception of intentional motion. *Perception, 23*, 253–268.

Dukas, R. (2008). Evolutionary biology and insect learning. *Annual Review of Entomology, 53*, 145–160.

Fox, R., & McDaniel, C. (1982). The perception of biological motion by human infants. *Science, 218*, 486–487.

Frankenhuis, W. E., Barrett, H. C., & Johnson, S. P. (2013). Developmental origins of biological motion perception. In K. L. Johnson & M. Shiffrar (Eds.), *People watching: Social, perceptual, and neurophysiological studies of body perception* (pp. 121–138). New York: Oxford University Press.

Frankenhuis, W. E., House, B., Barrett, H. C., & Johnson, S. P. (2013). Infants' perception of chasing. *Cognition, 126*, 224–233.

Frankenhuis, W. E., & Panchanathan, K. (2011). Balancing sampling and specialization: An adaptationist model of incremental development. *Proceedings of the Royal Society of London, Series B: Biological Sciences, 278*, 3558–3565.

Frankenhuis, W. E., & Ploeger, A. (2007). Evolutionary psychology versus Fodor: Arguments for and against the massive modularity hypothesis. *Philosophical Psychology, 20*, 687–710.

Gao, T., McCarthy, G., & Scholl, B. J. (2010). The wolfpack effect: Perception of animacy irresistibly influences interactive behavior. *Psychological Science, 21*, 1845–1853.

Gao, T., Newman, G. E., & Scholl, B. J. (2009). The psychophysics of chasing: A case study in the perception of animacy. *Cognitive Psychology, 59*, 154–179.

Gao, T., & Scholl, B. J. (2011). Chasing vs. stalking: Interrupting the perception of animacy. *Journal of Experimental Psychology: Human Perception and Performance, 37*, 669–684.

Gelman, R. (1990). First principles organize attention and learning about relevant data: Number and the animate-inanimate distinction as examples. *Cognitive Science, 14*, 79–106.

Gergely, G. (2010). Kinds of agents: The origins of understanding instrumental and communicative agency. In U. Goshwami (Ed.), *Blackwell handbook of childhood development* (2nd ed., pp. 76–105). Oxford: Blackwell Publishers.

Gergely, G., Bekkering, H., & Király, I. (2002). Rational imitation in preverbal infants. *Nature, 415*, 755.

Gergely, G., & Csibra, G. (2003). Teleological reasoning in infancy: The naïve theory of rational action. *Trends in Cognitive Sciences, 7*, 287–292.

Gergely, G., Nádasdy, Z., Csibra, G., & Bíró, S. (1995). Taking the intentional stance at 12 months of age. *Cognition, 56*, 165–193.

Gould, J. L., & Marler, P. (1987). Learning by instinct. *Scientific American, 256*, 74–85.

Greenough, W., Black, J., & Wallace, C. (1987). Experience and brain development. *Child Development, 58*, 539–559.

Hamlin, J. K., Wynn, K., & Bloom, P. (2007). Social evaluation by preverbal infants. *Nature, 450*, 557–560.

Hare, B., Call, J., & Tomasello, M. (2001). Do chimpanzees know what conspecifics know? *Animal Behaviour, 61*, 139–151.

Harvell, C. D. (1990). The ecology and evolution of inducible defenses. *Quarterly Review of Biology, 65,* 323–340.

Heider, F., & Simmel, M. (1944). An experimental study of apparent behaviour. *American Journal of Psychology, 57,* 243–249.

Henrich, J., & Gil-White, F. (2001). The evolution of prestige: Freely conferred deference as a mechanism for enhancing the benefits of cultural transmission. *Evolution and Human Behavior, 22,* 165–196.

Jacobs, R. A. (1997). Nature, nurture, and the development of functional specializations: A computational approach. *Psychonomic Bulletin & Review, 4,* 299–309.

Kuhl, P. K. (2004). Early language acquisition: Cracking the speech code. *Nature Reviews: Neuroscience, 5,* 831–843.

Kuhlmeier, V., Wynn, K., & Bloom, P. (2003). Attribution of dispositional states by 12-month-olds. *Psychological Science, 14,* 402–408.

Lettvin, J. Y., Maturana, H. R., McCulloch, W. S., & Pitts, W. H. (1959). What the frog's eye tells the frog's brain. *Proceedings of the Institute of Radio Engineers, 47,* 1940–1951.

Mandler, J. M. (1992). How to build a baby: II. Conceptual primitives. *Psychological Review, 99,* 587–604.

Maynard Smith, J., & Harper, D. (2003). *Animal signals.* Oxford: Oxford University Press.

McAleer, P., & Pollick, F. E. (2008). Understanding intention from minimal displays of human activity. *Behavior Research Methods, 40,* 830–839.

Michotte, A. (1963). *The perception of causality.* Oxford: Basic Books.

Morris, M. W., & Peng, K. (1994). Culture and cause: American and Chinese attributions for social and physical events. *Journal of Personality and Social Psychology, 67,* 949–971.

Morton, J., & Johnson, M. H. (1991). CONSPEC and CONLERN: A two-process theory of infant face recognition. *Psychological Review, 98,* 164–181.

Nelson, C. A. (2001). The development and neural bases of face recognition. *Infant and Child Development, 10,* 3–18.

New, J., Cosmides, L., & Tooby, J. (2007). Category-specific attention for animals reflects ancestral priorities, not expertise. *Proceedings of the National Academy of Sciences of the United States of America, 104,* 16598–16603.

Onishi, K. H., & Baillargeon, R. (2005). Do 15-month-old infants understand false beliefs? *Science, 308,* 255–258.

Oyama, S., Griffiths, P. E., & Gray, R. D. (Eds.). (2001). *Cycles of contingency: Developmental systems and evolution*. Cambridge, MA: MIT Press.

Pascalis, O., de Haan, M., & Nelson, C. A. (2002). Is face processing species-specific during the first year of life? *Science, 296*, 1321–1323.

Pellegrini, A. D., Dupuis, D., & Smith, P. K. (2007). Play in evolution and development. *Developmental Review, 27*, 261–276.

Piaget, J., & Inhelder, B. (1951). *La genése de l'idée de hasard chez l'enfant*. [The origin of the idea of chance in the child.] Paris: Presses Universitaires de France.

Rakoczy, H., Warneken, F., & Tomasello, M. (2008). The sources of normativity: Young children's awareness of the normative structure of games. *Developmental Psychology, 44*, 875–881.

Rimé, B., Boulanger, B., Laubin, P., Richir, M., & Stroobants, K. (1985). The perception of interpersonal emotions originated by patterns of movement. *Motivation and Emotion, 9*, 241–260.

Rochat, P., Morgan, R., & Carpenter, M. (1997). Young infants' sensitivity to movement information specifying social causality. *Cognitive Development, 12*, 536–561.

Rochat, P., Striano, T., & Morgan, R. (2004). Who is doing what to whom? Young infants' developing sense of social causality in animated displays. *Perception, 33*, 355–369.

Scholl, B. J., & Tremoulet, P. D. (2000). Perceptual causality and animacy. *Trends in Cognitive Sciences, 4*, 299–309.

Seligman, M. E. P. (1970). On the generality of the laws of learning. *Psychological Review, 77*, 406–418.

Sherry, D., & Schachter, D. (1987). The evolution of multiple memory systems. *Psychological Review, 94*, 439–454.

Shor, R. E. (1957). Effect of preinformation upon human characteristics attributed to animated geometric figures. *Journal of Abnormal and Social Psychology, 54*, 124–126.

Simion, F., Regolin, L., & Bulf, H. (2008). A predisposition for biological motion in the newborn baby. *Proceedings of the National Academy of Sciences of the United States of America, 105*, 809–813.

Smith, P. K., & Lewis, K. (1985). Rough-and-tumble play, fighting, and chasing in nursery school children. *Ethology and Sociobiology, 6*, 175–181.

Southgate, V., Johnson, M. H., & Csibra, G. (2008). Infants attribute goals even to biomechanically impossible actions. *Cognition, 107*, 1059–1069.

Steen, F. F., & Owens, S. A. (2001). Evolution's pedagogy: An adaptationist model of pretense and entertainment. *Journal of Cognition and Culture, 1,* 289–321.

Surian, L., Caldi, S., & Sperber, D. (2007). Attribution of beliefs by 13-month-old infants. *Psychological Science, 18,* 580–586.

Szego, P. A., & Rutherford, M. D. (2007). Actual and illusory differences in constant speed influence the perception of animacy similarly. *Journal of Vision, 7,* 1–7.

Szego, P. A., & Rutherford, M. D. (2008). Dissociating the perception of speed and the perception of animacy: A functional approach. *Evolution and Human Behavior, 29,* 335–342.

Thomsen, L., Frankenhuis, W. E., Ingold-Smith, M., & Carey, S. (2011). Big and mighty: Preverbal infants mentally represent social dominance. *Science, 331,* 477–480.

Tremoulet, P. D., & Feldman, J. (2000). Perception of animacy from the motion of a single object. *Perception, 29,* 943–951.

Tronick, E. (1972). Stimulus control and the growth of the infant's effective visual field. *Perception & Psychophysics, 11,* 373–376.

Wagner, L., & Carey, S. (2005). 12-month-old infants represent probable endings of motion events. *Infancy, 7,* 73–83.

Woodward, A. L. (2009). Infants' grasp of others' intentions. *Current Directions in Psychological Science, 18,* 53–57.

Woodward, A., & Sommerville, J. A. (2000). Twelve-month-old infants interpret action in context. *Psychological Science, 11,* 73–77.

Zacks, J. M. (2004). Using movement and intentions to understand simple events. *Cognitive Science, 28,* 979–1008.

9 Perceiving Animacy and Intentionality: Visual Processing or Higher-Level Judgment?

Brian J. Scholl and Tao Gao

Abstract

We can identify social agents in our environment not only on the basis of how they look, but also on the basis of how they move—and even simple geometric shapes can give rise to rich percepts of animacy and intentionality based on their motion patterns. But why should we think that such phenomena truly reflect visual processing, as opposed to higher-level judgment and categorization based on visual input? This chapter explores five lines of evidence: (1) The phenomenology of visual experience, (2) dramatic dependence on subtle visual display details, (3) implicit influences on visual performance, (4) activation of visual brain areas, and (5) interactions with other visual processes. Collectively, this evidence provides compelling support for the idea that visual processing itself traffics in animacy and intentionality.

Introduction

Identifying social agents in our environment and predicting their behavior is of obvious and extreme importance to our fitness, and cognitive scientists have accordingly devoted considerable attention to several varieties of social perception. The most obvious way to identify agents is on the basis of what they look like (regardless of how they may be moving). Agents such as people and animals tend to have distinctive visual features (e.g., faces and eyes) compared to nonagents such as vehicles and plants. But we can also identify agents on the basis of how they move (regardless of what they look like) such that even simple geometric shapes can give rise to rich percepts of animacy and intentionality based on their motion patterns. This phenomenon has fascinated researchers for many decades, largely based on the possibility that it represents actual visual processing.

But why should we think that such phenomena truly reflect perception, as opposed to higher-level judgment and categorization based on visual input? The task of marshaling such evidence is severely complicated by the fact that the dependent measures that are almost universally used in this field—viz. verbal descriptions and categorical ratings—are highly ill-suited to isolating visual processing (or to isolating any underlying mechanisms, for that matter). Recent advances, in contrast, have allowed for the study of perceived animacy via several implicit measures of visual performance. This chapter explores five relevant lines of evidence that can be derived from this newer work: (1) The phenomenology of visual experience, (2) dramatic dependence on subtle visual display details, (3) implicit influences on visual performance, (4) activation of visual brain areas, and (5) interactions with other visual processes. Collectively, this evidence provides compelling support for the idea that visual processing itself traffics in animacy and intentionality, in rich and interesting ways.

Detecting Agents

There is too much information in our local visual environments at any given moment to fully process all of it, and so vision is necessarily selective. This selection is realized in at least two ways. First, visual processing during online perception is selective via the operation of attention (driven both voluntarily and involuntarily) such that some visual objects (or regions, or features) are processed more than others (Chun et al., 2011; Treisman, 2006). This can lead to better acuity for some objects (e.g., Carrasco et al., 2004), and in the extreme it can lead observers to completely fail to consciously perceive even very salient but unattended objects (e.g., Most et al., 2005). And given the importance of agents in our local environment to our immediate fitness, it is perhaps no surprise that they are automatically prioritized via such attentional selection. For example, when viewing natural scenes, attention is automatically drawn to animate agents such as people and animals relative to inanimate objects such as vehicles and plants, even while carefully controlling for lower-level stimulus differences (New et al., 2007). And indeed, this type of attentional prioritization seems to be so fundamental that it is spared even in syndromes, such as autism spectrum disorder, in which other aspects of social perception are impaired (New et al., 2010). In addition, attention may also prioritize some specific types of especially fitness-relevant agents even more than others—for example, spiders (e.g., Rakison & Derringer, 2008) and snakes (e.g., Öhman & Mineka, 2003).

Selection in vision is also realized in a second way: specialized visual processing, independent of attentional selection, may be devoted to some types of stimuli relative to others. And other agents in our local environment (and their features) also seem to trigger this type of selection. For example, faces may automatically receive specialized processing that elbows do not, even when both are equally well attended—perhaps because the mind has specialized routines for processing faces in particular ways, whereas some other body parts may receive only domain-general visual processing (Kanwisher, 2010; Sinha et al., 2006; Sugita, 2009; Wilmer et al., 2010). And this form of selectivity may also have an even finer grain, as when subparts of faces such as the eyes receive even more specialized processing (e.g., Emery, 2000).

Perceiving Animacy from Motion Cues

The examples of selective prioritization for animacy discussed in the previous section all operate on the basis of visual surface features—that is, on the basis of the ways in which certain visual stimuli *look like* agents. And much of this detection is based on static features, perhaps so that people and animals can be detected even in motionless scenes. But another tradition in social perception research has identified a parallel way in which agents are identified and prioritized on the basis of how they move—even if their static visual features are as simple as can be. This sort of cue to animacy has fascinated psychologists ever since the seminal demonstrations of Fritz Heider (Heider & Simmel, 1944) and Albert Michotte (1950), who showed that even simple geometric shapes were automatically perceived as animate and goal-directed on the basis of how they moved. In a dynamic display such as that depicted in figure 9.1, for example, observers might reflexively see the larger triangle *chasing* the smaller shapes, with the goal of catching one or both of them. Beyond detecting animacy per se, observers may also identify specific intentions and goals; for example, they may perceive that a shape is trying to reach a specific place (e.g., Csibra, 2008; Dasser et al., 1989) or catch another shape (e.g., Gao et al., 2009; Rochat et al., 2004). And the contents of such percepts may also extend to a wide variety of other types of social interactions beyond pursuit, such as fighting and guarding (e.g., Barrett et al., 2005; Blythe et al., 1999). In all of these cases, the relevant displays may only involve simple geometric shapes, yet they may give rise to richly animate percepts that nearly all observers can immediately appreciate.

A large and interesting literature on such phenomena has been developed since the initial demonstrations of Michotte and Heider. This literature variously refers to the phenomena in terms of the perception of animacy (e.g., Gelman et al., 1995), intentionality (e.g., Dasser et al., 1989),

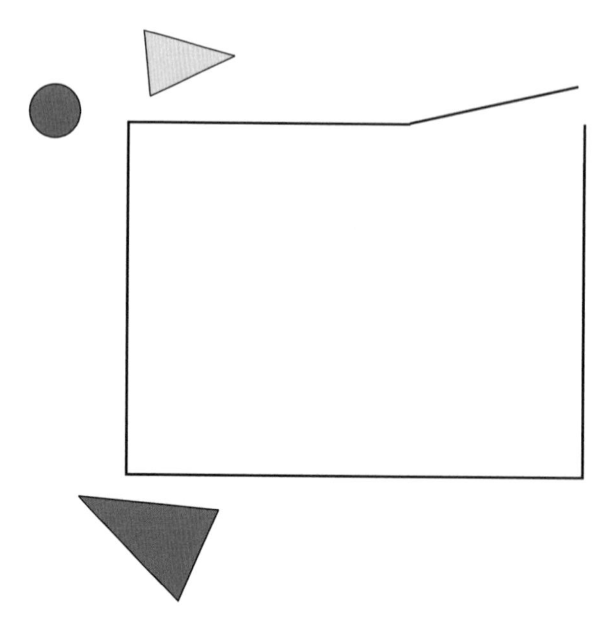

Figure 9.1
A sample static "snapshot" frame based on Heider and Simmel's (1944) classic animation, in which the large triangle is seen to chase the two smaller shapes around the box.

goal-directedness (e.g., Csibra, 2008), social causality (e.g., Rochat et al., 1997), action understanding (e.g., Baker et al., 2009), or social meaning (Tavares et al., 2008). Sometimes these distinctions are important, as when researchers wish to attribute one property but not another—for example, goal-directedness without other aspects of mental-state reasoning (e.g., Gergely & Csibra, 2003) or intentions without animacy (e.g., Gao, New, & Scholl, under review; Gao, Scholl, & McCarthy, 2012). But such differences will not be important for the present discussion, and so we will typically refer throughout this chapter only to the perception of animacy.

This body of research has distilled a host of specific cues that seem to trigger such percepts. These cues include relatively low-level properties such as self-propulsion (e.g., Dasser et al., 1989), synchronous movements (e.g., Bassili, 1976), sudden changes in speed and/or heading (e.g., Tremoulet & Feldman, 2000), certain patterns of approach or avoidance (e.g., Dittrich & Lea, 1994; Gao et al., 2009), coordinated orientation cues (e.g., Gao et al., 2010; Gao, Scholl, & McCarthy, under review), and others (for reviews see Scholl & Gao, to appear; Scholl & Tremoulet, 2000). These cues have then collectively been explained by appeal to more general principles such as the apparent violation of Newtonian mechanics (e.g., Gelman et al., 1995; Tremoulet & Feldman, 2006) and the apparent violation of a rationality principle—the notion that agents will move in the most efficient way possible in order to satisfy their goals (e.g., Baker et al., 2009; Csibra, 2008; Gao & Scholl, 2011).[1]

What Do We Mean by "Perception" of Animacy?

We have been repeatedly referring to the *perception* of animacy, but in what sense do these phenomena truly reflect perception as opposed to higher-level categorization and judgment? This is a key question—and perhaps even *the* key question—for this research area, in part because visual processing and notions of animacy and intentionality seem to be rather divorced. Certainly this is true in the wider field of cognitive science, where agency and intentionality are most often studied in the context of so-called Theory of Mind—the ability to perceive, predict, and explain the behavior of others in terms of their underlying mental states (for reviews, see Leslie et al., 2004; Saxe et al., 2004). Such definitions of Theory of Mind may sometimes invoke perception (e.g., Scholl & Leslie, 2001), but of course we typically think of perception as being involved primarily (or only) with the reconstruction of the local physical environment, not via mental state attributions, but via properties such as color, shape, and orientation. How to bridge this gulf?

One possibility is that the term "perception" should be read loosely, or even metaphorically, in this context. Phenomena of "perceived" animacy may simply show that we judge or interpret some stimuli as involving animacy. This would still involve perception, but only insofar as these interpretations derive from visual stimuli. But this form of higher-level attribution of animacy seems rather pedestrian and unworthy of such a rich scientific literature. After all, we can choose to interpret (visually observed) objects in all sorts of rich ways based on their visual characteristics. For example, we might talk of perceiving irony or sarcasm when viewing a piece of modern art, but in the present context that seems like loose talk insofar as we don't actually think that visual processing itself is trafficking in such notions. Indeed, if such examples are akin to the perception of animacy, then perhaps we just need to be more careful—to talk not of perceiving animacy, per se, but rather of *judging animacy on the basis of visual percepts*, where those percepts do not themselves involve any sense of animacy at all.[2]

Another possibility, though, is that this allusion to perception of animacy is not loose or metaphorical at all: these phenomena may reflect actual visual processing that is specialized for the extraction of animacy from visual motion (see also Rutherford, this volume), just as other types of visual processing are specialized for the extraction of depth or color from visual scenes. On this interpretation, vision itself may traffic in animacy as well as physical features such as shape and orientation. This is an exciting possibility, suggesting (as with other forms of social perception) that the purpose of vision is (in part) not only to recover the physical structure of the local environment, but also to recover its causal and social structure (Adams et al., 2010). The goal of this chapter is to review some of the findings that support this "social vision" interpretation.

Perception versus Cognition

If our goal is to determine whether the "perception of animacy" reflects perception or cognition, then we ought to say something about what the difference is. This general task, of course, is notoriously difficult, and is rather far beyond the scope of this chapter (see Pylyshyn, 1999). But this is also the sort of question where we can proceed without a full solution to this more general problem: We may not know exactly how and where to draw the line between perception and cognition, but we can nevertheless know that such a line exists, we can appreciate that the difference matters, and we can point to some of the characteristic properties of perception vs. cognition at some remove from the line. The need for drawing this distinction in some form, in any case, should be clear from the preceding discussion: There is one (cognitive) interpretation of the phenomena in question that seems downright boring (since it is no big news that we can choose to interpret a visual stimulus as animate) and one (perceptual) interpretation that seems relatively exciting (since animacy is not typically characterized as a visual feature).

Perception ought to be distinguished from cognition because there are wide and powerful generalizations that seem to hold of perceptual processes but not cognitive ones, and vice versa. (And of course the business of trying to identify and explain generalizations is what science is all about.) In general, "perception" refers in this context to a family of processes that is relatively automatic and irresistible, and that operates without the ability to consciously introspect its nature. For example, the perception of depth from stereopsis is not something that we try to do, and it is not something that we can avoid (at least when our eyes are open, in sufficient light); the results of stereoscopic vision (i.e., percepts of depth) seem immediate in our experience, with no hint of how they were derived. Note that a good heuristic in this context is that a percept may conflict with your actual *beliefs* about the nature of the world—which in turn cannot readily influence the percept itself. Thus, you might viscerally *see* depth while viewing a random-dot stereogram, while simultaneously *knowing* that there isn't *really* any depth.

The initial gloss of perception outlined in the previous paragraph is not nearly sufficient, though, since there are plenty of automatic, irresistible, and nonintrospectable mental processes that have nothing to do with visual input—semantic priming, for example (e.g., McNamara, 2005). These cognitive processes may also conflict with explicit beliefs, as in the case of the operation of implicit stereotypes (e.g., Dovidio et al., 2001). So for the heuristic purposes of this chapter, we will simply add the requirement that for a process to count as perceptual, it needs to be strongly and directly controlled by specific and subtle features of the visual input itself. This is true, obviously, of stereoscopic vision: When viewing vivid depth from a random-dot stereogram, for example, the cues are certainly specific, visual, and subtle. (Without knowing about how such images work, you might stare at a stereogram for a long time indeed—all the while experiencing vivid depth—without having any notion of what the relevant visual cues actually are.) And in contrast, there are many, many forms of higher-level judgment (and also many forms of implicit/automatic cognition) that are not controlled by specific and subtle visual details.

The Trouble with Asking People Questions

Distinguishing perception and cognition has traditionally been especially difficult in the context of perceived animacy; this is due to some troublesome aspects of the methods that have been most commonly used to assess it. These methods have essentially involved showing displays to observers and then asking the observers about what they see. And both parts of this—the showing and the asking—have created deep difficulties (Gao et al., 2009).

In terms of stimulus presentation, many (but certainly not all) studies have used haphazardly created stimuli, with the shapes' trajectories defined, not by motion algorithms, but by the intuitions of the "director" who scripted each event by hand in an animation program. This process has the great advantage of resulting in especially rich and cinematic stories told by the motions of the shapes—stories evoking mental states such as fear, mocking, and coaxing (e.g., Abell et al., 2000; Castelli et al., 2000; Heider & Simmel, 1944). But it also has a great disadvantage, which is that you can't evaluate the role of subtle visual characteristics if you're not able to vary them carefully and systematically (see also Pantelis & Feldman, 2012).

This difficulty has been further compounded by the questions that are subsequently asked of observers as the primary source of data. Such questions have come in several varieties. For example, (1) sometimes observers are just asked directly to freely describe what they see (as in the original studies by Heider & Simmel, 1944); (2) sometimes the resulting descriptions are formally coded, in various elaborate ways, for the use of social words (e.g., Heberlein & Adolphs, 2004; Klin, 2000); (3) sometimes observers are asked to assign explicit ratings of perceived animacy (e.g., rating animacy on a seven-point scale, from 1 = "definitely not alive" to 7 = "definitely alive"; Tremoulet & Feldman, 2000); and (4) sometimes ratings are solicited via a more qualitative scale, such as a four-point scale that consists of "physical," "rather physical," "rather personal," and "personal" (e.g., Santos et al., 2008). All of these methods are really just ways of asking observers to reports their percepts, though, and the problem with such explicit reports is that they cannot readily implicate perception (or any other underlying process) directly. Take, for example, a free report such as "it looks alive." This could reflect a direct visual experience, analogous to "it looks red." But such descriptions can be naturally influenced—or, more to the point, *contaminated*—by observers' higher-level beliefs and expectations, analogous to a report of "it looks mysterious" when viewing a painting. In particular, such reports can be directly influenced by inferences about how reliable observers *think* the manipulations should be for diagnosing animacy, beyond the percepts (or lack thereof) themselves.

A Tricky Example

A concrete example may help here. Tremoulet and Feldman (2000; see also Tremoulet & Feldman, 2006) reported an attempt to isolate some of the simplest direct cues to perceived animacy. Their stimuli were all systematically and algorithmically generated: In the simplest case, a single dot moved on an otherwise-empty screen and, at some point during its motion, changed its direction and/or its speed. To determine their influence on perceived animacy, the direction and speed-change magnitudes were systematically varied. This was an especially interesting and rigorous study (and many of our own experiments reported herein were directly inspired by it). But from our perspective, their use of ratings of animacy hamstrung the study in serious ways. Observers provided animacy ratings of each display (on a seven-point scale from "definitely not alive" to "definitely alive"), with no other guidance other than the further instruction that they should give low ratings to those dots whose motion seemed to be "artificial, mechanical, or strange" (Tremoulet & Feldman, 2000, p. 945). The simplest results from these conditions were ... well, they were exactly what you'd expect (and that's the problem): The more extreme the angle and/or speed change, the higher the animacy ratings.

But since such influences are so intuitive, it remains possible that observers were structuring their many hundreds of ratings not on the basis of what they *saw*, but rather on what kinds of changes they *judged* to be a reliable signal of an animate agent. Thus "definitely alive" might simply reflect the higher-level judgment that such changes should matter for animacy—so that "definitely alive" might really arise less from a sense that a shape definitely *looked* alive and more from a sense that a shape definitely seemed like its motion should probably be explained by appeal to animacy. And indeed, in a way it would seem difficult *not* to receive such ratings from observers, regardless of their percepts or lack thereof. After all, the observers have to vary their responses somehow across the several hundreds of displays (since to do otherwise would contravene an especially strong task demand), and speed/direction were the only variables that ever varied in these initial experiments—so what other type of data *could* be generated, even in principle? In fact, we would expect similar results from an experiment without any visual stimulation at all: If the displays were only *described* to the observers—along with the relevant manipulations and the magnitudes of those manipulations in each trial—we suspect that they would continue to predict that greater angle and speed changes would likely correspond to more animate entities. We can summarize this problem provocatively by proposing that this study does not imply anything at all about perception, but that the results are entirely fueled by higher-level reasoning (see also endnote 3). This is thus an example where the question of perception vs. cognition looms large; in fact, we think that this sort of concern permeates most of the extant studies of perceived animacy.

Implicating Perception

Despite the sorts of concerns reviewed in the preceding sections, we believe that the "perception of animacy" is aptly named—that it reflects true specialized processing in the visual system rather than (or in addition to) higher-level cognitive attributions. In other words, we think that the perception of animacy is more akin to the perception of depth and color in a painting than to the perception of sarcasm or irony in a painting.

Of course, in making such claims we do not mean to suggest that the relevant displays *actually* have any animate properties (they don't!) or that such percepts cannot be evoked by inanimate stimuli (they can!). But this is the situation with nearly all aspects of perception: Almost all visual processing operates in a heuristic manner, so that, for example, we can have real but incorrect percepts of depth, and depth percepts can also be evoked in depthless displays in various ways. In the case of depth, this is because the light that enters the eyes is insufficient to deductively infer depth without the additional use of unconscious inferences about the source of the visual stimulation (e.g., Gregory, 1980; Rock, 1983)—often in the form of heuristic assumptions about the world. Similar unconscious inferences may fuel the perception of animacy, and for exactly the same reasons: The light that enters the eyes may be insufficient to deductively infer the presence of agents, but heuristic assumptions (wired into the operation of the visual system) may lead certain spatiotemporal patterns to automatically trigger very real visual percepts of animacy.

How can such a view be tested? With other phenomena, that task can sometimes be straightforward, insofar as there are some hallmark processes of perception that seem clear-cut and uncontroversial. This is true, for example, in the context of the related phenomenon of causal perception, in which certain spatiotemporal patterns (akin to one billiard ball launching another) give rise to rich percepts of causality (e.g., Choi & Scholl, 2004, 2006; Newman et al., 2008; Wagemans et al., 2006). Despite the universal phenomenological impact of such displays (no pun intended, and on which more below), a few brave souls have occasionally suggested that such phenomena do not actually reflect perception in any substantive sense (e.g., Boyle, 1972; Joynson, 1971; Rips, 2011). But these sorts of displays seem to admit of relatively straightforward empirical tests. For example, one recent study showed that causal perception was subject to visual adaptation that was both (1) retinotopically specific and (2) not explainable by appeal to adaptation of any lower-level visual features (Rolfs et al., 2013). This seems impossible to explain without implicating visual processing; we know of no higher-level form of cognition or judgment that has ever evoked retinotopically specific processing of any kind. Unfortunately, this sort of strategy does not seem to be available in the case of the perception of animacy, in part because the cues to animacy in this domain are so necessarily dynamic in both space and time. So how then can we implicate perception?

We see at least five reasons that may support a true "social vision" interpretation of perceived animacy, as outlined in the following sections.

Phenomenology

In the first place, it is worth noting explicitly that the rich phenomenology of perceived animacy is consistent with an interpretation in terms of visual processing. Indeed, the phenomenology elicited by such displays is surely the driving force behind this research program as a whole. These phenomena often operate as fantastic demonstrations: observers simply see animacy and intentionality when viewing the displays, effortlessly and automatically, and without any instructions or preparation. (This is no doubt why students today still view digitized versions of the original Heider and Simmel animation, almost seventy years after its creation.) And such percepts are interesting and notable, since of course observers know that such properties are in fact absent. Moreover, such percepts seem irresistible. You can no more fail to see animacy simply by trying than you can fail to see color simply by trying; as long as you meet the precondition of having your eyes open and focused, the percepts seem to be ballistically generated. (A nice demonstration of this type of seeming encapsulation from higher-level knowledge involves explicitly instructing observers to describe such displays *without* using any mental-state terms. When under a heavy processing load, most observers fail at this task in occasionally comical ways: "Okay, now the square is moving up to the left, and the triangle is trying to ... oh, damn, sorry: it's not *trying* to do anything....")

The power and seeming universality of such phenomenology is apparent when exploring free-report descriptions of the relevant displays. For example, in one recent study (to be described in more detail below), observers viewed an array of oriented dart shapes that moved through a display along with a moving green disc (Gao, McCarthy, & Scholl, 2010). In all cases, the darts moved completely randomly and independently of the disc's motion. In one condition (figure 9.2a), the darts' orientations were continually adjusted during the motion so that they were always pointing directly at the disc. This condition generated what we termed the *wolfpack effect*, which leads observers to misperceive the darts' random motions in terms of animate behavior such as chasing (presumably because the way that an agent is 'facing' is, and always has been, a reliable indicator of its intentions and future behavior). A perpendicular control condition was nearly identical, except that the darts' orientations were now continually adjusted during the motion so that they were always pointing 90 degrees away from the disc (figure 9.2b). This condition completely eliminated the wolfpack effect, despite the fact that the two conditions had identical (1) motion trajectories, (2) rotational motion in the darts, and (3) degrees of correlation between the rotational motion and the disc's movement. In

(a)

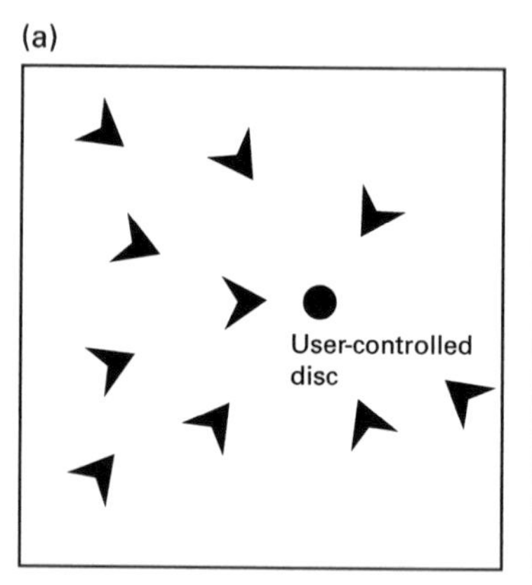

(b)

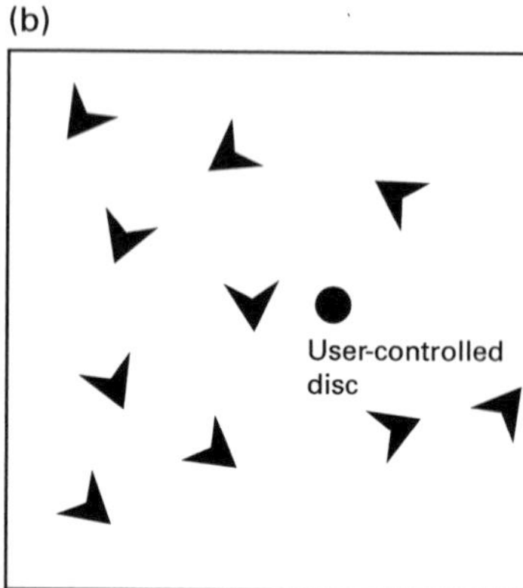

Figure 9.2
Sample displays from the study of the wolfpack effect by Gao, McCarthy, and Scholl (2010). (a) A static snapshot of a Wolfpack trial, wherein the darts are all "facing" the disc. (b) A static snapshot of a perpendicular control trial, wherein the darts are all "facing" 90 degrees away from the disc.

short, the two displays were well matched in terms of all lower-level visual properties, yet one evoked a rich percept of animacy and intentionality and the other did not. This was clear from the resulting phenomenology (as readers can experience for themselves via the demos at http://www.yale.edu/perception/wolfpack/), but it was also clear in terms of free reports. When asked to freely describe the wolfpack displays, nearly every observer invoked notions of animacy, such as "many white arrows were chasing after the green dot" and "the triangles ... follow the green dot wherever it goes." In contrast, no observer of the Perpendicular control displays ever used such language—they instead used descriptions such as "arrows that went in random directions," "chaotically floating white chevrons," and "a bunch of white snow flakes or jacks swirling tumultuously around my green circle" (Gao et al., 2010).

This difference in free reports of animacy seemed universal in the context of the wolfpack effect, and other studies (using different displays) with much more varied subject populations are consistent with the possibility that such percepts of animacy are cross-cultural universals (Barrett et al., 2005). Indeed, perhaps the only human observers who do not readily and even irresistibly describe such displays in animate terms are those with neuropsychological disorders of social perception, as in autism spectrum disorder (Abell, Happé, & Frith, 2000; Klin, 2000; Rutherford, Pennington, & Rogers, 2006) or amygdala damage (Heberlein & Adolphs, 2004). In short, the phenomenology of perceived animacy seems just as consistent, universal, and reflexive as does the phenomenology of perceived depth or color.[3]

Of course, phenomenology alone cannot settle the question at hand (cf. Rips, 2011). Indeed, perhaps the central lesson of cognitive science is that our introspective intuitions about how the mind works are fantastically poor guides to how the mind actually works. But it still seems worth noting that the perception of animacy appears to be no different from other forms of perception in this way and is very much in contrast to the phenomenology of many other forms of slow, deliberative, intentional, and eminently resistible decisions.

Dramatic Dependence on Subtle Visual Display Details

We suggested above that a hallmark feature of perception (vs. cognition) is its strict dependence on subtle visual display details; percepts seem to be irresistibly controlled by the nuances of the visual input regardless of our knowledge, intentions, or decisions. This is certainly true in the context of perceived animacy, as we will illustrate here with a recent study of perceived chasing. We have focused on chasing in several studies (e.g., Gao et al., 2009; Gao & Scholl, 2011; Gao, New, & Scholl, under review; Gao, Scholl, & McCarthy, 2012) as part of an attempt to break the complex cinematic movies of the Heider and Simmel variety into their component cues, which can then be systematically varied. Rather than animacy being picked up by some form of monolithic animacy detector, we suspect that it is detected via a number of independent cues, just as depth is detected via many different cues (e.g., binocular disparity, motion parallax, occlusion, etc.). And just as with depth, the best way to understand the underlying processing is likely to involve studying individual cues in relative isolation.

We have explored chasing in particular for both methodological and theoretical reasons (see also Frankenhuis & Barrett, this volume). Methodologically, it seems relatively easy to rigorously and systematically manipulate the variables that underlie chasing (cf. Nahin, 2007). And theoretically, chasing seems to have especially acute adaptive importance: Certainly for our ancestors, whether they were the chasers or the ones being chased, the outcome of chases would often have direct implications for fitness and survival. (And researchers who study dreaming report that, even in today's world, dreams that we are being chased are one of the most commonly remembered dream types; e.g., Garfield, 2001; Revonsuo, 2000.)

In our initial studies of perceived chasing (Gao et al., 2009), observers watched simple geometric shapes move about a computer display, with one shape (the "wolf") pursuing another (the "sheep"; Dittrich & Lea, 1994) in a crowded display containing many other shapes. Critically, the ability to detect chasing was assessed not (only) via free reports and ratings, but also via a variety of performance-based tasks. In the *Don't-Get-Caught* task, for example, subjects controlled the sheep's movement themselves during each trial by moving the computer mouse. The explicit task was to move the sheep about the display in order to avoid getting caught (i.e., touched) by the wolf. However, because each display contained many different objects (as depicted in figure 9.3), subjects could not avoid them all. And while the sheep itself was always highlighted (by being drawn in a different color), the wolf always looked identical to the many other distractors. Each trial ended when either the user-controlled sheep got caught by the wolf (a failure) or after a certain period of time had passed (a successful trial)—with performance assessed in terms of the percentage of trials in which subjects avoided being caught. As a result, this was really a chasing-detection task: Subjects had to detect the wolf's presence and position purely on the basis of its spatiotemporal behavior, after which they could

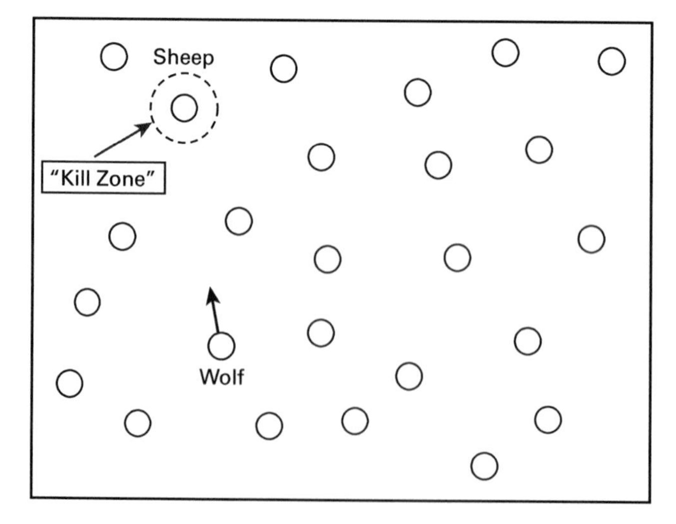

Figure 9.3
A sample screenshot from the dynamic display in the Don't-Get-Caught task from Gao, Newman, and Scholl (2009). The subject must use the mouse to move the sheep around the crowded display so that the moving wolf that is chasing it never enters its "kill zone."

escape from it (since the sheep could move faster than the wolf). Crucially, this ability to detect chasing could not be explained by appeal to any lower-level form of perception such as correlated motion or proximity (Gao et al., 2009; Gao & Scholl, 2011).

A key variable in our initial studies of chasing (Gao et al., 2009) was termed *Chasing Subtlety*. This was essentially a measure of the maximal angular deviation of the wolf's heading compared to perfect heat seeking. When the Chasing Subtlety was 0° (figure 9.4a), the wolf always headed directly toward the sheep. When the Chasing Subtlety was 30°, though, the wolf always headed in the general direction of the sheep, but not perfectly; instead, it could move in any direction within a 60° window, with the window always centered on the moving sheep. In other words, in this condition, the next displacement of the wolf could be in any direction within the shaded area in figure 9.4b. With 90° of Chasing Subtlety, the

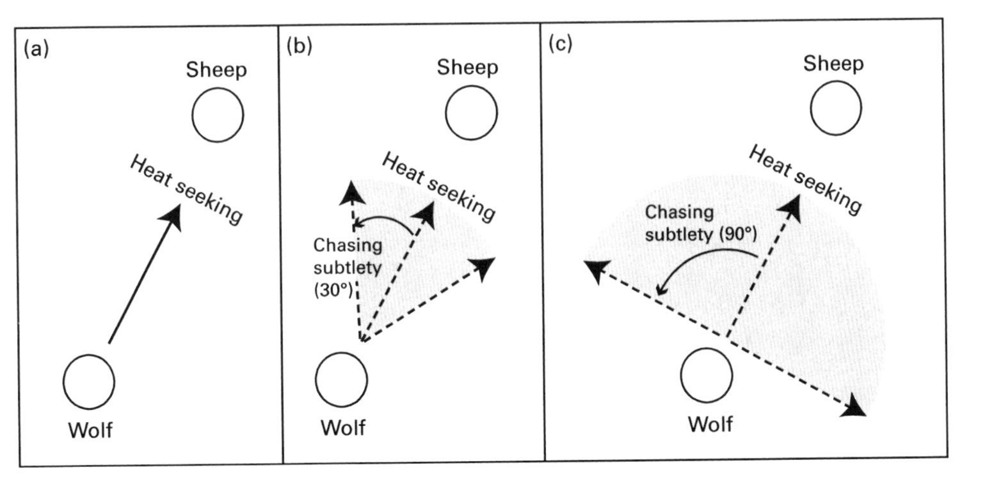

Figure 9.4
An illustration of the *Chasing Subtlety* manipulation used in Gao, Newman, and Scholl (2009). (a) When the Chasing Subtlety is 0 degrees, the wolf always heads directly toward the (moving) sheep, in a "heat-seeking" manner. (b) When the Chasing Subtlety is 30 degrees, the wolf is always heading in the general direction of the sheep, but is not perfectly heat-seeking: instead, it can move in any direction within a 60 degree window, with the window always centered on the (moving) sheep. (c) When the Chasing Subtlety is 90 degrees, the wolf's direction of movement is even less constrained: now the wolf may head in an orthogonal direction to the (moving) sheep, but can still never be heading *away* from it. The gray areas in (b) and (c) indicate the angular zone which constrains the wolf's direction of movement on that given frame of motion.

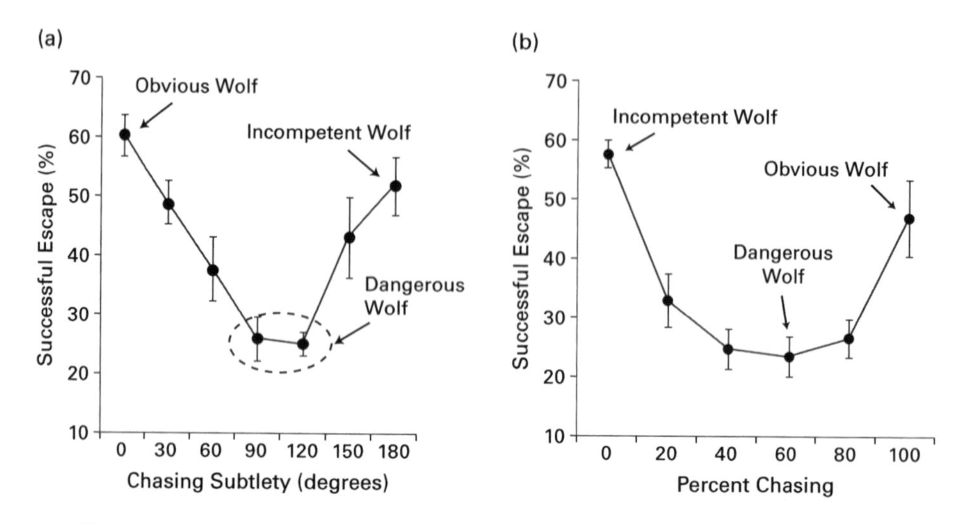

Figure 9.5
Sample results from experiments on the psychophysics of chasing. (a) Results from the Don't-Get-Caught task of Gao, Newman, and Scholl (2009): The percentage of trials in which the subjects successfully avoided being caught by the wolf, as a function of chasing subtlety. (b) Results from the analogous task of Gao and Scholl (2011): The percentage of trials in which the subjects successfully avoided being caught by the wolf, as a function of the percentage of chasing vs. random motion.

wolf could head orthogonal to the sheep (being displaced in any direction within the shaded region of figure 9.4c), but still could never move directly *away* from it. We varied Chasing Subtlety in a continuous manner across trials and measured the resulting ability to detect and avoid the wolf. As depicted in figure 9.5a, performance fell along a U-shaped curve. (In such graphs, the absolute value of performance is not meaningful, since it can be pushed around via global variables such as the speed of the movement or the number of distractors. In these experiments, we thus set such variables so as to get middling performance with the goal of seeing how that level of performance was then influenced by Chasing Subtlety and other factors.)

This U-shaped performance function had three notable aspects. First, performance with little or no Chasing Subtlety was quite good indeed, because the wolf's identity seemed to "pop out" to observers such that they could immediately detect chasing and then proceed to avoid the wolf (for online demonstrations, see http://www.yale.edu/perception/chasing/). Second, performance with near-maximal Chasing Subtlety was also good, but for a very different reason: Here, there was not much actual chasing to begin with, so the wolf tended to catch the sheep only by chance. Finally, and most interesting, performance with middling levels of Chasing Subtlety led to large declines in performance (by up to 40%). These deficits reveal a dissociation between the presence of *actual* chasing and the ability to *perceive* chasing: The wolf itself was still highly effective and tended to get reliably closer and closer to the sheep over time, but observers simply had trouble detecting this.

In a follow-up study, we generalized these results to the temporal domain via a manipulation of *interrupted* chasing (Gao & Scholl, 2011). Chasing Subtlety was now fixed at a small value, and what varied across trials was how often chasing was occurring. Rather than chasing the sheep continually during each trial, the wolf's chasing was frequently and periodically interrupted by periods of random motion. The extent of this random motion was then systematically varied. Specifically, the display always alternated between chasing and random motion, but the relative percentage of each type of motion during each short interval was varied from 0 to 100 percent (for online demonstrations, see http://www.yale.edu/perception/Interrupted-Chasing/). We again observed a U-shaped performance function (figure 9.5b) with a similar interpretation. Small percentages of random motion (on the right side of the graph) led to relatively successful performance, since the wolf was relatively easy to detect and avoid. And large percentages of random motion (on the left side the graph) also led to relatively successful avoidance, but only because the chasing was less effective. But in the middle, chasing was still highly effective, but chasing *detection* was greatly impaired.[4]

The point of reviewing these studies here is to note that chasing detection (as a form of perceived animacy) is influenced in systematic ways by rather subtle display parameters, in the form of a psychophysical function (and in ways that do not seem readily explainable by appeal to higher-level judgment). We may have some categorical intuitions about the factors that *should* count as animate, but we surely do not have intuitions about the difference between 30° and 60° of chasing subtlety or about the difference between 20 percent and 40 percent of random motion. (Indeed, it is not always even possible to tell the difference between such displays during free viewing, and it is difficult to make a conscious decision about animacy on the basis of information that you are not even aware of.) Yet these subtle variations in the visual input mattered great deal. Subjects could not simply decide to interpret any set of parameters as chasing; instead, the ability to detect chasing seemed confined to only a relatively narrow range of temporal and spatial parameters. Additionally, even within this narrow range, the ability to detect chasing took the form of a continuous function operating over a rather specialized type of input. This type of strict dependence on specific and subtle visual display details, coupled with a large degree of independence from overt intentions, is a hallmark of specialized visual processing, but is uncharacteristic of higher-level inference.

Beyond Phenomenology: Irresistible Influences on Visual Performance

The studies of perceived chasing reviewed in the previous section also support a social vision interpretation of perceived chasing in an even more direct way. Beyond the compelling phenomenology of the displays themselves, the data reported in these studies are all measures of visuomotor *performance* rather than explicit reports or ratings. This is an important distinction since overt decisions about what should and should not count as animacy can directly influence reports and ratings (as discussed above) but have no way to directly influence visual performance of the type studied here.

This is especially salient in the studies of perceived chasing because chasing detection seems subject to inflexible limitations. The most natural way to think of a higher-level decision is that you can decide whether to make it or not, and you can also decide just when to make it. But the results of the chasing studies reveal stark limits on the ability of observers to simply "decide" what counts as chasing. Rather, the implicit performance measures seem to be tapping into an underlying ability whose limits cannot be influenced merely by decisions about what features should matter for detecting animacy; rather, only those factors that actually *do* matter will facilitate detection and avoidance. To take a particular example, it would have been helpful in these studies if subjects could have decided to treat 90° of Chasing Subtlety as indicating chasing (as in fact it did in terms of the underlying motion patterns!), but apparently they were unable to do so. Indeed, this limitation is especially noteworthy in the studies of interrupted chasing (Gao & Scholl, 2011), since, before each experiment, subjects were carefully and explicitly informed about the precise nature of the wolf's motions. Thus, for example, subjects knew all about the random-motion interruptions, and had every incentive to discount them—but they could not do so in all cases.

This property is perhaps even starker in the case of the wolfpack effect. Beyond the phenomenology of the wolfpack displays, the associated experiments also measured the impact of the wolfpack configuration on several varieties of implicit visuomotor performance (Gao, McCarthy, & Scholl, 2010). In what was perhaps the strongest case, the wolfpack display was

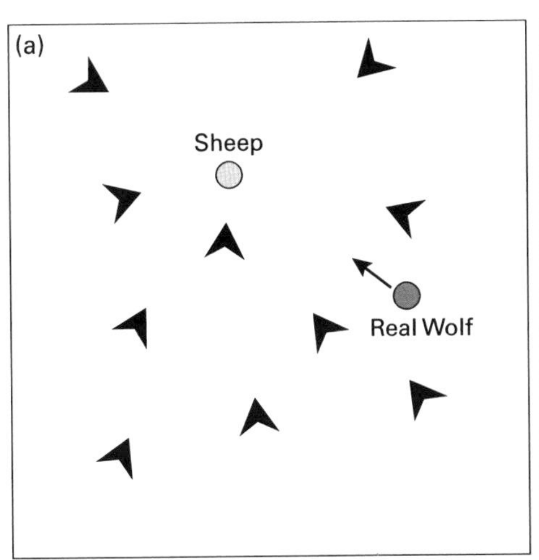

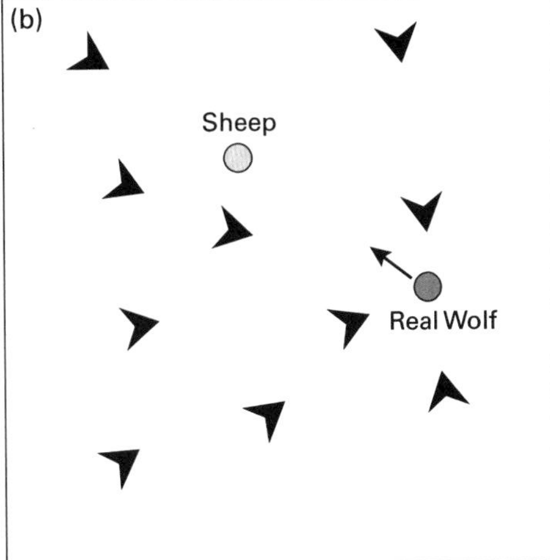

Figure 9.6
Screenshots from the interactive display used in a study of the wolfpack effect by Gao, McCarthy, and Scholl (2010). Subjects had to use the mouse to move the green disc across the display in order to avoid touching the display border, the darts, or a red "wolf" disc. (a) A screenshot from a dynamic wolfpack-to-sheep trial. Each dart was always oriented towards the user-controlled green disc. (b) A screenshot from a dynamic wolfpack-to-wolf trial. Each dart was always oriented toward the red "wolf" disc.

combined with the chasing task in a novel way: Subjects moved a sheep disc about a dart-filled display and attempted not to contact either (1) any of the darts or (2) a wolf disc that continually chased the sheep (figure 9.6a). (This was not a chasing detection task, however: subjects knew that the darts could never chase them, and the wolf—which was the only other disc in the display—was always colored bright red in order to make it especially salient.) This combination of the wolfpack plus the chasing wolf served two purposes. First, it forced subjects to keep moving constantly, even in a display that was considerably sparser than some of the initial chasing displays. Second, and more importantly, the "real" wolf could also now serve on some trials as the target of the wolfpack (such that the darts all continually pointed to the wolf rather than to the subject-controlled sheep disc). This manipulation (depicted in figure 9.6b) allowed us to essentially vary the social significance of the wolfpack while perfectly equating all visual factors.

The results of our study (Experiment 5 of Gao et al., 2010) were especially clear: Subjects were significantly worse at this task—by more than

20 percent!—in the wolfpack condition compared to the perpendicular control condition, but this effect was greatly attenuated when the wolfpack instead "faced" the wolf rather than the subject-controlled sheep (thus ruling out any explanation based on grouping or any other lower-level visual factor). As with the chasing results, these results seem impossible to explain by appeal to explicit higher-level decisions, because what they show is how impotent such decisions can be. After all, subjects in this experiment had every reason to simply ignore the wolfpack (and not to treat it as animate at all) since it was irrelevant or even disruptive to their overt task. But, despite these powerful task demands, they could not simply decide not to treat the wolfpack as animate. And this limitation is especially noteworthy here, since subjects were fully informed prior to the experiment about the irrelevance of the darts' orientations. These sorts of results strongly suggest that the resulting data reflect some properties and constraints of automatic perceptual processing rather than higher-level decisions that subjects are overtly making about the contents of the displays.

Activation of Visual Brain Areas

Another fairly direct way of asking about the broad categorical nature of the processes underlying a phenomenon is to ask about the brain regions that selectively support it. This sort of strategy requires several assumptions, and will only be helpful (among other constraints) when (1) there *are* brain regions that are selective for the process in question, (2) experiments can be devised to isolate the relevant processes in practice, and (3) the resulting regions actually tell you something categorical (e.g., about perception vs. cognition).

The perception of animacy, like every other variety of perception and cognition, has certainly received its share of neuroscientific studies aimed primarily at identifying its neural correlates (e.g., Blakemore et al., 2003; Castelli et al., 2000; Martin & Weisberg, 2003; Schultz et al., 2004, 2005; for a review see Heberlein, 2008; see also McAleer & Love, this volume). From the perspective of the present question about perception vs. higher-level judgment, these studies all face at least two major challenges. First, the stimuli used in previous studies have made it difficult or impossible to isolate animacy, per se, relative to other lower-level visual factors such as stimulus energy and motion correlation. This is in large part because prior stimuli did not impose a tight degree of control between animate vs. inanimate displays—either by constructing such displays entirely haphazardly (e.g., Castelli et al., 2000) or by using notably different motion trajectories in animate vs. inanimate displays (e.g., Blakemore et al., 2003; Schultz et al., 2005). As a result, it is not possible to definitively tie the resulting brain activity to perceived animacy. And of course this challenge is especially dire if researchers want to explore brain areas involved in visual processing, since (for example) motion regions (including MT+/V5) would be expected to be especially sensitive to differences in motion trajectories.

A second major challenge in this area from the perspective of the present chapter is that the brain areas that do result from such studies do not decisively implicate either perception or cognition. Perhaps the most common finding in this literature is selective activation for animate displays in the posterior superior temporal sulcus (pSTS). On the one hand, this brain area has previously been implicated in other forms of visual processing such as biological motion perception (Allison, Puce, & McCarthy, 2000) as well as attention shifting (e.g., Corbetta & Shulman, 2002; Serences et al., 2005). But on the other hand, this region has also been implicated in theory of mind more generally, in studies that report selective activation from stimuli such as verbal sentences, with no visual stimulation whatsoever (e.g., Saxe, in press; Saxe & Kanwisher, 2003). And more generally, the functional properties of this brain region are simply confusing and controversial in several respects (e.g., Mitchell, 2008), with one recent review of the controversy defining the STS (in the very first sentence of its abstract!) as "the chameleon of the brain" (Hein & Knight, 2008, p. 2125).

Recently, however, we have collaborated with Greg McCarthy on an attempt to explore the neural bases of one specific form of perceived animacy in a study that (1) explores some visual brain regions directly and (2) avoids both of these major challenges. The primary design of the experiment (Gao, Scholl, & McCarthy, under review; see also McCarthy, Gao, & Scholl, 2009) was straightforward: We utilized the especially well-controlled contrast between the wolfpack condition and the perpendicular condition from the wolfpack effect in a subtraction design to explore the brain regions that were selective for animacy in wolfpack blocks vs. perpendicular blocks. Subjects again completed the Don't-Get-Caught task while in the scanner; in order to directly explore the role of attention, we also included a separate control condition in which the shapes repeatedly flashed on and off.

The results of this study were rich and varied, and they implicated regions such as the pSTS that (as described earlier in this section) are not helpful for the goals of the present chapter. But the resulting activation patterns also implicated brain regions that are strongly associated with

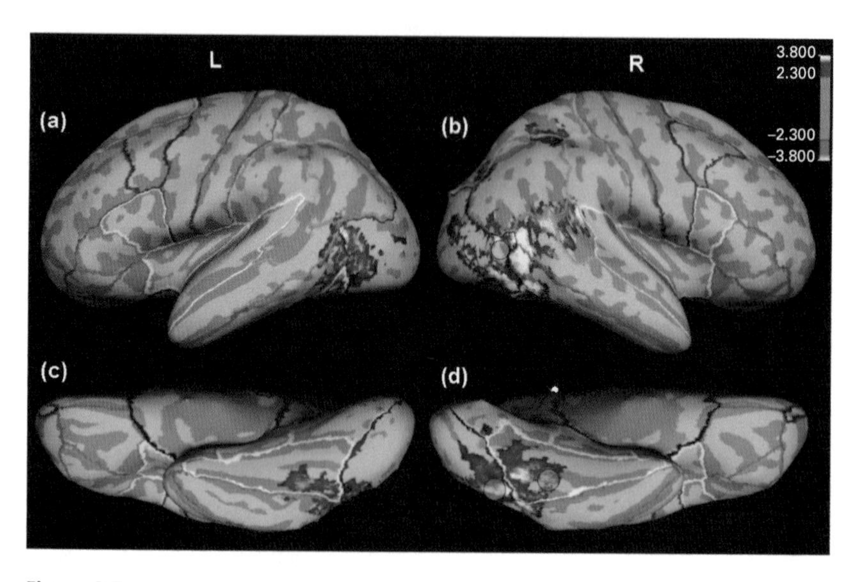

Figure 9.7
The cluster-corrected contrast for wolfpack > perpendicular blocks on the inflated MNI152 brain in the study of Gao, Scholl, and McCarthy (under review). The single isolated blue circle in (c) indicates the peak coordinates of area MT derived from each individual's motion localizer. The additional blue circles in (d) indicate the peak coordinates of areas FFA and OFA derived from each individual's face-scene localizer.

visual processing, such as the fusiform face area, and the motion-selective region MT+. This last result is especially exciting insofar as (1) the contrast controlled perfectly for lower-level motion parameters, and (2) this brain region is especially strongly identified with visual processing per se. Indeed, despite considerable prior research, this is the first study to our knowledge that has directly implicated this brain region as being correlated with perceived animacy while controlling for lower-level motion (cf. Castelli et al., 2000). This selective activation is depicted directly in figure 9.7; the semi-transparent blue circle illustrates the peak coordinates of area MT+ as derived from an independent motion localizer task. And the selectivity for animacy in this area is especially well illustrated in the timecourse activation depicted in figure 9.8, with a considerably greater percent signal change for the wolfpack blocks at several timepoints relative to both the perpendicular control blocks and the flashing control blocks. Though this is only a single study exploring a single cue to perceived animacy, these results nevertheless constitute a compelling case study suggesting that the

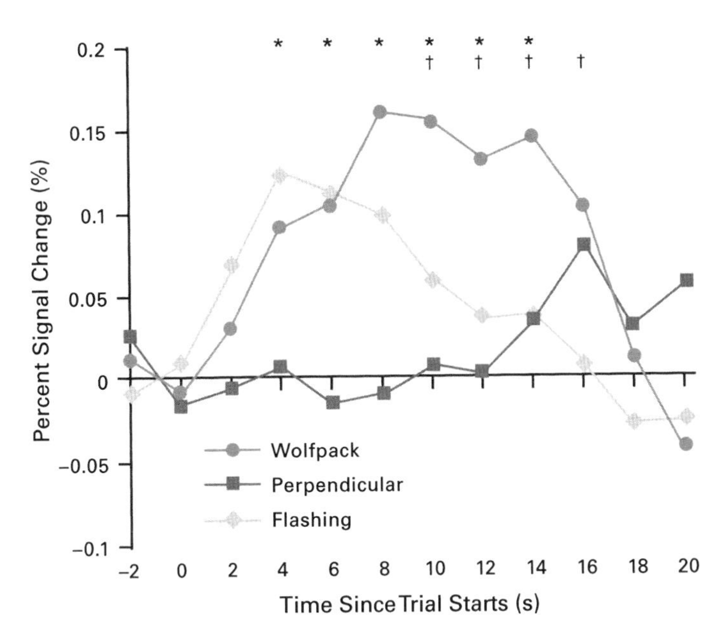

Figure 9.8
Time courses of the average BOLD signal change from the region of interest consisting of the peak coordinates of area MT derived from each individual's motion localizer, in the study of Gao, Scholl, and McCarthy (under review).

perception of animacy pervades several levels of visual processing in the human brain. These results suggest that MT+ is either sensitive to configural cues for animacy detection per se, or that, once animacy is detected, this area is selectively activated through feedback from higher processing regions.

Rich Interaction with Other Visual Processes

A final way of implicating visual processing is to explore the ways in which perceived animacy interacts with other visual processes. As a case study of this kind of interaction, we will now come full circle, returning to the discussion of selectivity in vision from the beginning of this chapter. We noted in that initial discussion of detecting agents that there are at least two primary forms of selective processing. One is that specialized visual processing may be devoted to some types of stimuli relative to others, and this possibility is consistent with our previous discussion of the types of subtle and specific cues that drive perceived chasing. But the more obvious form of selection in vision involves selective attention in online visual processing. In a recent study of perceived chasing, we directly explored the connections between animacy and attention and focused in particular on the role of perceived intentionality (Gao, New, & Scholl, under review). (This is part of a larger attempt to look at perceived animacy not only as a putative end product of perception, but also in terms of its downstream *effects* on other forms of perception and cognition.)

In these studies, observers again watched a wolf chase a sheep, but now in the context of a new *wavering wolf* display (see also Gao, Scholl, & McCarthy, 2012). Each animation contained only three discs, one of which was the wolf, and two of which were potential sheep. The potential sheep always moved randomly, while the wolf pursued one of the two potential sheep at any given time, but frequently shifted between the two sheep as pursuit targets. This basic display is depicted in figure 9.9 (with online animations available at http://www.yale.edu/perception/chasing-attention/). This display yields robust percepts (as revealed by free reports) not only of animacy and chasing (both of which are held constant throughout each animation), but also of perceived intentions that frequently change. It is immediately clear which disc is chasing which other disc, and the shifts to new chasing targets seem especially salient.

We then sought to explore the influence of this display on attention, taking advantage of straightforward and well-studied experimental paradigms from attention research. In particular, to explore how attention is distributed across such dynamic displays during passive viewing (with no instructions or indeed any mention of animacy or chasing), we employed

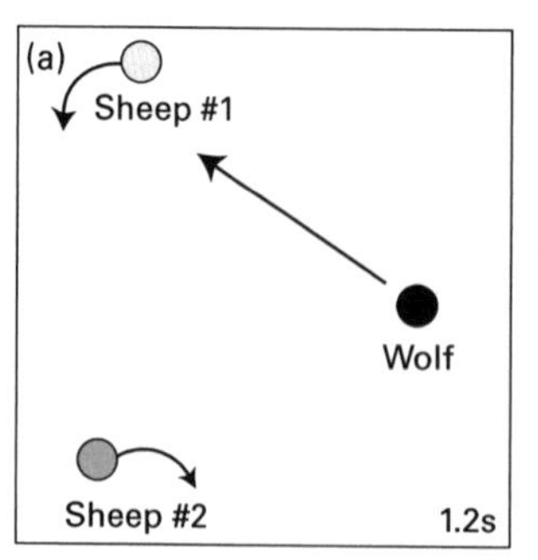

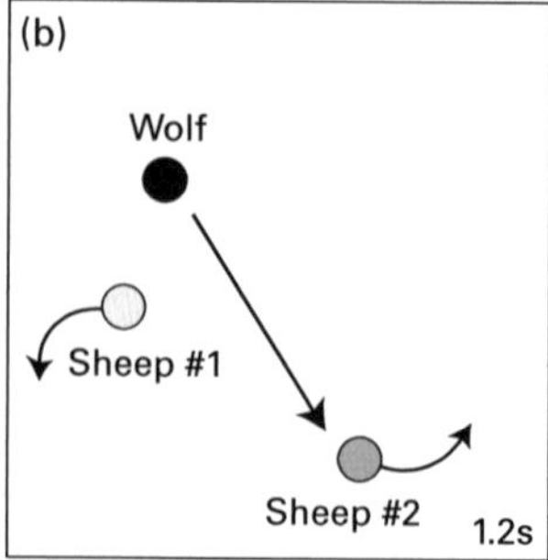

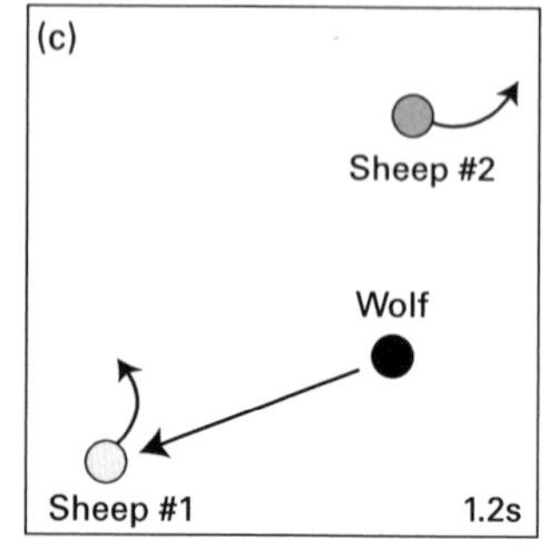

Figure 9.9
An illustration of the wavering wolf display used in the study of Gao, New, and Scholl (under review). The wolf chased continuously, but its goal (i.e., the sheep) switched among other two (randomly moving) shapes every 1.2 seconds.

a simple form of probe detection. Small probes were briefly and sporadically presented on each object and the resulting probe detection rates were taken as indices of the degree to which each object was currently attended. This method was initially used in some of the first studies of spatial attention (e.g., Posner et al., 1980); more recently, it has been used to explore how attention is distributed over space and time in dynamic multiobject displays that are not perceived in animate terms (e.g., Alvarez & Scholl, 2005; Flombaum, Scholl, & Pylyshyn, 2008).

The resulting patterns of attention revealed two primary results. First, observers selectively prioritized the wolf during chasing, even though no distinctions were made among the items in the instructions. Second, the chased object (i.e., the sheep) was also attentionally prioritized over the third shape in the display; this was the case even though they both moved according to the same underlying algorithm. Both effects depended on perceived intentions since additional control studies ruled out explanations based on lower-level properties such as proximity and common motion.

This close online control over attention is perhaps what would be expected by a visual process, but it would be difficult to explain by appeal to higher-level judgment and intentions (see also Pratt et al., 2010). This is true in a general way, but also in a more specific way: The prioritization in this study of the current sheep over the now-suddenly-ignored sheep occurred extremely quickly, in the course of less than a second—and such shifts occurred reliably over and over throughout the experiment. This is exactly what would be expected if perceived animacy was controlling attention involuntarily as a part of ongoing visual processing, but it would be surprising for such fast, reliable, and repetitive shifts to occur in the context of voluntary higher-level attentional control.

"Perception of Animacy": Aptly Named

The five lines of evidence reviewed in the preceding section all converge on the same conclusion: The "perception of animacy" is aptly named. This phenomenon really seems to reflect perception rather than higher-level judgment and inference—not just in a weak sense (because it begins with visual input) but in a strong and much more intriguing sense (as it seems to invoke specialized visual processing). We conclude the perception of animacy is just as "visual" as is the perception of other properties such as depth. In each case, the visual system engages in specialized unconscious inferences in order to detect the presence of (physical or social) information in the local environment that cannot be deductively inferred but is of critical importance to our fitness and survival.

Acknowledgments

For helpful conversation and/or comments on previous drafts, we thank Greg McCarthy, Josh New, George Newman, and the organizers and attendees at the fantastic McMaster Workshop on Social Perception. Address correspondence and reprint requests to:

Brian Scholl
Department of Psychology
Yale University
Box 208205
New Haven, CT 06520–8205
Email: brian.scholl@yale.edu
Web: http://www.yale.edu/perception/
Phone: 203–432–4629

Notes

1. Percepts of animacy may also be driven by more specific types of motion cues, such as when the relative motions of many points in point-light displays are perceived in terms of particular forms of biological motion (for a review see section 1 of this volume and Blake & Shiffrar, 2007). We do not delve into such phenomena in the present chapter, however, since they are thought to reflect a distinct form of motion-based processing that integrates over multiple parts of an agent (e.g., multiple limbs or joints). Instead, we focus here only on percepts of animacy in which each agent is realized via the movement of a single shape.

2. This characterization may be accurate for other related research programs. For example, recent years have seen the growth of "mind perception" as a distinct area of study, with review articles and even entire laboratories so named (e.g., Gray et al., 2007; for a review see Waytz et al., 2010). But the connection to visual perception seems tenuous or nonexistent. Instead, this research program simply involves asking participants about their intuitions and judgments directly—e.g., asking participants a question such as "Does this entity have the capacity to feel pain?," with their answers indicated by numerical ratings. Referring to such data in terms of mind perception (rather than "mind judgments" or "mind ratings") sounds exciting, but visual processing is never actually invoked.

3. Consistent with this discussion, it is perhaps also worth noting that the displays of Tremoulet and Feldman (involving single moving dots that underwent a change

in direction and/or speed; Tremoulet & Feldman, 2000) do *not* give rise to rich phenomenological percepts of animacy. Indeed, as Gao, New, and Scholl (under review) have argued, rich percepts of animacy intrinsically require some form of perceived *interaction*. This flies in the face of many previous appeals to self-propulsion as a robust cue to perceived animacy (e.g., Dasser et al., 1989), but it is consistent with the resulting phenomenology. For example, the objects in multiple-object tracking (MOT) displays (see Scholl, 2009) almost always move in a self-propelled manner, and they change their speeds and angles frequently. Nevertheless, to our knowledge, nobody viewing a MOT display has ever exclaimed that the shapes appeared to be animate. From our perspective, this is because the objects in MOT displays move independently of each other (or at most are repulsed by each other), whereas interobject interactions may be a key to perceived animacy (Gao, New, and Scholl, under review).

4. In fact, the purpose of this study was to explore other questions, such as the degree to which perceived animacy can be explained by the operation of a "principle of rationality" in online perception, but we do not discuss such themes in this chapter.

References

Abell, F., Happé, F., & Frith, U. (2000). Do triangles play tricks? Attribution of mental states to animated shapes in normal and abnormal development. *Journal of Cognition and Development, 15,* 1–16.

Adams, R., Ambady, N., Nakayama, K., & Shimojo, S. (Eds.). (2010). *The science of social vision.* Oxford: Oxford University Press.

Allison, T., Puce, A., & McCarthy, G. (2000). Social perception from visual cues: Role of the STS region. *Trends in Cognitive Sciences, 4,* 267–278.

Alvarez, G. A., & Scholl, B. J. (2005). How does attention select and track spatially extended objects? New effects of attentional concentration and amplification. *Journal of Experimental Psychology: General, 134,* 461–476.

Baker, C., Saxe, R., & Tenenbaum, J. (2009). Action understanding as inverse planning. *Cognition, 113,* 329–349.

Barrett, H. C., Todd, P. M., Miller, G. F., & Blythe, P. (2005). Accurate judgments of intention from motion alone: A cross-cultural study. *Evolution and Human Behavior, 26,* 313–331.

Bassili, J. (1976). Temporal and spatial contingencies in the perception of social events. *Journal of Personality and Social Psychology, 33,* 680–685.

Blake, R., & Shiffrar, M. (2007). Perception of human motion. *Annual Review of Psychology, 58,* 47–73.

Blakemore, S. J., Boyer, P., Pachot-Clouard, M., Meltzoff, A., Segebarth, C., & Decety, J. (2003). The detection of contingency and animacy from simple animations in the human brain. *Cerebral Cortex, 13*, 837–844.

Blythe, P. W., Todd, P. M., & Miller, G. F. (1999). How motion reveals intention: Categorizing social interactions. In G. Gigerenzer, P. M. Todd, & the ABC Research Group (Eds.), *Simple heuristics that make us smart* (pp. 257–286). New York: Oxford University Press.

Boyle, D. G. (1972). Michotte's ideas. *Bulletin of the British Psychological Society, 25*, 89–91.

Carrasco, M., Ling, S., & Read, S. (2004). Attention alters appearance. *Nature Neuroscience, 7*, 308–313.

Castelli, F., Happé, F., Frith, U., & Frith, C. (2000). Movement and mind: A functional imaging study of perception and interpretation of complex intentional movement patterns. *NeuroImage, 12*, 314–325.

Choi, H., & Scholl, B. J. (2004). Effects of grouping and attention on the perception of causality. *Perception & Psychophysics, 66*, 926–942.

Choi, H., & Scholl, B. J. (2006). Perceiving causality after the fact: Postdiction in the temporal dynamics of causal perception. *Perception, 35*, 385–399.

Chun, M., Golomb, J., & Turk-Browne, N. (2011). A taxonomy of external and internal attention. *Annual Review of Psychology, 62*, 73–101.

Corbetta, M., & Shulman, G. L. (2002). Control of goal-directed and stimulus-driven attention in the brain. *Nature Reviews: Neuroscience, 3*, 201–215.

Csibra, G. (2008). Goal attribution to inanimate agents by 6.5-month-old infants. *Cognition, 107*, 705–717.

Dasser, V., Ulbaek, I., & Premack, D. (1989). The perception of intention. *Science, 243*, 365–367.

Dittrich, W., & Lea, S. (1994). Visual perception of intentional motion. *Perception, 23*, 253–268.

Dovidio, J. F., Kawakami, K., & Beach, K. R. (2001). Implicit and explicit attitudes: Examination of the relationship between measures of intergroup bias. In R. Brown & S. L. Gaertner (Eds.), *Blackwell handbook of social psychology* (Vol. 4, *Intergroup relations*, pp. 175–197). Oxford: Blackwell.

Emery, N. (2000). The eyes have it: The neuroethology, function, and evolution of social gaze. *Neuroscience and Biobehavioral Reviews, 24*, 581–604.

Flombaum, J. I., Scholl, B. J., & Pylyshyn, Z. W. (2008). Attentional resources in tracking through occlusion: The high-beams effect. *Cognition, 107*, 904–931.

Gao, T., McCarthy, G., & Scholl, B. J. (2010). The wolfpack effect: Perception of animacy irresistibly influences interactive behavior. *Psychological Science, 21*, 1845–1853.

Gao, T., New, J. J., & Scholl, B. J. (under review). The wavering wolf: Perceived intentionality controls attentive tracking. Manuscript submitted for publication.

Gao, T., Newman, G. E., & Scholl, B. J. (2009). The psychophysics of chasing: A case study in the perception of animacy. *Cognitive Psychology, 59*, 154–179.

Gao, T., & Scholl, B. J. (2011). Chasing vs. stalking: Interrupting the perception of animacy. *Journal of Experimental Psychology: Human Perception and Performance, 37*, 669–684.

Gao, T., Scholl, B. J., & McCarthy, G. (under review). The perception of animacy pervades visual processing: Selective engagement of cortical regions associated with faces and motion. Manuscript submitted for publication.

Gao, T., Scholl, B. J., & McCarthy, G. (2012). Dissociating the detection of intentionality from animacy in the right posterior superior temporal sulcus. *Journal of Neuroscience, 32*, 14276–14280.

Garfield, P. (2001). *Universal dream key: The 12 most common dream themes around the world*. New York: HarperCollins.

Gelman, R., Durgin, F., & Kaufman, L. (1995). Distinguishing between animates and inanimates: Not by motion alone. In D. Sperber, D. Premack, & A. J. Premack (Eds.), *Causal cognition: A multidisciplinary debate* (pp. 150–184). Oxford: Clarendon Press.

Gergely, G., & Csibra, G. (2003). Teleological reasoning in infancy: The naïve theory of rational action. *Trends in Cognitive Sciences, 7*, 287–292.

Gray, H. M., Gray, K., & Wegner, D. M. (2007). Dimensions of mind perception. *Science, 315*, 619.

Gregory, R. L. (1980). Perceptions as hypotheses. *Philosophical Transactions of the Royal Society of London, Series B: Biological Sciences, 290*, 181–197.

Heberlein, A. (2008). Animacy and intention in the brain: Neuroscience of social event perception. In T. Shipley & J. Zacks (Eds.), *Understanding events: From perception to action* (pp. 363–388). Oxford: Oxford University Press.

Heberlein, A. S., & Adolphs, R. (2004). Impaired spontaneous anthropomorphizing despite intact perception and social knowledge. *Proceedings of the National Academy of Sciences of the United States of America, 101*, 7487–7491.

Heider, F., & Simmel, M. (1944). An experimental study of apparent behavior. *American Journal of Psychology, 57*, 243–259.

Hein, G., & Knight, R. T. (2008). Superior temporal sulcus—it's my area: Or is it? *Journal of Cognitive Neuroscience, 20*, 2125–2136.

Joynson, R. B. (1971). Michotte's experimental methods. *British Journal of Psychology, 62*, 293–302.

Kanwisher, N. (2010). Functional specificity in the human brain: A window into the functional architecture of the mind. *Proceedings of the National Academy of Sciences of the United States of America, 107*, 11163–11170.

Klin, A. (2000). Attributing social meaning to ambiguous visual stimuli in higher functioning autism and Asperger syndrome: The social attribution task. *Journal of Child Psychology and Psychiatry, and Allied Disciplines, 41*, 831–846.

Leslie, A. M., Friedman, O., & German, T. P. (2004). Core mechanisms in "theory of mind." *Trends in Cognitive Sciences, 8*, 528–533.

Martin, A., & Weisberg, J. (2003). Neural foundations for understanding social and mechanical concepts. *Cognitive Neuropsychology, 20*, 575–587.

McCarthy, G., Gao, T., & Scholl, B. J. (2009). Processing animacy in the posterior superior temporal sulcus [abstract]. *Journal of Vision, 9*(8), 775a. http://journalofvision.org/9/8/775/.

McNamara, T. (Ed.). (2005). *Semantic priming: Perspectives from memory and word recognition*. New York: Psychology Press.

Michotte, A. (1950). The emotions regarded as functional connections. In M. Reymert (Ed.), *Feelings and emotions: The Mooseheart symposium* (pp. 114–125). New York: McGraw-Hill. Reprinted in Thinès, G., Costall, A., and Butterworth, G. (Eds.), *Michotte's experimental phenomenology of perception* (pp. 103–116). Hillsdale, NJ: Erlbaum, 1991.

Mitchell, J. P. (2008). Activity in right temporo-parietal junction is not selective for theory-of-mind. *Cerebral Cortex, 18*, 262–271.

Most, S. B., Scholl, B. J., Clifford, E., & Simons, D. J. (2005). What you see is what you set: Sustained inattentional blindness and the capture of awareness. *Psychological Review, 112*, 217–242.

Nahin, P. J. (2007). *Chases and escapes: The mathematics of pursuit and evasion*. Princeton: Princeton University Press.

New, J. J., Cosmides, L., & Tooby, J. (2007). Category-specific attention for animals reflects ancestral priorities not expertise. *Proceedings of the National Academy of Sciences of the United States of America, 104*, 16598–16603.

New, J. J., Schultz, R. T., Wolf, J., Niehaus, J. L., Klin, A., German, T., et al. (2010). The scope of social attention deficits in autism: Prioritized orienting to people and animals in static natural scenes. *Neuropsychologia, 48*, 51–59.

Newman, G. E., Choi, H., Wynn, K., & Scholl, B. J. (2008). The origins of causal perception: Evidence from postdictive processing in infancy. *Cognitive Psychology, 57*, 262–291.

Öhman, A., & Mineka, S. (2003). The malicious serpent: Snakes as a prototypical stimulus for an evolved module of fear. *Current Directions in Psychological Science, 12,* 5–8.

Pantelis, P. C., & Feldman, J. (2012). Exploring the mental space of autonomous intentional agents. *Attention, Perception & Psychophysics, 74,* 239–249.

Posner, M. I., Davidson, B. J., & Snyder, C. R. R. (1980). Attention and the detection of signals. *Journal of Experimental Psychology: General, 109,* 160–174.

Pratt, J., Radulescu, P., Guo, R. M., & Abrams, R. A. (2010). It's alive! Animate motion captures visual attention. *Psychological Science, 21,* 1724–1730.

Pylyshyn, Z. W. (1999). Is vision continuous with cognition? The case for cognitive impenetrability of visual perception. *Behavioral and Brain Sciences, 22,* 341–423.

Rakison, D. H., & Derringer, J. L. (2008). Do infants possess an evolved spider-detection mechanism? *Cognition, 107,* 381–393.

Revonsuo, A. (2000). The reinterpretation of dreams: An evolutionary hypothesis of the function of dreaming. *Behavioral and Brain Sciences, 23,* 877–1121.

Rips, L. (2011). Causation from perception. *Perspectives on Psychological Science, 6,* 77–97.

Rochat, P., Morgan, R., & Carpenter, M. (1997). Young infants' sensitivity to movement information specifying social causality. *Cognitive Development, 12,* 537–561.

Rochat, P., Striano, T., & Morgan, R. (2004). Who is doing what to whom? Young infants' developing sense of social causality in animated displays. *Perception, 33,* 355–369.

Rock, I. (1983). *The logic of perception.* Cambridge, MA: MIT Press.

Rolfs, M., Dambacher, M., & Cavanagh, P. (2013). Visual adaptation of the perception of causality. *Current Biology, 23,* 250–254.

Rutherford, M. D., Pennington, B. F., & Rogers, S. J. (2006). The perception of animacy in young children with autism. *Journal of Autism and Developmental Disorders, 36,* 983–992.

Santos, N. S., David, N., Bente, G., & Vogeley, K. (2008). Parametric induction of animacy experience. *Consciousness and Cognition, 17,* 425–437.

Saxe, R. (in press). The right temporo-parietal junction: A specific brain region for thinking about thoughts. In A. Leslie & T. German (Eds.), *Handbook of theory of mind.* Hillsdale, NJ: Erlbaum.

Saxe, R., Carey, S., & Kanwisher, N. (2004). Understanding other minds: Linking developmental psychology and functional neuroimaging. *Annual Review of Psychology, 55,* 87–124.

Saxe, R., & Kanwisher, N. (2003). People thinking about thinking people: The role of the temporo-parietal junction in "theory of mind." *NeuroImage, 19*, 1835–1842.

Scholl, B. J. (2009). What have we learned about attention from multiple object tracking (and vice versa)? In D. Dedrick & L. Trick (Eds.), *Computation, cognition, and Pylyshyn* (pp. 49–78). Cambridge, MA: MIT Press.

Scholl, B. J., & Gao, T. (to appear). Seeing minds in motion: The perception of animacy and intentionality. *Annual Review of Psychology*.

Scholl, B. J., & Leslie, A. M. (2001). Minds, modules, and meta-analysis. *Child Development, 72*, 696–701.

Scholl, B. J., & Tremoulet, P. (2000). Perceptual causality and animacy. *Trends in Cognitive Sciences, 4*, 299–309.

Schultz, J., Friston, K. J., O'Doherty, J., Wolpert, D. M., & Frith, C. D. (2005). Activation in posterior superior temporal sulcus parallels parameter inducing the percept of animacy. *Neuron, 45*, 147–156.

Schultz, J., Imamizu, H., Kawato, M., & Frith, C. D. (2004). Activation of the human superior temporal gyrus during observation of goal attribution by intentional objects. *Journal of Cognitive Neuroscience, 16*, 1695–1705.

Serences, J., Shomstein, S., Leber, A., Golay, X., Egeth, H., & Yantis, S. (2005). Coordination of voluntary and stimulus-driven attentional control in human cortex. *Psychological Science, 16*, 114–122.

Sinha, P., Balas, B., Ostrovsky, Y., & Russell, R. (2006). Face recognition by humans: Nineteen results all computer vision researchers should know about. *Proceedings of the IEEE, 94*, 1948–1962.

Sugita, Y. (2009). Innate face processing. *Current Opinion in Neurobiology, 19*, 39–44.

Tavares, P., Lawrence, A., & Barnard, P. (2008). Paying attention to social meaning: An fMRI study. *Cerebral Cortex, 18*, 1876–1885.

Treisman, A. (2006). How the deployment of attention determines what we see. *Visual Cognition, 14*, 411–443.

Tremoulet, P. D., & Feldman, J. (2000). Perception of animacy from the motion of a single object. *Perception, 29*, 943–951.

Tremoulet, P. D., & Feldman, J. (2006). The influence of spatial context and the role of intentionality in the interpretation of animacy from motion. *Perception & Psychophysics, 68*, 1047–1058.

Wagemans, J., Van Lier, R., & Scholl, B. J. (2006). Introduction to Michotte's heritage in perception and cognition research. *Acta Psychologica, 123,* 1–19.

Waytz, A., Gray, K., Epley, N., & Wegner, D. M. (2010). Causes and consequences of mind perception. *Trends in Cognitive Sciences, 14,* 383–388.

Wilmer, J., Germaine, L., Chabris, C., Chatterjee, G., Williams, M., Loken, E., et al. (2010). Human face recognition ability is specific and highly heritable. *Proceedings of the National Academy of Sciences of the United States of America, 107,* 5238–5241.

10 How Are the Actions of Triangles and People Processed in the Human Brain?

Antonia F. de C. Hamilton and Richard Ramsey

Abstract

Comprehension of actions is a core social skill. Here we provide a critical review of the dominant mirror neuron theory of action comprehension. Recent data demonstrate that parts of the mirror system respond to actions performed by nonhuman shapes and are insensitive to actor identity. Regions beyond the mirror system are also important for action comprehension. We suggest this data is not compatible with a strong mirror system hypothesis, and we outline alternative theories.

1 Introduction

Humans invest a large proportion of their time acting and interacting with other people and, as such, are social animals. A central feature of social interaction is reciprocal nonverbal understanding: Individuals must make sense of each other's actions in order to interact appropriately. For example, if a mother and child are building a sand castle, each must interpret the other's actions of scooping sand or tapping the bucket in order to coordinate their own actions and achieve their joint goal. The ability to understand other people's actions is a key feature of human social cognition. The present paper focuses on the cognitive and brain mechanisms that underpin how adults make sense of other people's actions and goals.

Making sense of others' actions is not a unitary cognitive process—it is multifaceted. An observed action encompasses features at multiple different levels of description, including kinematics (e.g., grip type), target objects (e.g., the type of object grasped), and broader motivations (e.g., deceit). Although it is important to keep track of all these levels during social interaction, the major theme of the current paper is the relationship between actions and target objects, which we refer to as object-goals. As we carry out our daily activities, we frequently observe other people performing object-directed actions, which commonly achieve specific goals. Different types of objects are normally associated with different goals. For example, a person who grasps a banana is likely to want to eat, while one who grasps a hammer is more likely to hit a nail and not likely to eat. Thus, interpretation of even simple object-grasping actions can help us to (1) predict future actions, (2) learn from them, and (3) interact appropriately. The present paper reviews a series of neuroimaging studies that examine the cognitive and brain processes that allow us to interpret other people's goal-directed actions.

This chapter is organized into three parts. In the first part, we outline an account of goal understanding—one that has dominated the literature for the past fifteen years—that is based on directly matching observed actions onto one's own motor system; this process is linked to the human mirror neuron system (MNS) (Rizzolatti, Fogassi, & Gallese 2001). We further consider some limitations of both this framework and past research into brain systems for goal understanding.

In the second part, we review a series of studies from our laboratory that take inspiration from developmental psychology and are designed to examine how actions and goals are processed in the adult human brain. We consider how different brain systems may represent two features of observed actions: (1) The target object of an action (object-goal) and (2) the identity of the agent performing an action. In particular, we suggest that one specific brain region within the MNS—the anterior intraparietal sulcus (aIPS)—is sensitive to the target object of an action (food vs. tool) but does not distinguish agent identity. Responses in the aIPS are the same for two different human agents and even for a human agent as compared to an animated geometric shape that is devoid of humanlike form and motion.

Finally, we consider the implications of these findings in relation to claims that (1) the MNS is tuned only to human actions and (2) actions are understood by a direct-matching mechanism within the MNS (Rizzolatti et al., 2001). We suggest that (1) activation of the MNS is driven by the behavior and interpretation of an animate actor rather than by its form, and (2) strong direct-matching theories cannot account for these results. Further, we suggest that brain systems beyond the MNS are required to interpret who is performing an action and to understand unusual actions; this implies that the MNS is not the sole brain network used for action comprehension. We further consider some future directions for research into goal understanding. We suggest a multidisciplinary approach comprising developmental, social, and cognitive psychology, with the addition of neuroscience methods, as a good starting point.

2 Background

2.1 Brain Systems for Action Understanding

A dominant view in the action cognition literature is that we understand other people's actions by matching observed actions onto our own motor repertoire (Rizzolatti et al., 2001). Evidence for this proposal has mainly been provided by the discovery of mirror neurons in ventral premotor cortex (di Pellegrino, Fadiga, Fogassi, Gallese, & Rizzolatti, 1992; Gallese, Fadiga, Fogassi, & Rizzolatti, 1996) and inferior parietal lobule (Fogassi et al., 2005). These neurons respond when a monkey performs a specific action, such as a precision grip to a piece of food, or observes the same action performed by another monkey or a human.

Since the discovery of mirror neurons in the monkey brain, cognitive neuroscientists have used neuroimaging techniques, such as functional magnetic resonance imaging (fMRI), to examine whether analogous regions in the human brain respond in a similar manner. In general, results have been consistent with findings in the monkey brain: The inferior frontal gyrus, as well as the adjacent ventral premotor cortex and the inferior parietal lobule, respond to performed and observed actions in a similar manner (for meta-analyses, see Grèzes & Decety, 2001; Caspers, Zilles, Laird, & Eickhoff, 2010). Based on these findings, it has been claimed that the activity in the observer's brain "mirrors" that of the performer's brain. Further, it is this matching or mirroring process that results in comprehension of the observed action, which is a primary component of social interaction (Rizzolatti et al., 2001; Gallese, Keysers, & Rizzolatti, 2004).

One limitation is that human neuroimaging and monkey neurophysiology experiments have used a variety of definitions for what it is to "understand" an action (Hickok, 2009). When one sees a child grasp an apple, the brain is able to rapidly process many different types of action information, including the kinematic features of the action (the speed of the hand and shape of the grasp), object-goal information (the hand grasps an apple), and broader motivations (the child wants the reddest apple on the tree) (Grafton & Hamilton, 2007). In a conventional fMRI experiment, which employs a subtraction design, it is not easy to separate these different components and determine whether a particular brain region responds to action kinematics, object-goals, broader motivations, or even some combination of these.

One study manipulated context in an attempt to identify brain regions that are sensitive to the goals or intentions underlying observed actions. Iacoboni and colleagues (2005) showed a hand as it grasped a teacup from either a messy or tidy table. The authors reasoned that, in the former case, the action was suggestive of the goal "cleaning," whereas in the latter case, the action was suggestive of the goal "drinking." When participants observed equivalent actions both within a context and with no context, stronger responses for the former were found in the inferior frontal gyrus (IFG) and the adjacent ventral premotor cortex (PMv). Furthermore, the authors of this study reported stronger activity in this region for drinking actions than for cleaning actions. Thus, the authors argued that the inferior frontal node of the MNS performed a context-sensitive mirroring process to understand the goals behind other people's actions (Iacoboni et al., 2005). Making such claims from these data, however, is problematic for several reasons. First, differences in context covary with other features such as visual scene complexity, task difficulty, movement kinematics, and saliency (Grafton & Hamilton, 2007). Second, there is no reason to predict that an observation of drinking should lead to stronger brain activity than an observation of cleaning. And finally, the process by which one could mirror a context has not, as yet, been clearly specified (Uithol, van Rooij, Bekkering, & Haselager, 2011a). Considering these limitations, it is difficult to claim that the observed response in IFG is a mirroring process responsible for goal inference.

A second study attempted to localize brain systems that are sensitive to the goals and intentions of other people's actions by comparing brain responses when observing an object-directed action to responses when observing an identical action toward empty space (Pelphrey, Morris, & McCarthy, 2004). This study found that the posterior superior temporal sulcus (pSTS) showed stronger responses when the action was directed toward an empty space rather than an object. The authors suggest that reaching to an empty location (when a target object is available) violates our expectations about how intentional agents will behave and pSTS is sensitive to the intentionality of observed actions. Although it is possible that this pSTS response reflects the interpretation of an unusual action, it is difficult to argue that this region encodes object-goals or intentions during actions that we typically observe, such as those directed toward objects.

One common feature of these two early attempts to localize brain regions that are sensitive to the goals served by actions is that they both used subtraction as an experimental design. Subtraction designs in neuroimaging have been criticized on broad theoretical grounds (Friston et al., 1996). More specifically, in the case of action-goal perception, it does not seem to be possible to create an optimal subtraction between a stimulus with a goal and a matched stimulus without a goal to reveal "goal" alone.

As previously mentioned, these two conditions also typically vary in other features, such as context, visual scene complexity, kinematics, or similar confounds (Grafton & Hamilton, 2007). To circumvent this issue, the studies we describe from our laboratory have used a different approach, which is inspired by work in developmental psychology.

2.2 Repetition Suppression

In a seminal infant study of action comprehension (Woodward, 1998), Amanda Woodward showed that six-month-olds are sensitive to the object-goals of other people's actions. She first habituated infants to the sight of an actor reaching toward a ball on the left—thereby ignoring a teddy bear on the right. At test, the infants looked longer when the actor reached toward the teddy bear, even when it was located on the left, and did not look longer at reaches toward the ball, even when the ball was located on the right. Thus, infants were sensitive to the identity of the object-goal but not to the direction of the hand motion. For the purposes of neuroimaging, the critical feature of this experimental design is that testing how an infant responds to a change in the target object—from ball to teddy bear—can reveal infant sensitivity to an object-goal.

Similarly, many regions of the human cortex are highly sensitive to changes in a sequence of stimuli. For example, if a participant in an fMRI study observes the same stimulus twice in a row, the blood oxygen level dependent (BOLD) response in brain regions that encode that stimulus is weaker for the second trial, but recovers when a novel stimulus is shown. This phenomenon is termed repetition suppression (RS), but also known as fMRI adaptation (Grill-Spector & Malach, 2001; Naccache & Dehaene, 2001). RS has been shown to be a general property of the human cortex across many domains, including the processing of objects (Grill-Spector et al., 1999), faces (Winston, Henson, Fine-Goulden, & Dolan, 2004), numbers (Naccache & Dehaene, 2001), and syntax (Noppeney & Price, 2004), as well as memory (Buckner et al., 1998) and motor performance tasks (Hamilton & Grafton, 2008b). Although there is some debate about the neural mechanism underlying RS (Grill-Spector, Henson, & Martin, 2006), there is agreement that the presentation of a repeated stimulus feature results in suppression of the BOLD signal, whereas presentation of novel a stimulus feature results in a release from suppression.

There are a number of advantages of RS paradigms over conventional subtraction designs. First, RS is sensitive to specific features of a stimulus and permits independent analysis of each feature. For example, a movie of an everyday action typically includes both object-goals (e.g., take cookie)

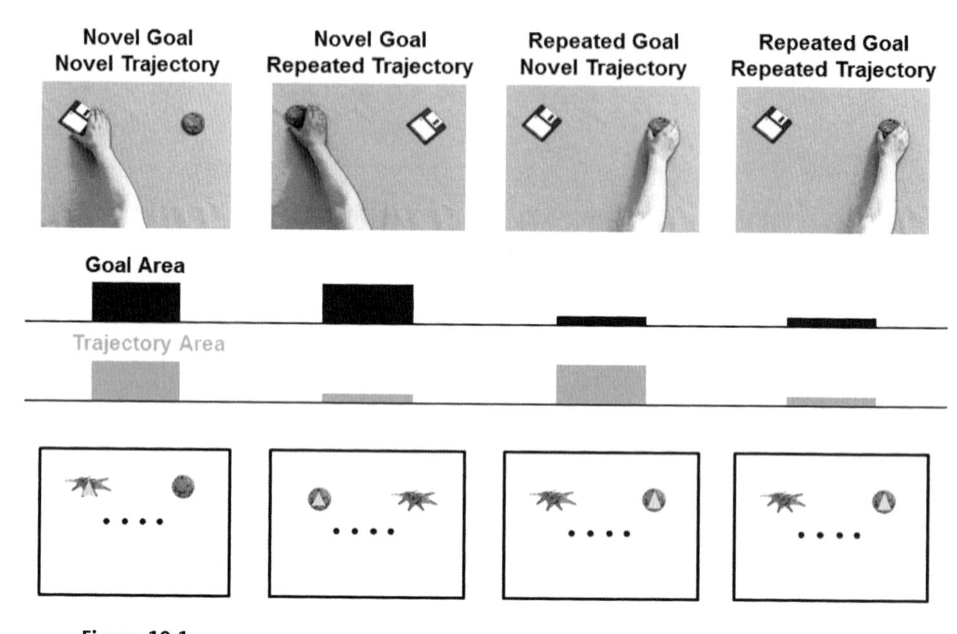

Figure 10.1
Stimulus sequences for repetition suppression studies of goal-directed actions performed by a human hand or an animated triangle. Top row—sample sequence of movies for the hand action study. Middle rows—predicted BOLD signal in brain regions encoding goal and trajectory. Bottom row—sample sequence of movies for the animated triangles study.

and kinematic features (e.g., move left). In a sequence of stimuli for an RS experiment, we can control the repetition of each feature independently (figure 10.1, top row). Consider the trial sequence [1. take-cookie-right] [2. take-disk-right] [3. take-disk-left]. In trial 2, the goal of the action (disk) is novel relative to trial 1, but the direction of movement (right) is repeated. Thus, we would predict RS in any brain regions sensitive to movement direction, but no suppression in any brain regions sensitive to goal. On trial 3, the goal (disk) is now repeated relative to trial 2, but the direction of movement (left) is novel. Thus, we would predict RS in brain regions encoding goal and a robust response in regions encoding movement direction. By presenting different sequences of stimuli, we can test for sensitivity to repeated kinematic features or repeated goal features independently; we can also distinguish these features within the brain. (figure 10.1, middle rows).

Second, RS designs are well balanced for attention and low-level features. All of our studies use a one-back RS paradigm whereby each video is coded as novel or repeated (in terms of a specific stimulus feature) relative to the previous video that was presented (figure 10.1). One-back RS designs differ from other work on RS or adaptation that show multiple repeats in order to adapt participants to a certain stimulus feature, such as leftward gaze (Calder et al., 2007). Specifically, within the one-back RS design, the same stimulus movie can appear in different repetition contexts, which means that conditions are balanced for the precise details of the movies. While observing videos, participants typically perform an incidental task, such as answering simple questions about the videos they have been watching. This means that the participants' task during scanning is always constant; they are unaware of the structure of novel and repeated stimulus features.

Third, RS studies can potentially be interpreted in terms of the tuning of neuronal populations within specific brain regions. The dominant neurophysiological explanation of RS suggests that populations of neurons within a brain region encode particular stimulus features. When a feature is repeated, the neuronal population that encodes that feature may respond with different timing, a sharper response tuning curve, or a weaker overall firing rate (Grill-Spector et al., 2006; but see Sawamura, Orban, & Vogels, 2006). These changes in the population response are believed to cause the weaker BOLD signal on repeated trials. Thus, finding RS for a particular stimulus feature in a particular brain region implies that distinct populations of neurons encode that feature within the same brain region (Naccache & Dehaene, 2001). Consequently, compared to conventional subtraction fMRI designs, RS can reveal the populations of neurons that are coding within a brain area and not just between brain areas.

Fourth, as described above, the RS method has conceptual parallels with the habituation methods commonly used in developmental psychology. Experimental designs and stimuli that have been developed for infant habituation can often be used for adult RS studies and vice versa; this allows for parallel study of both the adult and developing cognitive systems. It is not yet clear how the neural and cognitive mechanisms underlying changes in looking time in infants relate to the brain responses seen in adults. Nevertheless, the possibility of using equivalent paradigms in infants and adults can enhance the links between these two often-unrelated literatures.

In the following section we turn to empirical evidence. Several studies will be outlined that have used RS-fMRI to examine the neural foundations of human action understanding.

3 RS-fMRI Studies of Action Understanding

3.1 Perception of Human Action Kinematics, Object-Goals, and Outcomes

In an initial study using the RS method, Hamilton and Grafton (2006) adapted the work of Woodward (1998) to investigate which regions of the human brain are sensitive to the object-goals of observed hand actions (Hamilton & Grafton, 2006). Participants saw short videos in which two distinct objects are visible on a table. A human hand reaches toward one of the objects, grasps it, and moves it back to the starting location. The objects in the video are matched for size and shape in order to afford a similar grasp configuration, but they are from distinct object categories: tool (e.g., a hammer) and food (e.g., a banana). This means that the two objects have distinct semantic associations and likely future actions. The goal of the action was defined in terms of the target object grasped by the actor (tool vs. food) and could be manipulated independent of the reach trajectory of the action (left vs. right) (figure 10.1).

Brain regions showing RS for the object-goal were those regions that showed suppression when the same object was grasped repeatedly, but release from suppression when a novel object was grasped (novel > repeated object-goal). One brain region, the left aIPS, showed this response pattern at cluster-corrected significance levels (figure 10.2a). By contrast, RS for trajectory was found in the lateral occipital and superior frontal brain areas. This result demonstrates that the left aIPS is sensitive to the object-goal of an observed reaching action irrespective of reach trajectory. It also served as a proof-of-principle by demonstrating that RS could be used with fMRI to separate different features of observed actions independently.

Two follow-up studies applied the same experimental design to separate other features of perceived actions, such as handgrip (fingertip vs. whole hand), action means (push vs. pull), and action outcome (open box vs. closed box). First, Hamilton and Grafton (2007) showed participants movies of an actor grasping a wine bottle or a dumbbell. Each object could be grasped with a whole-hand grip or a fingertip grip. RS for the object-goal of the action was found in the left aIPS, which replicates the previous result (Hamilton & Grafton, 2006). RS was also evaluated for the two different grip configurations. When novel grips were compared to repeated grips (regardless of object-goal), RS for grip was found in the occipital regions, IFG, supplementary motor area, middle frontal gyrus, and middle IPS. There was no response in the aIPS or IPL for the grip contrast. These data are consistent with other RS studies that show sensitivity to both perceived

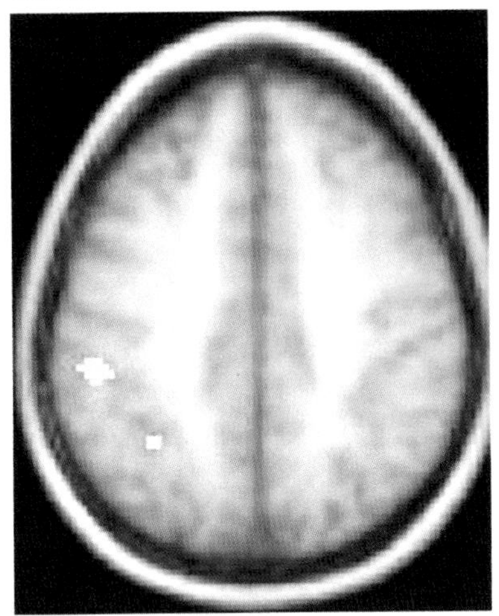

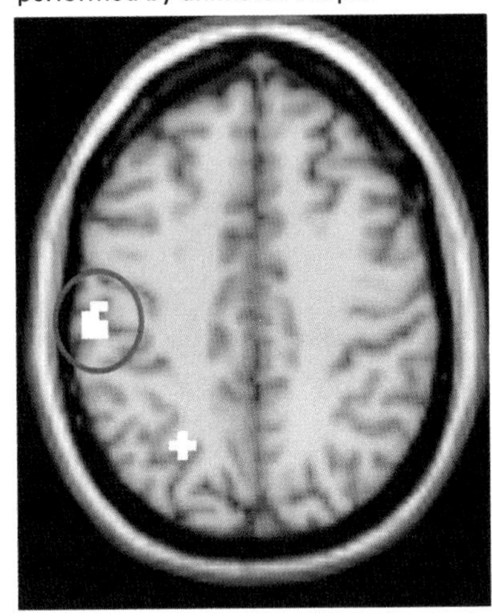

Figure 10.2
Brain regions showing RS for goal-directed hand actions (left) and for goal-directed actions performed by animated shapes (right). Both studies found engagement of left anterior intraparietal sulcus (aIPS) in this contrast.

hand kinematics (Kilner, Neal, Weiskopf, Friston, & Frith, 2009) and the relationship between handgrip and object type (Johnson-Frey et al., 2003) in the IFG adjacent to the PMv.

In a second study, the outcome of observed actions was separated from the means used to achieve the outcome (Hamilton & Grafton, 2008a). For example, opening or closing a box (outcome) could be achieved by pulling the lid with the fingers or pushing with the thumb (means). Sensitivity to action outcome was found in the right IFG, right IPL, and left aIPS. By contrast, RS for means was observed in the lateral occipital cortex, STS, and middle IPS. This result suggests that when actions are more complex than just taking a single object, the right IFG and right IPL are more engaged. Furthermore, across all three experiments, visual areas in occipitotemporal cortex provide an analysis of trajectory, grip, and means, thus supporting a role for visual brain systems in action perception.

The studies reviewed thus far have demonstrated how different regions of the human brain are specifically sensitive to different features of observed actions, including kinematics (IFG), object-goals (left aIPS), and outcomes (left aIPS and right IFG and IPL). The next challenge is to explore the cognitive processes that underlie these responses. A key prediction of the direct-matching hypothesis is that it should only be possible to employ direct matching for stimuli that have human features that relate to one's own motor system, such as human form and motion (Kilner, Paulignan, & Blakemore, 2003; Press, 2011). In the next section, we outline an experiment that tests if human features are necessary for object-goal sensitivity in the aIPS by using simple, geometric shapes as agents instead of human hands (Ramsey & Hamilton, 2010c).

3.2 Triangles Have Goals Too

Developmental studies suggest that in some but not all circumstances, infants can attribute goals to nonhuman agents (Gergely, Nadasdy, Csibra, & Bíró, 1995; Gergely & Csibra, 2003; Csibra, 2008). For example, twelve-month-old infants look longer (i.e., they're surprised) when a circle violates the most efficient means of achieving a desired goal (Gergely et al., 1995). The authors interpret these findings as evidence that infants at around one year of age are able to treat circles as intentional agents that have goals (but see Paulus, Hunnius, van Wijngaarden et al., 2011). In contrast, Woodward (1998) showed that five-, six- and nine-month-old infants treat a human hand as goal directed, but they did not respond in the same way if an inanimate rod or claw performed the same actions. This suggests that not all movements toward an object are equivalent for infants, but that contextual features or the form of the actor impact the interpretation of the action as goal directed. However, it is not clear from these infant studies whether the same cognitive and brain mechanisms are used to interpret the goal-directed actions of human and nonhuman agents. Neuroimaging studies allow us to address this question with adult participants.

Studies of human brain responses to observation of animated shapes have often shown overlap between the processing of shapes and the processing of human stimuli. The superior temporal sulcus (STS) is known to respond to biological motion (Blake & Shiffrar, 2007)—walking human figures and eye/head movements in particular (Allison, Puce, & McCarthy, 2000). This region is also engaged when participants observe interacting spheres with increasing animacy (Schultz, Friston, O'Doherty, Wolpert, & Frith, 2005) and when they observe shapes moving in a context that makes them seem human (Wheatley, Milleville, & Martin, 2007). When the behavior of animated shapes is more complex, such that typical observers attribute mental states to the shapes (Heider & Simmel, 1944), activation is seen in the temporoparietal junction (TPJ) and the medial prefrontal cortex (mPFC) (Castelli, Happé, Frith, & Frith, 2000). This same brain network is engaged when participants consider the mental states of other people (Frith & Frith, 2003).

In all these studies, the same brain systems respond to humans and shapes if participants interpret the action they see as animate or if participants engage in mentalizing. Thus, brain responses seem to be determined by the participants' interpretation of the stimuli rather than the form of the agent. In contrast, it is widely believed that responses of the MNS are specifically tuned to human actions (Press, 2011). These human actions have particular low-level features—namely the shape of the hand and the characteristic biological motion trajectory—that seem to engage the MNS. For example, stronger MNS engagement has been reported for actions with biological (rather than linear) movement trajectories (Shimada, 2010), for actions that obey the two-thirds power law (Casile et al., 2010), and for observation of a human hand rather than a robot hand (Perani et al., 2001; Tai, Scherfler, Brooks, Sawamoto, & Castiello, 2004). These and other findings have been used to argue that (1) MNS regions are specifically tuned to human actions and (2) this tuning reflects a direct-matching process.

Our study aimed to determine which brain systems respond to the perception of simple shapes performing object-directed actions (Ramsey & Hamilton, 2010c). The design of the study was the same as used previously to study the perception of human goal-directed actions (Hamilton & Grafton, 2006). Short videos were sequenced to systematically manipulate goal (target object: tool vs. food) and trajectory (left vs. right). Importantly, instead of using a human hand as the agent, geometric shapes (a triangle, star, and diamond) were animated to act as agents (figure 10.1, bottom row). The shapes did not possess two key features of human stimuli: biological form and motion. Specifically, the shapes *did not* look like human hands, nor did they move with the minimum-jerk trajectory that is typical of human hand movements (Hogan, 1984); instead, they moved according to a linear velocity profile.

Despite this lack of low-level human features, the shapes did behave in a manner that induces the percept of animacy. First, their motion was object directed, which acts a potent cue to animacy (Opfer, 2002). Second, the shapes appeared self-propelled and would grow or shrink as they made contact with the target objects, as well as when they returned to the start location (Tremoulet & Feldman, 2000). Third, a barrier (four red circles) was placed between the animated shape and the two target objects (cookie and keys) because moving shapes appear more goal-directed when they negotiate barriers (Csibra, 2008). Overall, the shape stimuli behaved as if "alive" but did not display any low-level perceptual cues of human action (e.g., a handlike form or handlike biological motion). This means that kinematic features of the stimuli, such as form and motion, could not be directly matched onto the observer's own motor system.

Twenty-eight participants observed the movies depicting animated shapes during fMRI. As before, we searched for brain regions showing a stronger response to novel object-goals compared to repeated object-goals. This pattern of response was found in the left aIPS (figure 10.2b). There was no evidence for object-goal sensitivity in other brain regions of specific interest, such as the IFG adjacent to the PMv. These findings suggest that the left aIPS distinguishes the object-goals of actions performed by simple geometric shapes. The pattern and location of this response closely matches that observed previously with human hand actions (Hamilton & Grafton, 2006, 2007) and suggests that the aIPS shows similar sensitivity to object-goals independent of the agent's form.

One possible limitation of our study is that the participants did not observe any videos of human hand actions; thus, our conclusion that the same brain region processes both human and shape actions is based on a comparison across studies rather than a comparison within participants. However, it is also an advantage that participants did not see human-action videos, because it means that these participants were not primed within the experiment to make analogies between the movement of the shapes and the behavior of a human hand.

The results of this study raise two critical questions. First, how can we interpret these data in relation to the large number of previous studies that claim the MNS is specifically tuned to human biological motion? Second, if the response of the aIPS is the same for people and triangles, what brain systems distinguish who is acting? Recent data allow us to consider each of these questions in turn.

3.3 A Human-Specific MNS?

Numerous studies suggest that the response of the human MNS is stronger when observing human actions (Press, 2011)—that is, actions performed by a human body and using typical human movement profiles, such as minimum-jerk trajectories for reaching actions and the two-thirds power law for curved movements. The results of the triangles study suggests that brain regions that encode object-goals of human actions can also encode object-goals when the actor is a triangle and lacks low-level human features. This demonstrates that goal encoding is not specific to human actors. However, our results do not allow us to test if the response to human goal-directed actions is greater than the response to goal-directed actions performed by nonhuman actors, because participants did not view both types of action within a single experiment.

A recent study by Cross and colleagues did directly address this question (Cross et al., forthcoming). Using a conventional factorial design, Cross and colleagues tested how the human brain responds to seeing a person or a robotic figure dance in a smooth human style or a jerky robotic style. Surprisingly, responses in both the parietal and premotor MNS brain regions were stronger when participants observed the robotic dance style. There were no differences in these regions when viewing a real human form compared to a humanoid robot. These results run counter to the dominant claim that the MNS is tuned only to natural human motion (Press, 2011) and suggest that responses within these brain systems are more flexible than previously considered.

One possible account of these results focuses on how participants interpret or categorize a stimulus rather than its low-level features. As hinted above, in other regions of the social brain, activation seems to be determined by how a stimulus is interpreted. Thus, seeing a variety of animate, moving agents engaged the MTG and STS regardless of the specific form of the agent (Schultz et al., 2005; Wheatley et al., 2007). Similarly, when the actions of an agent can be interpreted in terms of mental states, the TPJ and mPFC are engaged (Castelli et al., 2000). This response can even be seen when the agent's behavior does not change at all; for example, the TPJ and mPFC are active when participants believe they are interacting with a person compared to an identical condition where participants believe they are interacting with a computer (Gallagher, Jack, Roepstorff, & Frith, 2002).

The same principle might apply across the MNS. That is, if an action is perceived as directed toward an object-goal, the aIPS is engaged regardless of whether the actor is a human or a triangle. Similarly, if a figure is perceived as dancing, the MNS is engaged regardless of whether the figure has human or robotic form or motion. Under this model, responses of the MNS are not tuned specifically to human or biological features, but are driven more by the top-down interpretation of the stimulus. Further experiments will be required to test if this idea is valid.

3.4 Who Is Acting?

If responses of the MNS to observed actions do not distinguish whether the actor is a human, an animated triangle (Ramsey & Hamilton, 2010c), or a robot (Cross et al., forthcoming), which cognitive and brain systems distinguish *who* is acting? Addressing this question is critical for many social interactions. For example, when another person takes ten dollars from your hand, it matters if they are a shopkeeper or a robber. That is, the meaning of a simple, goal-directed action can vary depending on the identity of the actors involved. We recently used an RS paradigm to separate the brain systems that code actor identity from the action goal performed (Ramsey & Hamilton, 2010b). Participants watched video clips of two different actors with two different object-goals, arranged in an RS design during fMRI. We calculated RS for repeated, compared to novel, actor identity as well as object-goal.

Our results demonstrated that the observation of the same actor repeatedly performing an object-directed action suppresses the BOLD response in the fusiform gyrus and occipitotemporal cortex, but observation of a novel actor performing the action results in a release from suppression in these regions. In contrast, brain regions within the IFG, IPL, and MTG showed RS for the object-goal of the performed action. Previous work on person identity most commonly examined the BOLD response in the fusiform gyrus and occipitotemporal cortex using static images of motionless faces or body parts (Kanwisher, McDermott, & Chun, 1997; Downing, Jiang, Shuman, & Kanwisher, 2001). Our data suggest that similar cortical regions that have previously been associated with person identity are also recruited in more social and dynamic contexts; they distinguish between two intentional agents who are acting in a goal-directed fashion. These results demonstrate that regions beyond the MNS are critical for distinguishing between different actors in a social scene and thus understanding actions that occur in everyday social situations.

These data have relevance for the problem of understanding the *who* of a social situation. Previously, it has been suggested that if performed and observed actions are represented in the same brain systems, then an additional "who" system is needed to distinguish between the self and others (Georgieff & Jeannerod, 1998). Discussion of this "who" system has been limited to the problem of deciding if I am acting or another person is acting (Georgieff & Jeannerod, 1998; de Vignemont & Fourneret, 2004). The current experiment considers the problem of distinguishing between two distinct other people; it shows that the MNS does not discriminate between two other agents. Rather, a "who" system encoding the identity of different actors might be needed; our results implicate the fusiform and occipitotemporal brain regions in this process. These results mean that the MNS is not the sole brain system responsible for comprehending actions, but that other brain networks respond to the broader social context of action.

3.5 Summary of Empirical Evidence

In sum, we have presented a series of fMRI experiments using RS that have examined how different brain systems process others' actions in a social context. We have shown that the left aIPS is sensitive to the object-goal of an action regardless of whether the actor is a human hand or a simple geometric shape that is devoid of human form and motion. Further studies have shown that the MNS responds to the actions of both human and robotic figures, and that it does not distinguish between two different human actors. These results are compatible with the claim that the MNS encodes actions but is not tuned to particular actors or agents. We suggest that other brain systems are required to encode actor identity. In the next section we outline and evaluate several interpretations of this finding in the aIPS and discuss the resulting implications for cognitive theories of goal understanding.

4 Theoretical Implications

The new findings outlined in the previous section lead us to consider two questions. First, what aspect of an action or observed action is represented in the aIPS? And second, how is the visual image on the retina transformed into this representation?

4.1 What Is Represented in the aIPS?

Traditional accounts of aIPS focused on the role of this region in encoding performed hand grasps. Single neurons in the AIP region of the macaque brain encode different hand shapes (Sakata, Taira, Murata, & Mine, 1995; Murata, Gallese, Luppino, Kaseda, & Sakata, 2000), and sensitivity to hand shape has also been detected in the human aIPS (Króliczak, Quinlan, McAdam, & Culham, 2006). However, these studies tend to use meaningless objects as stimuli, which are differentiated only by their shape. The data from our series of studies show that when object shape is matched but object identity changes, aIPS is sensitive to the identity of the object-goal (Hamilton & Grafton, 2006; 2007). This encoding is independent of hand shape information, because the same pattern of response was seen when the action was performed by an animated triangle with no human grasp or biological motion trajectory.

Other studies also suggest that the aIPS encodes a more abstract representation of action than just hand grasp. Jastorff and colleagues showed that the aIPS is sensitive to the direction that other people move objects (i.e., toward vs. away from the body) irrespective of whether the person performed the action with their hand, foot, or mouth (Jastorff, Clavagnier, Gergely, & Orban, 2010). They suggest that for actions typically performed by the hand, such as moving small objects, we may process these actions in hand-centered space, even when a different effector, such as a foot or mouth, is being used. This result is consistent with our data showing actor independence in the aIPS (see also Sommerville & Loucks, this volume).

The parietal cortex is also sensitive to the difference between typical and unusual person-object relationships (Newman-Norlund, van Schie, van Hoek, Cuijpers, & Bekkering, 2010). Specifically, bilateral parts of the IPL distinguished between typical effector-object relations (a hand touching a phone) compared to atypical effector-object relations (a foot touching a phone). Finally, a study using multivoxel pattern analysis to search for overlapping neuronal representations of goal-directed actions performed by the self and another person found that the left aIPS was one of the few regions encoding the goal of actions for both self and the other person (Oosterhof, Wiggett, Diedrichsen, Tipper, & Downing, 2010).

These results could be considered within a framework of intentional relations (Barresi & Moore, 1996). This framework defines an intentional relation as a three-way link between an agent, a directed activity, and an object. Such relationships can involve real objects, such as observing somebody grasping an apple, and more complex mental relations, which involve beliefs and desires directed toward imaginary objects or world states. Barresi and Moore (1996) proposed that these intentional relationships form the basis of how social interactions are understood and processed.

The response profile we have demonstrated in the aIPS could be interpreted as a simple type of intentional relationship. The aIPS is sensitive to the object (Hamilton & Grafton, 2006; Ramsey & Hamilton, 2010c) and also to the type of action (Jastorff et al., 2010; Newman-Norlund et al., 2010; Oosterhof et al. 2010). While this region does not discriminate between different actors (Ramsey & Hamilton, 2010b; Ramsey & Hamilton, 2010c), it is likely that the presence of an animate agent is required to engage the aIPS. Further study will be needed to determine how the representation of actions and object-goals within the aIPS fits into broader frameworks for social cognition.

4.2 How Are Goal Representations Calculated?

If the aIPS represents the object-goal of an action, possibly in the form of a simple intentional relationship, it is then useful to know how this representation is achieved. How can the moving visual image on the retina be transformed into a representation of an action on an object? The dominant model of action understanding within the human brain is based on the idea of direct matching. This is the claim that an observed action can be "directly matched" onto a motor representation in the observer's own motor system. A strong version of this direct-matching hypothesis might require matching at the level of kinematics (Rizzolatti, Fadiga, Fogassi, & Gallese, 1999). However, more recent variants suggest direct matching could occur primarily at the level of goals (Gazzola, Rizzolatti, Wicker, & Keysers, 2007), or there could be two separate routes for direct matching and goal processing within the MNS (Rizzolatti & Sinigaglia, 2010).

An alternative to the direct-matching account is a teleological reasoning theory, which proposes that actions can be understood using visual inference alone, without the involvement of the motor system (Gergely & Csibra, 2003; Csibra, 2007). This mechanism considers the rationality of the action in relation to a desired goal-state and environmental constraints. The data described in section 3 above provide us with new insights into which of these mechanisms might lead to a goal representation.

First, our data allow us to rule out a strong version of the direct matching hypothesis. This is the idea that an observed action must be first matched to a kinematic motor representation of hand shape and movement in the MNS in order for a goal to be calculated (Rizzolatti et al., 2001). Such a direct-matching mechanism could contribute to the perception of goal-directed human hand actions (Hamilton & Grafton, 2006) and even of humanoid robots (Gazzola et al., 2007; Cross et al. forthcoming). However, a mechanism that matches biological form and motion cannot apply to the current findings because the shapes that served as actors had neither handlike body parts nor biological motion trajectories. Therefore, the present result demonstrates that goal representations in the aIPS can be achieved without a strong form of direct matching.

This result is also consistent with the idea that action comprehension can occur without access to biological form or motion (Csibra, 2007). In doing so, this empirical evidence raises a challenge to the MNS theories of action understanding, which has previously been raised on theoretical grounds (Uithol, van Rooij, Bekkering, & Haselager, 2011b)—namely, what does the mirror neuron system "mirror"?

More recent theories of direct matching do not make such strong claims; these theories suggest that the primary role of the MNS is to encode goals rather than kinematics (Gazzola et al., 2007). Under this account, any intentional agent, independent of form and motion, can be incorporated into the MNS. Once an observed agent is deemed intentional and object-directed, the aIPS could subsequently treat it *as if* it were a human hand

and process object-goals in a similar manner. In this sense the shapes in our study may have been treated as hands in a functional sense, in that they can manipulate objects. It is important to note that we did not present hands and shapes in the same study in order to reduce the likelihood that participants would interpret the shapes as hands, but we cannot rule out that participants did this spontaneously.

However, these direct-matching accounts, by abstracting away from the motor-kinematic features of an action and suggesting matching of goals, lose a lot of the power of the original mirror hypothesis. A goal is not necessarily a motor representation, so a visual input could be matched to a goal by means of visual processing, without any contribution from the motor system. Thus, these variants on the direct-matching hypothesis do not have much unique predictive power and can be hard to distinguish from visual or teleological hypotheses.

The core claim of the teleological reasoning hypothesis is that actions are interpreted with reference to whether an action is rational (Gergely & Csibra, 2003). Under a teleological approach, a goal is assigned to an action by evaluating the efficiency of the agent's action with respect to environmental constraints on goal attainment (Gergely & Csibra, 2003). Evidence for teleological processing of actions can be found in both infant and adult research. Typical infants look longer at irrational actions (Gergely et al., 1995; Csibra, Gergely, Bíró, Koós, & Brockbank, 1999) and imitate actions rationally (Gergely, Bekkering, & Kiraly, 2002; but see Paulus, Hunnius, Vissers, & Bekkering, 2011). In the adult brain, the TPJ and mPFC distinguish rational from irrational actions (Brass, Schmitt, Spengler, & Gergely, 2007; Marsh & Hamilton, 2011). Because motor or kinematic representations are not required for teleological processing, the teleological account can explain the response of the aIPS to actions performed by triangles.

Distinguishing between the teleological and direct-matching accounts of goal understanding will be an important focus in future research. Critical differences between the models emerge when we consider the impact of both experience and rationality on each. The direct-matching model predicts that the ability to understand a goal should be closely linked to one's experience of performing that goal-directed action (Kanakogi & Itakura, 2010; Paulus, Hunnius, Vissers, & Bekkering, 2011). In contrast, the teleological model predicts that understanding of goals should be related to judgments of whether a goal is rational given current environmental constraints (Csibra, 2003). Current data do not conclusively distinguish between these models yet.

5 Broader Implications and Future Directions

5.1 The Variety of Goal

The majority of this chapter has discussed evidence for one simple type of action goal, an object-goal, which is defined by the identity of the object manipulated by an agent. But there are many other types of goals that can be served by action (Jacob & Jeannerod, 2005; Uithol et al., 2011a). These range along a continuum from concrete, tangible actions, which can be performed over a relatively short period of time (seconds and minutes), such as grasping a teapot and pouring tea, to longer-term, intangible goals, such as planning a holiday for next summer or striving for a promotion. However, the relationship between goal inference from specific concrete actions and goal inference from the processing of longer-term thoughts, desires, and beliefs is not yet clearly specified.

Some studies claim that the motor system, specifically the MNS, plays a more sophisticated role in the perception of action than merely processing kinematic features and simple goals; it does so by showing sensitivity to the context surrounding an action (Iacoboni et al. 2005; Liepelt & Brass, 2009). For example, Liepelt and Brass (2009) showed that motor-related readiness potentials were modulated if participants observed a finger action imposed by a mechanical clamp instead of under free control. Similarly, Iacoboni et al. (2005) suggest that responses in the IFG to observed action are modulated by the context of the action.

Other work has suggested that the social competence of the MNS is limited and that, instead, a growing body of fMRI studies implicate brain regions associated with mentalizing/Theory of Mind in the perception of action (Grèzes, Frith, & Passingham, 2004a, 2004b; Brass et al., 2007; Ramsey & Hamilton, 2010a; Spunt, Satpute, & Lieberman, 2010; Marsh & Hamilton, 2011). Recent fMRI studies have shown that the mentalizing network is sensitive to the perception of irrational actions (Brass et al., 2007; Marsh & Hamilton, 2011) as well as to instances where the observed actor's knowledge state is different from the perceiver's knowledge state (Ramsey & Hamilton, 2010a). In these studies, no instructions were given to consider the mental states of the observed actors, but the mentalizing network still showed sensitivity to aspects of observed actions.

Taken together, the work described in this section suggests that the diverse range of goals that action can serve is processed by a distributed brain network comprising, but not restricted to, brain areas associated with the MNS and mentalizing network (Keysers & Gazzola, 2007; Uddin,

Iacoboni, Lange, & Keenan, 2007). We suggest that future work should aim to further delineate the contributions to goal understanding made by the MNS, the mentalizing network, and other social brain systems. Specifically, it will be pertinent to examine how these systems work together during social interactions. Approaches from motor control, developmental, cognitive, and social psychology, as well as neuroscience methods, will be needed to fully examine human goal understanding and its neural substrates.

5.2 Linking Actions and Actors

To make sense of other people's actions, it is not sufficient to process only action features such as kinematics, objects-goals, and broader motivations. It is also important to link the identity of the agent to these action features. Our data suggest that the MNS regions, specifically the aIPS, are agent neutral. These regions do not discriminate between different human actors (Ramsey & Hamilton, 2010b) and show similar responses to human and nonhuman actors (Ramsey & Hamilton, 2010c). This insensitivity to actor form is consistent with brain-imaging work using animated shapes during biological motion and mentalizing tasks (Castelli et al., 2000; Schultz et al., 2005). In all these studies, the BOLD response is determined by the type of social processing engaged, such as mental state reasoning, animacy perception, or object-goal perception, rather than the form of the actor.

However, in real social interactions, it matters immensely who you are interacting with, because each different actor is likely to have his or her own beliefs, desires, and motivations. Initial work has shown differing brain responses to individuals associated with positive and negative behavior (Singer, Kiebel, Winston, Dolan, & Frith, 2004) and various social stereotypes (Krendl, Macrae, Kelley, Fugelsang, & Heatherton, 2006). The question of how these types of actor information are linked to, and integrated with, information about an individual's current goal-directed behavior will be an important area for future research.

Conclusion

This chapter reviewed evidence that the human MNS (1) is finely tuned to goal-directed hand actions and (2) encodes the identity of the object that an actor reaches for. In contrast, these brain systems are insensitive to actor identity and show equivalent responses to different people and animated shapes. These results imply that a strong direct-matching account of action comprehension is implausible, and they highlight the importance of future research into comprehension of different types of goals and actors as well as the links between them.

References

Allison, T., Puce, A., & McCarthy, G. (2000). Social perception from visual cues: Role of the STS region. *Trends in Cognitive Sciences, 4*(7), 267–278.

Barresi, J., & Moore, C. (1996). Intentional relations and social understanding. *Behavioral and Brain Sciences, 19,* 107–122.

Blake, R., & Shiffrar, M. (2007). Perception of human motion. *Annual Review of Psychology, 58,* 47–73.

Brass, M., Schmitt, R. M., Spengler, S., & Gergely, G. (2007). Investigating action understanding: Inferential processes versus action simulation. *Current Biology, 17*(24), 2117–2121.

Buckner, R. L., Goodman, J., Burock, M., Rotte, M., Koutstaal, W., Schacter, D., et al. (1998). Functional-anatomic correlates of object priming in humans revealed by rapid presentation event-related fMRI. *Neuron, 20*(2), 285–296.

Calder, A. J., Beaver, J. D., Winston, J. S., Dolan, R. J., Jenkins, R., Eger, E., et al. (2007). Separate coding of different gaze directions in the superior temporal sulcus and inferior parietal lobule. *Current Biology, 17*(1), 20–25.

Casile, A., Dayan, E., Caggiano, V., Hendler, T., Flash, T., & Giese, M. A. (2010). Neuronal encoding of human kinematic invariants during action observation. *Cerebral Cortex, 20*(7), 1647–1655.

Caspers, S., Zilles, K., Laird, A. R., & Eickhoff, S. B. (2010). ALE meta-analysis of action observation and imitation in the human brain. *NeuroImage, 50,* 1148–1167.

Castelli, F., Happé, F., Frith, U., & Frith, C. (2000). Movement and mind: A functional imaging study of perception and interpretation of complex intentional movement patterns. *NeuroImage, 12*(3), 314–325.

Cross, E. S., Liepelt, R., Hamilton, A. F., Parkinson, J., Ramsey, R., Stadler, W., et al. (2012). Robotic actions preferentially engage the human mirror system. *Human Brain Mapping, 33*(9), 2238–2254.

Csibra, G. (2003). Teleological and referential understanding of action in infancy. *Philosophical Transactions of the Royal Society of London, Series B: Biological Sciences, 358*(1431), 447–458.

Csibra, G. (2007). Action mirroring and action understanding: An alternative account. In P. Haggard, Y. Rosetti, & M. Kawato (Eds.), *Sensorimotor foundations of higher cognition: Attention and performance XXII.* Oxford: Oxford University Press.

Csibra, G. (2008). Goal attribution to inanimate agents by 6.5-month-old infants. *Cognition, 107*(2), 705–717.

Csibra, G., Gergely, G., Bíró, S., Koós, O., & Brockbank, M. (1999). Goal attribution without agency cues: The perception of "pure reason" in infancy. *Cognition, 72*(3), 237–267.

de Vignemont, F., & Fourneret, P. (2004). The sense of agency: A philosophical and empirical review of the "Who" system. *Consciousness and Cognition, 13*(1), 1–19.

di Pellegrino, G., Fadiga, L., Fogassi, L., Gallese, V., & Rizzolatti, G. (1992). Understanding motor events: A neurophysiological study. *Experimental Brain Research, 91*(1), 176–180.

Downing, P. E., Jiang, Y., Shuman, M., & Kanwisher, N. (2001). A cortical area selective for visual processing of the human body. *Science, 293*(5539), 2470–2473.

Fogassi, L., Ferrari, P. F., Gesierich, B., Rozzi, S., Chersi, F., & Rizzolatti, G. (2005). Parietal lobe: From action organization to intention understanding. *Science, 308*(5722), 662–667.

Friston, K. J., Price, C. J., Fletcher, P., Moore, C., Frackowiak, R. S., & Dolan, R. J. (1996). The trouble with cognitive subtraction. *NeuroImage, 4*(2), 97–104.

Frith, U., & Frith, C. D. (2003). Development and neurophysiology of mentalizing. *Philosophical Transactions of the Royal Society of London, Series B: Biological Sciences, 358*(1431), 459–473.

Gallagher, H. L., Jack, A. I., Roepstorff, A., & Frith, C. D. (2002). Imaging the intentional stance in a competitive game. *NeuroImage, 16*(3 Pt 1), 814–821.

Gallese, V., Fadiga, L., Fogassi, L., & Rizzolatti, G. (1996). Action recognition in the premotor cortex. *Brain, 119*(Pt 2), 593–609.

Gallese, V., Keysers, C., & Rizzolatti, G. (2004). A unifying view of the basis of social cognition. *Trends in Cognitive Sciences, 8*(9), 396–403.

Gazzola, V., Rizzolatti, G., Wicker, B., & Keysers, C. (2007). The anthropomorphic brain: The mirror neuron system responds to human and robotic actions. *NeuroImage, 35*(4), 1674–1684.

Georgieff, N., & Jeannerod, M. (1998). Beyond consciousness of external reality: A "who" system for consciousness of action and self-consciousness. *Consciousness and Cognition, 7*(3), 465–477.

Gergely, G., Bekkering, H., & Kiraly, I. (2002). Rational imitation in preverbal infants. *Nature, 415*(6873), 755.

Gergely, G., & Csibra, G. (2003). Teleological reasoning in infancy: The naive theory of rational action. *Trends in Cognitive Sciences, 7*(7), 287–292.

Gergely, G., Nadasdy, Z., Csibra, G., & Bíró, S. (1995). Taking the intentional stance at 12 months of age. *Cognition, 56*(2), 165–193.

Grafton, S. T., & Hamilton, A. F. (2007). Evidence for a distributed hierarchy of action representation in the brain. *Human Movement Science, 26*(4), 590–616.

Grèzes, J., & Decety, J. (2001). Functional anatomy of execution, mental simulation, observation, and verb generation of actions: A meta-analysis. *Human Brain Mapping, 12*(1), 1–19.

Grèzes, J., Frith, C., & Passingham, R. E. (2004a). Brain mechanisms for inferring deceit in the actions of others. *Journal of Neuroscience, 24*(24), 5500–5505.

Grèzes, J., Frith, C. D., & Passingham, R. E. (2004b). Inferring false beliefs from the actions of oneself and others: An fMRI study. *NeuroImage, 21*(2), 744–750.

Grill-Spector, K., Henson, R., & Martin, A. (2006). Repetition and the brain: Neural models of stimulus-specific effects. *Trends in Cognitive Sciences, 10*(1), 14–23.

Grill-Spector, K., Kushnir, T., Edelman, S., Avidan, G., Itzchak, Y., & Malach, R. (1999). Differential processing of objects under various viewing conditions in the human lateral occipital complex. *Neuron, 24*(1), 187–203.

Grill-Spector, K., & Malach, R. (2001). fMR-adaptation: A tool for studying the functional properties of human cortical neurons. *Acta Psychologica, 107*(1–3), 293–321.

Hamilton, A. F., & Grafton, S. T. (2006). Goal representation in human anterior intraparietal sulcus. *Journal of Neuroscience, 26*(4), 1133–1137.

Hamilton, A. F., & Grafton, S. T. (2007). The motor hierarchy: from kinematics to goals and intentions. In P. Haggard, Y. Rosetti, & M. Kawato (Eds.), *Sensorimotor foundations of higher cognition: Attention and performance XXII*. Oxford: Oxford University Press.

Hamilton, A. F., & Grafton, S. T. (2008a). Action outcomes are represented in human inferior frontoparietal cortex. *Cerebral Cortex, 18*(5), 1160–1168.

Hamilton, A. F., & Grafton, S. T. (2008b). Repetition suppression for performed hand gestures revealed by fMRI. *Human Brain Mapping, 30*(9), 2898–2906.

Heider, F., & Simmel, M. (1944). An experimental study of apparent behavior. *American Journal of Psychology, 57*(2), 243–259.

Hickok, G. (2009). Eight problems for the mirror neuron theory of action understanding in monkeys and humans. *Journal of Cognitive Neuroscience, 21*(7), 1229–1243.

Hogan, N. (1984). An organizing principle for a class of voluntary movements. *Journal of Neuroscience, 4*(11), 2745–2754.

Iacoboni, M., Molnar-Szakacs, I., Gallese, V., Buccino, G., Mazziotta, J. C., & Rizzolatti, G. (2005). Grasping the intentions of others with one's own mirror neuron system. *PLoS Biology, 3*(3), e79.

Jacob, P., & Jeannerod, M. (2005). The motor theory of social cognition: A critique. *Trends in Cognitive Sciences, 9*(1), 21–25.

Jastorff, J., Clavagnier, S., Gergely, G., & Orban, G. A. (2010). Neural mechanisms of understanding rational actions: Middle temporal gyrus activation by contextual violation. *Cerebral Cortex, 21*(2), 318–329.

Johnson-Frey, S. H., Maloof, F. R., Newman-Norlund, R., Farrer, C., Inati, S., & Grafton, S. T. (2003). Actions or hand-object interactions? Human inferior frontal cortex and action observation. *Neuron, 39*(6), 1053–1058.

Kanakogi, Y., & Itakura, S. (2010). Developmental correspondence between action prediction and motor ability in early infancy. *Nature Communications, 2*, 341.

Kanwisher, N., McDermott, J., & Chun, M. M. (1997). The fusiform face area: A module in human extrastriate cortex specialized for face perception. *Journal of Neuroscience, 17*(11), 4302–4311.

Keysers, C., & Gazzola, V. (2007). Integrating simulation and theory of mind: From self to social cognition. *Trends in Cognitive Sciences, 11*(5), 194–196.

Kilner, J. M., Neal, A., Weiskopf, N., Friston, K. J., & Frith, C. D. (2009). Evidence of mirror neurons in human inferior frontal gyrus. *Journal of Neuroscience, 29*(32), 10153–10159.

Kilner, J. M., Paulignan, Y., & Blakemore, S. J. (2003). An interference effect of observed biological movement on action. *Current Biology, 13*(6), 522–525.

Krendl, A. C., Macrae, C. N., Kelley, W. M., Fugelsang, J. A., & Heatherton, T. F. (2006). The good, the bad, and the ugly: An fMRI investigation of the functional anatomic correlates of stigma. *Social Neuroscience, 1*(1), 5–15.

Króliczak, G., Quinlan, D. J., McAdam, T. D., & Culham, J. C. (2006). AIP shows grasp-specific fMRI adaptation for real actions. Talk presented at the Society for Neuroscience, Atlanta, Georgia.

Liepelt, R., & Brass, M. (2009). Top-down modulation of motor priming by belief about animacy. *Experimental Psychology, 57*(3), 221–227.

Marsh, L. E., & Hamilton, A. F. C. (2011). Dissociation of mirroring and mentalising systems in autism. *NeuroImage, 56*(3), 1511–1519.

Murata, A., Gallese, V., Luppino, G., Kaseda, M., & Sakata, H. (2000). Selectivity for the shape, size, and orientation of objects for grasping in neurons of monkey parietal area AIP. *Journal of Neurophysiology, 83*(5), 2580–2601.

Naccache, L., & Dehaene, S. (2001). The priming method: Imaging unconscious repetition priming reveals an abstract representation of number in the parietal lobes. *Cerebral Cortex, 11*(10), 966–974.

Newman-Norlund, R., van Schie, H. T., van Hoek, M. E., Cuijpers, R. H., & Bekkering, H. (2010). The role of inferior frontal and parietal areas in differentiating meaningful and meaningless object-directed actions. *Brain Research, 1315*, 63–74.

Noppeney, U., & Price, C. J. (2004). An FMRI study of syntactic adaptation. *Journal of Cognitive Neuroscience, 16*(4), 702–713.

Oosterhof, N. N., Wiggett, A. J., Diedrichsen, J., Tipper, S. P., & Downing, P. E. (2010). Surface-based information mapping reveals crossmodal vision-action representations in human parietal and occipitotemporal cortex. *Journal of Neurophysiology, 104*, 1077–1089.

Opfer, J. E. (2002). Identifying living and sentient kinds from dynamic information: The case of goal-directed versus aimless autonomous movement in conceptual change. *Cognition, 86*(2), 97–122.

Paulus, M., Hunnius, S., van Wijngaarden, C., Vrins, S., van Rooij, I., & Bekkering, H. (2011). The role of frequency information and teleological reasoning in infants' and adults' action prediction. *Developmental Psychology, 47*(4), 976–983.

Paulus, M., Hunnius, S., Vissers, M., & Bekkering, H. (2011). Imitation in infancy: Rational or motor resonance? *Child Development, 82*(4), 1047–1057.

Pelphrey, K. A., Morris, J. P., & McCarthy, G. (2004). Grasping the intentions of others: The perceived intentionality of an action influences activity in the superior temporal sulcus during social perception. *Journal of Cognitive Neuroscience, 16*(10), 1706–1716.

Perani, D., Fazio, F., Borghese, N. A., Tettamanti, M., Ferrari, S., Decety, J., et al. (2001). Different brain correlates for watching real and virtual hand actions. *NeuroImage, 14*(3), 749–758.

Press, C. (2011). Action observation and robotic agents: Learning and anthropomorphism. *Neuroscience and Biobehavioral Reviews, 35*(6), 1410–1418.

Ramsey, R., & Hamilton, A. F. (2010a). How does your own knowledge influence the perception of another person's action in the human brain? *Social Cognitive and Affective Neuroscience*. doi:10.1093/scan/nsq102.

Ramsey, R., & Hamilton, A. F. (2010b). Understanding actors and object-goals in the human brain. *NeuroImage, 50*(3), 1142–1147.

Ramsey, R., & Hamilton, F. C. (2010c). Triangles have goals too: understanding action representation in left aIPS. *Neuropsychologia, 48*(9), 2773–2776.

Rizzolatti, G., Fadiga, L., Fogassi, L., & Gallese, V. (1999). Resonance behaviors and mirror neurons. *Archives Italiennes de Biologie, 137*(2–3), 85–100.

Rizzolatti, G., Fogassi, L., & Gallese, V. (2001). Neurophysiological mechanisms underlying the understanding and imitation of action. *Nature Reviews: Neuroscience, 2*(9), 661–670.

Rizzolatti, G., & Sinigaglia, C. (2010). The functional role of the parieto-frontal mirror circuit: Interpretations and misinterpretations. *Nature Reviews: Neuroscience, 11*(4), 264–274.

Sakata, H., Taira, M., Murata, A., & Mine, S. (1995). Neural mechanisms of visual guidance of hand action in the parietal cortex of the monkey. *Cerebral Cortex, 5*(5), 429–438.

Sawamura, H., Orban, G. A., & Vogels, R. (2006). Selectivity of neuronal adaptation does not match response selectivity: A single-cell study of the FMRI adaptation paradigm. *Neuron, 49*(2), 307–318.

Schultz, J., Friston, K. J., O'Doherty, J., Wolpert, D. M., & Frith, C. D. (2005). Activation in posterior superior temporal sulcus parallels parameter inducing the percept of animacy. *Neuron, 45*(4), 625–635.

Shimada, S. (2010). Deactivation in the sensorimotor area during observation of a human agent performing robotic actions. *Brain and Cognition, 72*(3), 394–399.

Singer, T., Kiebel, S. J., Winston, J. S., Dolan, R. J., & Frith, C. D. (2004). Brain responses to the acquired moral status of faces. *Neuron, 41*(4), 653–662.

Spunt, R. P., Satpute, A. B., & Lieberman, M. D. (2010). Identifying the what, why, and how of an observed action: An fMRI study of mentalizing and mechanizing during action observation. *Journal of Cognitive Neuroscience, 23*(1), 63–74.

Tai, Y. F., Scherfler, C., Brooks, D. J., Sawamoto, N., & Castiello, U. (2004). The human premotor cortex is "mirror" only for biological actions. *Current Biology, 14*(2), 117–120.

Tremoulet, P. D., & Feldman, J. (2000). Perception of animacy from the motion of a single object. *Perception, 29*(8), 943–951.

Uddin, L. Q., Iacoboni, M., Lange, C., & Keenan, J. P. (2007). The self and social cognition: The role of cortical midline structures and mirror neurons. *Trends in Cognitive Sciences, 11*(4), 153–157.

Uithol, S., van Rooij, I., Bekkering, H., & Haselager, P. (2011a). Understanding motor resonance. *Social Neuroscience, 6*(4), 388–397.

Uithol, S., van Rooij, I., Bekkering, H., & Haselager, P. (2011b). What do mirror neurons mirror? *Philosophical Psychology, 24*(5), 607–623.

Wheatley, T., Milleville, S. C., & Martin, A. (2007). Understanding animate agents: Distinct roles for the social network and mirror system. *Psychological Science, 18*(6), 469–474.

Winston, J. S., Henson, R. N., Fine-Goulden, M. R., & Dolan, R. J. (2004). fMRI-adaptation reveals dissociable neural representations of identity and expression in face perception. *Journal of Neurophysiology, 92*(3), 1830–1839.

Woodward, A. L. (1998). Infants selectively encode the goal object of an actor's reach. *Cognition, 69*(1), 1–34.

11 Infants Attribute Mental States to Nonhuman Agents

Yuyan Luo and You-jung Choi

Abstract

Recent research shows that from an early age, infants appear to attribute mental states such as goals, dispositions, perceptions, and beliefs to agents—whether human or nonhuman. Agents are entities that can detect their environment and exert control over their actions. In the present chapter, we review evidence on infants' understanding about nonhuman agents. We describe results highlighting certain featural information and behavioral cues infants may use to identify agents. In light of the empirical results, we discuss our theoretical account and speculate on the learning mechanisms involved in early psychological understanding.

When interacting with others, we make sense of their actions by referring to mental states such as goals, disposition (e.g., preferences), perceptions, and beliefs (Carey & Spelke, 1994; Leslie, 1994; Premack & Premack, 1995; Tomasello, Carpenter, Call, Behne, & Moll, 2005; Wellman, 1991). These "others" to which we are willing to attribute mental states are termed agents (Leslie, 1994, 1995) and can be people, other animals, or even inanimate objects (Heider & Simmel, 1944). For example, when we see a vacuum robot moving about, we assume that it can detect its way and navigate around the house to clean the dirt.

What is the origin of such understanding? Do infants also attribute mental states to both human and nonhuman agents? To what extent is infants' psychological understanding different from, or similar to, that of adults? What accounts for the development of their understanding? In the present chapter, we attempt to answer these questions by focusing on infants' understanding of nonhuman agents. We first present evidence showing that infants also appear to attribute mental states to nonhuman agents. We then consider what types of information infants use to identify agents. These empirical results lead us to the following propositions. We suggest that infants are born equipped with an innate notion of agency, similar to a built-in language learning mechanism. This notion may be vague or primitive and says little about what an agent does and what an agent looks like. However, this innate notion, acting as a sort of placeholder (Medin & Ortony, 1989), enables infants to implicitly believe that certain entities have minds to guide their behavior. Learning and experience directs infants' attention to specific kinds of agents (e.g., humans) and certain behavioral information (e.g., biological motion) that they use to identify nonhuman entities as agents. The presence of an agent elicits infants' attributions of various mental states as the underlying reasons for the agent's behavior. Our theoretical account thus paints a clear picture of the starting points of psychological understanding. In the final section of the chapter, we explore what drives this development by looking at the general learning mechanisms involved. By examining the scope of early psychological understanding (e.g., what an agent is to infants, what mental states infants attribute to an agent), we can better explain how an infant becomes a social being and understands others as social beings.

Young Infants Act as if They Reason about the Intentional Actions of Nonhuman Agents

Research on infants' understanding of nonhuman agents has been informed by what has been learned regarding their understanding of human agents. Beginning in the first year of life, infants attempt to make sense of human agents' intentional actions (e.g., Woodward, 2009). When watching an agent act on objects, for example, infants as young as three months of age act as if they attribute *goals* and *dispositions* to the agent that help explain and predict the agent's actions. A goal is defined as a particular outcome that an agent wants to achieve. Infants have been found to detect human agents' simple goals, such as contacting an object (e.g., Luo & Johnson, 2009; Phillips & Wellman, 2005; Woodward, 1998), pointing to an object (Woodward & Guajardo, 2002), or looking at an object (Johnson, Ok, & Luo, 2007; Luo, 2010; Woodward, 2003). A disposition is defined as a tendency that explains why an agent chooses a particular goal. Infants have been found to attribute to human agents simple dispositions, such as a preference for one object over another (e.g., Luo & Baillargeon, 2007; Repacholi & Gopnik, 1997), a preference for a particular color (Luo & Beck, 2010), or a predilection to engage in a certain activity, such as sliding objects (e.g., Song & Baillargeon, 2007; Song, Baillargeon, & Fisher, 2005).

Do infants attribute goals and dispositions to *nonhuman* agents? To address this question, researchers have designed studies using stimuli that bear little resemblance to people, such as a self-propelled box (Luo, 2011b; Luo & Baillargeon, 2005), a blob that beeps and moves itself (Johnson, 2003; Johnson, Shimizu, & Ok, 2007), and computer-animated geometric shapes (Csibra, Bíró, Koós, & Gergely, 2003; Gergely, Nádasdy, Csibra, & Bíró, 1995; Kuhlmeier, Wynn, & Bloom, 2003).

For example, in one experiment modeled after Woodward (1998), five-month-old infants were assigned to one of two conditions (Luo & Baillargeon, 2005). The infants in one condition first watched orientation trials in which a self-propelled box moved in the center of an apparatus. Next, two objects were introduced, one in each corner of the apparatus (two-object condition). Infants then received familiarization trials in which the box approached and contacted object-A as opposed to object-B. Finally, the positions of the two objects were reversed, and infants received test trials in which the box approached either object-A (old-goal event) or object-B (new-goal event). A second group of infants received similar trials except that, during the familiarization phase, only object-A was present (one-object condition). The infants in the two-object condition looked reliably longer at the new-goal event rather than the old-goal event. This suggests that they (1) attributed to the self-propelled box a preference for A over B (a dispositional state that explains why an agent chooses a particular goal-object in the presence of another option), (2) expected the box to approach A in the test trials, and hence (3) responded with heightened interest to the new-goal event (i.e., when B was approached). Infants therefore behaved as though they were able to reason about the intentional actions of a nonhuman agent. In the one-object condition, when object-B was absent during the familiarization trials, infants looked equally at the new- and old-goal test events. In this condition, infants had no information about which object the agent might choose during the test (i.e., the agent may again approach object-A, or it may now approach object-B), and they therefore looked equally at the two test events. These results have also been reached in studies involving three-month-old infants (Luo, 2011b)—the youngest age yet tested in goal-related studies involving human agents (Sommerville, Woodward, & Needham, 2005).

Infants also appear to attribute dispositions to agents in the one-object conditions (Bíró, Verschoor, & Coenen, forthcoming; Luo, 2011b). Just as we might attribute a fondness for chocolates to an individual who goes to great lengths to retrieve the last piece of chocolate in the house, three-month-olds seemed to attribute a preference for object-A to the box agent, even when object-B was absent, if (1) object-A occupied different positions in the familiarization trials, and (2) the box consistently adjusted its actions so as to approach object-A, thus demonstrating equifinal variations of its actions to achieve the same goal (Luo, 2011b).

Furthermore, infants seem to judge nonhuman agents' choices of goal-directed actions based on how efficient or rational these actions are within given situations—just as they do with human agents (Carpenter, Call, & Tomasello, 2005; Gergely, Bekkering, & Király, 2002). For example, in Gergely et al. (1995), twelve-month-old infants were habituated to the following computer-animated event. A small circle and a large circle first lay on the opposite sides of a barrier. They then engaged in a turn-taking interaction in which the large circle expanded, then contracted, followed by the small circle performing the same sequence. Next, the small circle approached the barrier, retreated, approached it again, jumped over it, and finally stopped against the large circle. The small circle approached the large one from left or right equally often, demonstrating equifinality in its actions. During the test, the barrier was removed and the small circle either still jumped (old-action event) or moved in a straight line (new-action event) to approach the large circle. The infants looked reliably longer at the old- than at the new-action test event. These and control results were consistent with the conclusion that infants (1) identified the small circle's goal of approaching the large circle and (2) realized that it had to jump over the barrier to achieve this goal during habituation. They thus expected the small circle to adapt to the change in the environment when the barrier was removed and choose a more efficient path to approach the large circle during the test. Hence, infants responded with increased attention when the small circle failed to do so in the old-action event. These results were later extended to younger (nine-month-old) infants (Csibra, Gergely, Bíró, Koós, & Brockbank, 1999).

Finally, just as adults sometimes judge people by whether or not they are helpful to others, infants appear to do so as well, even when those others are geometric shapes (Hamlin, Wynn, & Bloom, 2007; Kuhlmeier et al., 2003). For example, in Hamlin et al. (2007), based on the work of Kuhlmeier and colleagues (Kuhlmeier & Wynn, 2003; Kuhlmeier et al., 2003; see chapter 12 in the present volume), ten-month-old infants were habituated to events in which a circle that tried to climb a hill was helped by an agent-A (e.g., a triangle pushed the circle up) or hindered by an agent-B (e.g., a square pushed the circle down). During the test, the circle approached the helper or the hinderer; infants looked reliably longer when the circle approached the hinderer. These and follow-up conditions suggested that the infants formed expectations as to the circle's dispositions toward the helper and the hinderer based on their encounters in a previous context. They seemed to appreciate that the circle should show a positive disposition toward its helper but not the hinderer. Interestingly, after watching the habituations events depicting the interactions among these nonhuman agents, infants themselves also approached the helper (Hamlin et al., 2007). Three-month-olds were later found to also avoid looking at the hinderer (Hamlin, Wynn, & Bloom, 2010).

Infants Act as if They Consider Nonhuman Agents' Perceptions and Beliefs When Interpreting Their Intentional Actions

The evidence that infants behave as if they can reason about the intentional actions of nonhuman agents supports a *system-based view* of early psychological reasoning. According to this view, infants are born equipped with a psychological-reasoning system that provides them with a skeletal causal framework for interpreting and predicting the intentional actions of any entity they identify as an agent, whether human or nonhuman (e.g., Gergely & Csibra, 2003; Johnson, 2005; Leslie, 1995). This is not to say, of course, that experience and learning play little role in infants' understanding. We return to this issue in the next section.

What is the nature of the skeletal causal framework? We have proposed a *mentalistic* account of early psychological understanding (Luo & Baillargeon, 2010), which suggests that, similar to adults, infants use a coherent construal of agents' various mental states to make sense of their actions. That is, agents' acts in pursuit of goals are guided by their perceptions, knowledge, and/or beliefs about certain aspects of the environment—for example, where a goal object is in a scene. One key aspect of the mentalistic account is that infants' interpretations of an agent's intentional actions are based on the environment *as represented by the agent*, which can differ from reality. There is now evidence that infants recognize that an agent may have incomplete perceptions of a scene or may hold a false belief (e.g., Kovacs, Teglas, & Endress, 2010; Luo, 2011a; Luo & Baillargeon, 2007; Luo & Beck, 2010; Luo & Johnson, 2009; Onishi & Baillargeon, 2005; Sodian, Thoermer, & Metz, 2007; Song & Baillargeon, 2008; Surian, Caldi, & Sperber, 2007; Tomasello & Haberl, 2003). These studies involve both human and nonhuman agents. Here we focus on results with nonhuman agents.

In one experiment (Choi, Luo, & Baillargeon, 2011; fig. 11.1), for example, five-month-old infants received familiarization trials in which, as in Luo and Baillargeon (2005), the self-propelled box agent consistently moved toward and rested against object-A but not object-B. In one (hidden-object) condition, both objects were visible to the infant, but only object-A was "visible" to the agent—a tall screen hid object-B from it. In two other conditions, both object-A and object-B were perceptible to the agent: Either the screen was shorter so that object-B protruded above it (short-screen condition), or object-B itself was taller so that it protruded above the tall screen (tall-object condition). Following the familiarization trials, the objects' positions were reversed, the screen was removed, and the agent approached either object-A (old-goal event) or object-B (new-goal event). As expected, in the short-screen and tall-object conditions, infants looked reliably longer at the new-goal event rather than at the old-goal event. In the hidden-object condition, in contrast, infants looked about equally at the two events. The infants thus behaved as though they had realized that the agent's repeated actions toward object-A during the familiarization trials could not be interpreted as revealing its preference for A over B if the agent could not perceive object-B. The hidden-object condition was essentially the same, from the agent's perspective, as the one-object condition in Luo and Baillargeon (2005). Ongoing experiments are examining how infants characterize the box agent's perceptions. For example, if the box had an antenna that towered above the tall screen, would the object behind the screen become perceptible to the box?

Infants seem to consider the box agent's *perceptions* of which objects are present (both object-A and object-B, or object-A only) when attributing preferences to it. When the agent cannot perceive all objects in a setting, infants act as if they acknowledge that its perception is *incomplete*. This is similar to the situation in which agents hold *false* beliefs because they have not witnessed all *events* that happen in the setting (Wellman, Cross, & Watson, 2001). For example, Onishi and Baillargeon (2005) found that fifteen-month-old infants act as if they expect a human agent to search for a toy according to where he or she *believes* it is located (i.e., location-A, where the agent had hidden the toy) as opposed to the toy's *actual location* (i.e., location-B, where the toy had moved during his or her absence). They seem to understand that if the agent did not see the relocation of the toy, he or she should falsely believe that the toy is still in its original location. Therefore, infants respond with heightened interest when the agent reaches for the toy in its current hiding location.

These results were extended to younger (thirteen-month-old) infants in situations involving a nonhuman agent. In Surian et al. (2007), infants first watched the following computer-animated events during the familiar-

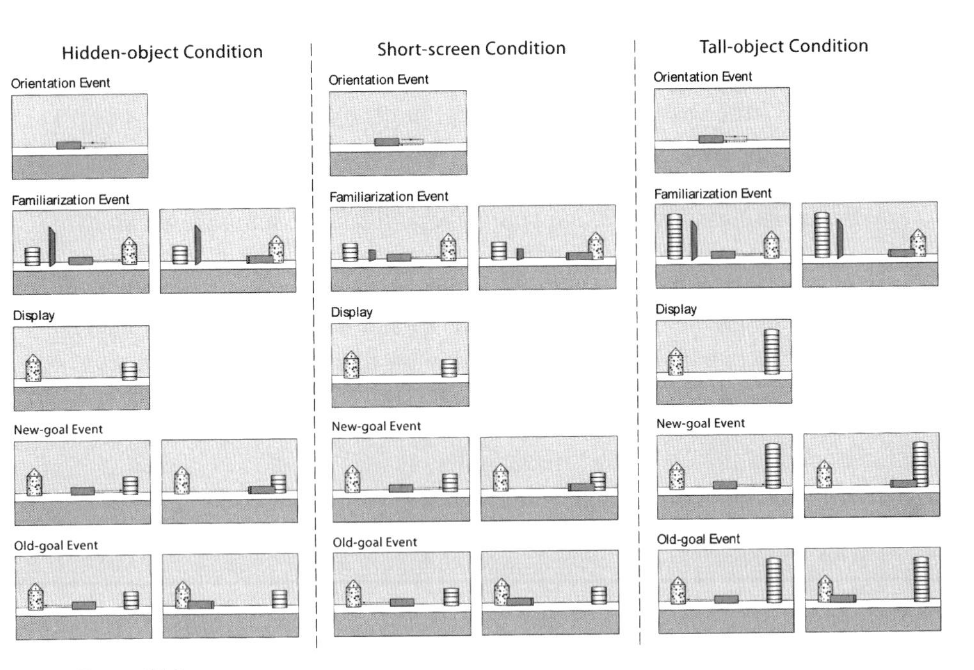

Figure 11.1
Schematic drawing of the events shown in the hidden-object, short-screen, and tall-object conditions of Choi, Luo, and Baillargeon (2011).

ization phase. To start, a caterpillar entered the scene and stopped between two screens. Next, the caterpillar watched an experimenter's hand place a piece of cheese behind one screen and an apple behind the other screen. Showing that it preferred the cheese to the apple, the caterpillar always moved behind the same screen to chew on the cheese. During the test, the caterpillar was removed and the hand switched the foods into reversed locations. When the caterpillar next came to the scene, it went behind either of the two screens to chew on the food. Infants looked reliably longer when the caterpillar went to the new location for the cheese. These and control results suggest that infants recognized that, since the caterpillar had not seen the hand switching the foods' hiding locations, it should falsely believe that the cheese was still in the original location.

Therefore, infants' understanding of agents, human or nonhuman, is consistent with a system-based, mentalistic account. The inborn psychological-reasoning system is triggered by the presence of an agent. Infants then behave as though they attribute to the agent a whole host of mental states—including goals, dispositions, perceptions, as well as beliefs—to explain its behavior.

This account makes sense because we adults are folk mentalistic psychologists. It is thus extremely difficult to imagine explaining others' behavior without positing mental states (Perner, 2010). Given that continuity and parallels have been found in infants' and adults' understanding about, for example, physical objects and numerosity (e.g., Baillargeon, 2008; Spelke & Kinzler, 2007), it is highly likely that such continuity also exists in yet another domain of folk knowledge: psychological understanding. Of course, this does not necessarily have to be the case. For example, for infant understanding of agents' false beliefs, it has been suggested that various behavioral rules, gathered from observations and experiences, are sufficient to explain their successes in the aforementioned situations without invoking mentalistic terms (e.g., Perner, 2010). Behavioral rules can presumably differ in different situations. Therefore, one challenge to the mentalistic account is to show infants' competencies in different situations at a given age (Perner, 2010). This challenge has been met in research on false-belief understanding in infancy, at least in the second year of life (for a review, see Baillargeon, Scott, & He, 2010). The same challenge can be applied to research on infants' understanding about nonhuman agents. With a focus primarily on the first year of life, the evidence reviewed so far in the present chapter shows infants' success in various situations involving nonhuman agents. Still, given the limited motor abilities of early infancy, the empirical evidence mostly comes from looking-time studies in which infants spontaneously respond to events that are either consistent or inconsistent with their expectations. Johnson and colleagues, however, have demonstrated that at least by twelve months of age, infants are willing to follow the "gaze" of a novel nonhuman agent (Johnson, Slaughter, & Carey, 1998) as they do with human agents. Although nonmentalistic accounts are admittedly difficult to rule out completely (Johnson, 2003), more studies such as these, especially with children in early infancy, would help move the field toward the mentalistic account.

Identifying Agents and Learning

Next, we further examine the following issues. First, what kind of entity is an infant willing to consider an agent? Could there be innate bases for understanding agents? Or, could it all be learned from the environment? Second, how does the psychological-reasoning system explain the learning processes and development? In the remainder of this chapter, we review the evidence regarding infants' identification of agents. We then argue that there may be an innate primitive concept of agents. Finally, we speculate on infants' learning mechanisms. In doing so, we point to future directions of research that will mark the starting point and chart the developmental trajectory of early psychological understanding.

How Do Infants Identify Agents?

We address this question by first comparing two studies with three-month-old infants, the youngest age to have been tested in studies about the psychology of agents. Sommerville et al. (2005) found that infants this young, who typically cannot reach for and grasp an object (Needham, Barrett, & Peterman, 2002), were unable to encode a *human* agent's arm and hand grasping one of two objects as being goal directed. In contrast to the five- and six-month-olds in Woodward (1998), the three-month-olds failed to respond to the change of the agent's goal when she grasped the new object during test. However, positive results were obtained if these infants first participated in an action task in which they were provided with an experience of grasping by wearing Velcro mittens they could use to swipe at objects. These results point to the importance of firsthand action experiences—even those acquired in a laboratory setting facilitate infants' attributions of goals to human agents. On the other hand, Luo (2011b) extended the task of Luo and Baillargeon (2005) to three-month-olds and reported that infants at this age also seem to attribute goals and preferences to a novel *nonhuman* agent—a self-propelled box.

Why did three-month-olds fail to perceive a human agent's arm and hand grasping an object as goal-directed in Sommerville et al. (2005) but readily attribute goals to a nonhuman agent contacting an object? We suggest that the situation used in Luo (2011b) was more supportive of young infants' agent detection than that of Sommerville et al. (2005) because the infants were provided with richer *featural* information as well as *behavioral* information about the agent. In earlier studies, six-month-old infants succeeded in encoding a person's bare hand reaching for and grasping an object as being goal directed (Woodward, 1998), but this was not the case when the hand had a glove on (Guajardo & Woodward, 2004); this finding speaks to the significance of featural information in infants' understanding about agents. In Sommerville et al. (2005), as in Woodward (1998), three-month-old infants only saw the agent's arm and hand, which was insufficient for them to infer the presence of the person acting as the intentional agent. (Interestingly, after the infants had experiences with the Velcro mittens in the action task, the agent in the looking-time task also wore a similar mitten.) In Luo (2011b), however, the infants always saw the entire box throughout the experiment. This analysis suggests that, when given sufficient featural information about a human agent—for example, being able to see the person's face and upper body—three-month-olds, when predicting and interpreting an agent's actions, act as if they take the agent's goals and dispositions into account. Preliminary positive results support this hypothesis (Choi & Luo, 2011).

Additionally, the infants in Luo (2011b) were given richer information about the agent's behavior than were those of Sommerville et al. (2005). In Sommerville et al. (2005), infants simply saw the agent's arm and hand reach toward one of two objects during the habituation phase. In Luo (2011b), however, infants first watched the box move back and forth in the center of the apparatus during the orientation trials. They then saw the box move to one corner of the apparatus to stop against object-A during the familiarization trials. Therefore, the box changed its behavior from orientation to familiarization. This behavioral cue suggested that the box was an agent that could detect its environment and control its actions to pursue its goal.

Support for the point outlined in the previous paragraph comes from a study by Johnson and colleagues (Johnson, Shimizu et al., 2007). In the study, twelve-month-old infants, during the habituation trials, saw a novel nonhuman agent—a self-propelled blob—move toward and contact object-A on the left rather than object-B on the right. Following habituation, the objects' positions were reversed, and the agent approached either object-A (old-goal event) or object-B (new-goal event) during the test. The infants looked reliably longer at the new-goal event rather than the old-goal event when the agent started each habituation trial facing forward and then turned to face object-A, but *not* when the agent was already aligned with object-A before moving toward the object. These results suggest that even a subtle behavioral change, such as a simple turn, signals the presence of an agent capable of acting to pursue its goal object.

In sum, infants need featural and/or behavioral information to identify agents (e.g., Bíró & Leslie, 2007; Johnson, 2003). Given infants' familiarity with people, human or animate features—such as a person's face, eyes, or bare hand—suggest to an infant the presence of a human agent. When these features are lacking and infants are faced with an inanimate object that looks nothing like a person, this object needs to give evidence that it can detect the environment and exert control over its actions to be perceived as an agent. Such evidence can be as simple as the turn shown in Johnson, Shimizu, et al. (2007) or as elaborate as the drama played out by the two circles in Gergely et al. (1995).

Csibra (2008) found another simple behavioral cue to agency: equifinality. During the habituation trials, six-and-a-half-month-old infants watched three-dimensional animated video events in which a box moved around an obstacle in the center of a room to reach its target. The box's detours around the obstacle were either one-way or two-way (i.e., from both left and right). During test, the obstacle was removed. The box moved as before or in a straight line to its target. Just as in Gergely et al. (1995), infants responded with heightened interest to the old-action event in only the two-way condition. These results and those from control conditions suggest that the equifinal variations—that is, moving in different ways to achieve the same goal—are important for infants in (1) identifying the box as an agent and (2) further evaluating the efficiency of its goal-directed actions. When the box moved in only one direction, it was unclear whether it did so to pursue its goal or whether it was only capable of this single movement. We obtained similar results with our self-propelled box (Choi et al., 2011).

Changes in behavior and equifinal variations of actions both suggest an agent's capacity for (1) perceiving the environment and (2) controlling its own actions. In other words, both cues convey an impression of choice, or free will, that signals to infants the presence of an agent (e.g., Bíró, Csibra, & Gergely, 2007; Bíró & Leslie, 2007; Caron, 2009; Csibra, 2008; Premack, 1990). It is important to note that as demonstrated by the negative results in Csibra (2008) and Johnson, Shimizu et al. (2007), self-propelled motion is an insufficient cue to agency; infants do not view the self-propelled object moving along a fixed path as an agent.

What sort of changes in behavior do infants use as cues to agency? So far we have seen infants' *successes* in agent detection in the following situations: a self-propelled box moves in one area and then moves in a different path to contact an object (three-to-five months of age, Luo, 2011b; Luo & Baillargeon, 2005); a self-propelled box moves around an obstacle in two directions to contact its target (six-and-a-half months of age, Csibra, 2008); a self-propelled blob turns to move toward its goal object (twelve months of age, Johnson, Shimizu et al., 2007). More research is needed to outline the developmental course of using behavioral agency cues. For instance, are older infants able to make use of fewer or simpler cues? Bíró and Leslie's (2007) study is the only one so far to address this question. In the study, infants were presented with a rod with a Velcro tip extending from outside of an apparatus. One end of the rod was hidden, so featural information about the rod was impoverished. However, nine- and twelve-month-olds, but not six-month-olds, perceived the rod as an agent if it moved randomly and then moved to pick up an object with its Velcro tip. Six-month-old infants needed one more behavioral cue (i.e., equifinal variations); this cue occurred when the rod each time touched the object at a different spot and then picked it up. Conversely, Shimizu and Johnson (2004) showed infants' *failures* in agent detection. Twelve-month-old infants did not view a self-propelled blob as an agent if it beeped and then moved to approach one of two toys. However, in their study, the toys were also present when the blob beeped. It was thus unclear why the blob first ignored them and then changed its "mind." It still remains to be seen whether the change from beeping to moving works for agency detection when the blob beeps with the toys absent and then moves itself to approach one of the toys.

The important line of research by Johnson and colleagues reveals yet another cue to agency: contingent interaction with another agent. After watching a novel inanimate blob beeping to a person as if engaging in a conversation (with the person speaking English) for one minute, twelve-month-old infants categorize the blob as an agent. When it turns as if to look at an object, infants follow its turn to see what catches its attention (Johnson et al., 2008; Johnson et al., 1998). In addition, infants attribute to the blob agent preferences for specific objects (Johnson, Shimizu et al., 2007). The cue of reciprocal conversation suggests that the agent can detect its environment and control its actions. The agent's behavior —engaging in conversation—is also quite similar to what a human agent can do. Indeed, all the behavioral cues reviewed so far arguably could be borrowed from what infants already know about human agents.

This claim is the essence of an experience-based view of early psychological understanding, in that infants' abilities to understand others come from their experiences with both themselves and other people. Many proponents of this empirical view single out infants' firsthand experiences with action as crucial to early intentional understanding. As infants become more and more accomplished in producing goal-directed actions, they are able to understand similar actions in others; this is in part due to an innate ability to pair their own actions and mental states with those of others (Meltzoff, 1995, 2005; Tomasello, 1999; Woodward, Sommerville, & Guajardo, 2001). For example, in Sommerville et al. (2005), when there was minimal featural and behavioral information about the agent (i.e., only the person's arm and hand were visible, and there was no variation in the hand's actions), the Velcro mitten experiences helped three-month-old infants encode the agent's goal. When infants are six months old and have had experiences with grasping objects, the same action is no longer ambiguous (Woodward, 1998).

The contrasts between Luo (2011b) and Sommerville et al. (2005) make clear that self-generated action experiences are not the precondition of infants' understanding of agents. At three months of age, infants also seem to attend to featural and behavioral information that signals an agent, which either is a person or looks nothing like a person but can act and react to its environment. This admittedly does not exclude the possibility that the behavioral agency cues infants use could be learned in their first three months of life, especially when empirical evidence from this time period is lacking. Assuming that many types of experiences, in addition to the hands-on action experiences, can be at work, we see the experience-based view complementing the system-based view. Therefore, infants' psychological understanding utilizes both (1) an innate predisposition to attend to agents and (2) prior experiences.

The Primitive Concept of Agents May Be Innate

We propose that, at the minimum, infants from early on possess a primitive concept of agents as being entities that have mental states to guide their behavior. This is what the psychological-reasoning system deals with, and it is distinct from domains of physical or numerical understanding that are concerned with, respectively, physical objects and numerosity (e.g., Carey & Spelke, 1994; Spelke & Kinzler, 2007). This innate concept of agents, we suppose, is similar to the hardwired language-learning mechanism that each infant comes into the world with. It may be vague and says little about what an agent should look like and what an agent should do. Just as the linguistic environment an infant is born into enables it to learn a specific language, learning about surroundings and experiences, of course primarily with self and other people, enables an infant to come to appreciate humanlike features and certain ways of behaving as essential for identifying agents.

Innate ideas are no doubt difficult to prove. However, three potential sources of empirical evidence could prove to be helpful in this endeavor. First, if infants from different cultural settings were found to identify not only people and other animals but also inanimate objects (e.g., the computer-animated geometric shapes) as agents, as adults do, this would be compelling evidence that the notion of agency might have roots in human nature. The available data so far comes mostly from infants in middle-class families in Western, industrialized cultures. Similar results extended to various cultures would constitute supporting evidence for innate bases of psychological understanding (Csibra & Gergely, 2009). Second, if newborn infants were found to attribute mental states to both human and nonhuman agents, this would strengthen the notion of innate competence. Moreover, it would show that the innate concept of agents is not vague and might include certain featural and/or behavioral information about agents. Third, nonhuman primates have been shown to engage in intentional reasoning about people and conspecifics (e.g., Call, Hare, Carpenter, & Tomasello, 2004; Call & Tomasello, 1998; Flombaum & Santos, 2005; Hare, Call, & Tomasello, 2001). Although these results are being debated, similar results with tasks involving inanimate agents would corroborate the argument for innate ideas from the evolutionary point of view. The idea that such evidence could come from studies involving newborn infants and nonhuman animals is not at all far-fetched. Newborns are found to (1) detect the eye gaze of others, (2) follow the gaze of others, and (3) imitate them (Farroni, Csibra, Simion, & Johnson, 2002; Farroni, Massaccesi, Pividori, Simion, & Johnson, 2004; Meltzoff & Moore, 1977). Two-day-old infants and newly hatched chicks are also found to be sensitive to biological motion and self-propelled objects (Mascalzoni, Regolin, & Vallortigara, 2010; Simion, Regolin, & Bulf, 2008; Vallortigara, Regolin, & Marconato, 2005). Interestingly, infant chimpanzees respond in a manner similar to that of human infants in a task adapted from Gergely et al. (1995) described earlier (Uller, 2004).

The data from studies of newborns arguably demonstrate their predisposition to treat people as more than just physical objects. In addition, from at least three months of age, infants appear to attribute mental states to both human and nonhuman agents. These abilities presumably rely on an innately based psychological-reasoning system with a notion of agency as its core. What learning mechanisms are employed by this domain-specific system? We list some candidate mechanisms in the following section. These mechanisms are domain general, meaning that they can work on input from various domains—be it physical objects and events, agents and their behavior, or numerocity. Note also that there has been evidence derived from studies of early infants, and even newborns, showing that they seem to engage in these types of learning.

Learning Mechanisms

The psychological-reasoning system in principle is a skeletal causal framework devoted to making sense of agents. When early infants look about and see agents acting in various ways, they presumably infer agents' mental states as reasons for their behavior. All the results reviewed in the present chapter suggest that infants engage in *causal learning* about agents. For example, when an agent consistently approaches object-A, but not object-B, during the familiarization phase of two-object conditions (Luo, 2011b; Luo & Baillargeon, 2005), infants come to the conclusion that (1) the agent has a preference for A over B, and (2) this preference is the cause of its behavior. Based on this causal analysis, infants expect the agent to continue approaching object-A during the test and hence respond with heightened interest when this expectation is violated as the agent approaches B. If the agent's actions toward object-A simply resulted in agent-object associations, infants might have no expectation about the agent's future actions (if the association has little predictive power), or infants might respond to the novel agent-object association when the agent approaches object-B. To rule out the latter possibility, all the components of the two-object conditions were preserved in the one-object conditions, except that, when the agent approached object-A during familiarization, object-B was absent. In these conditions, infants no longer responded with heightened interest when the agent approached object-B during the test; this finding argues against association-based explanations for the two-object conditions results. Many studies discussed in the present chapter included such control conditions, thus providing support for the claim that infants seem to reason about the psychological causes of agents' actions.

One prediction of causal learning is that, if provided with appropriate information, infants could possibly be taught to understand certain aspects of an agent's behavior at ages earlier than they normally would. To our knowledge, no such "teaching" studies have been conducted, since we do not yet know much about what infants learn or understand at different ages. Research on infants' understanding about physical objects and events, in contrast, has yielded much data in this regard and has culminated in an elaborate theoretical model and subsequent investigations, including teaching studies (Baillargeon, 2008; Wang & Baillargeon, 2008). To apply a similar logic here, we first need to compare infants' performances at different ages in similar tasks. For example, in our work, we find that at five-to-six months of age, infants seem to only consider an agent's perceptual access to objects when attributing preferences to it. When both object-A and object-B are "visible" to the agent, infants seem to attribute to it a preference for A over B if the agent consistently approaches A; this is the case regardless of the physical accessibility of object-B (Choi et al., 2011; Luo & Johnson, 2009). Object-B can be behind a screen, which makes it more difficult to access than object-A, as in the short-screen and tall-object conditions of Choi et al. (2011), and still a preference is assumed. At three months of age, however, infants seem to care about the physical accessibility of objects to the agent as well. When object-B is more difficult to reach for than object-A because it is farther away from the agent, infants no longer attribute preference to the agent (Choi & Luo, 2011). What information or experiences are obtained between three-and-five months of age that enable infants to realize that physical access does not matter much? If an agent preferred A to B, it would make an effort to obtain A even if it had to move around a screen or simply go a longer way to do so. Answers to this question could lead to experiments designed to teach three-month-old infants to disregard objects' physical accessibilities to agents in preference attributions.

Experiences that contribute to changes in infants' understanding about agents over time can certainly come from their observations of regularities and irregularities in agents' behavior—a phenomenon known as *statistical learning*. Based on given data (e.g., on frequencies), infants and children are found to make "informed" generalizations in a variety of areas such as speech perception, word learning, causal reasoning, and understanding about physical objects (e.g., Gopnik et al., 2004; Maye, Werker, & Gerken, 2002; Saffran, Aslin, & Newport, 1996; Sobel & Kirkham, 2007; Xu, 2007). For example, in the study by Xu and Garcia (2008), eight-month-old infants used information about a population (e.g., a box with mostly red and a few white Ping-Pong balls) to make inferences about a sample (e.g., an experimenter should randomly draw from the box four red balls and one white ball, as opposed to four white balls and one red ball). Conversely, infants could also make inferences about the population based on a random sample drawn from it.

Recently, researchers have started to ask how infants integrate their statistical learning skills with their understanding about agents. For example, in situations similar to those described in the previous paragraph, eleven-month-old infants no longer expected to see a sample of mostly red balls from a corresponding population when the experimenter had a specific goal of obtaining white but not red balls (Xu & Denison, 2009). We have examined a complementary situation: how infants use statistical information—specifically, frequencies of agents' actions toward objects—to make inferences about agents' goals and preferences (Luo, Markson, Fawcett, Mou, & vanMarle, 2010). We found, as in previous studies, that when a human agent consistently approaches object-A but not object-B, nine-month-old infants attribute to the agent a preference for A over B. However, when the agent does so inconsistently, approaching A in three familiarization trials and B in one familiarization trial (order counterbalanced), infants no longer view the agent's actions as evidence of a preference for A over B. These and control results suggest that infants seem to consider the consistency (or inconsistency) of an agent's actions toward objects when attributing preferences to that agent. Again, a developmental story needs to be told from further testing. For example, what is the consistency/inconsistency ratio for preference attributions at different ages?

In addition, *associative learning* has been suggested by many to be a powerful mechanism that helps infants to learn about static and dynamic features of objects (for a complete discussion, see Rakison & Lupyan, 2008). This learning mechanism may be important for infants' acquisition of featural and behavioral information about agents. Finally, infants learn about others, and how they act, by *imitation* (e.g., Meltzoff, 2005). Both this learning mechanism and *learning by analogy* (Chen, Sanchez, & Campbell, 1997; Goswami, 2001) may play key roles in infants' learning about different kinds of agents.

Concluding Remarks

In the present chapter, we have shown that infants seem to consider various mental states (e.g., goals, preferences, perceptions, and beliefs) as causes for agents' actions. Agents can be human or nonhuman. By focusing on infants' psychological understanding about *nonhuman* agents, we argue that infants may, at the very least, start out in life with a vague concept of agents as one kind of entity that has mental states; this concept does not specify agents' appearances or behaviors. Through learning and experiences, infants may come to single out certain featural and behavioral information that signals the presence of an agent. For example, an object may look like a person or a part of a person, or it may behave as if it can detect its environment and exert control over its actions. A variety of general learning mechanisms are possibly responsible for the development of infants' understanding about agents. Future research will concentrate on telling a complete developmental story.

Because we see the purpose for humans of understanding agents as being able to effectively interact with them, studying infants' psychological understanding about agents, human or nonhuman, can help us better explain how infants and children develop as social beings. In addition, by finding out to what extent the ability to reason about agents is built in and to what extent it is acquired through learning and experiences, we come to know more about the similarities and dissimilarities between human cognition and certain cognitive capacities found in other animal species.

References

Baillargeon, R. (2008). Innate ideas revisited: For a principle of persistence in infants' physical reasoning. *Perspectives on Psychological Science, 3,* 2–13.

Baillargeon, R., Scott, R. M., & He, Z. (2010). False-belief understanding in infants. *Trends in Cognitive Sciences, 14,* 110–118.

Bíró, S., Csibra, G., & Gergely, G. (2007). The role of behavioral cues in understanding goal-directed actions in infancy. *Progress in Brain Research, 164,* 303–323.

Bíró, S., & Leslie, A. M. (2007). Infants' perception of goal-directed actions: Development through cue-based bootstrapping. *Developmental Science, 10,* 379–398.

Bíró, S., Verschoor, S., & Coenen, L. (forthcoming). Evidence for a unitary goal concept in 12-month-old infants. *Developmental Science.*

Call, J., Hare, B., Carpenter, M., & Tomasello, M. (2004). "Unwilling" versus "unable": Chimpanzees' understanding of human intentional action. *Developmental Science, 7,* 488–498.

Call, J., & Tomasello, M. (1998). Distinguishing intentional action from accidental action in orangutans, chimpanzees, and human children. *Journal of Comparative Psychology, 112,* 192–206.

Carey, S., & Spelke, E. S. (1994). Domain-specific knowledge and conceptual change. In L. A. Hirschfeld & S. A. Gelman (Eds.), *Mapping the mind: Domain specificity in cognition and culture* (pp. 169–200). New York: Cambridge University Press.

Caron, A. J. (2009). Comprehension of the representational mind in infancy. *Developmental Review, 29,* 69–95.

Carpenter, M., Call, J., & Tomasello, M. (2005). Twelve- and 18-month-olds copy actions in terms of goals. *Developmental Science, 8,* F13–F20.

Chen, Z., Sanchez, R. P., & Campbell, T. (1997). From beyond to within their grasp: Analogical problem solving in 10- and 13-month-olds. *Developmental Psychology, 33,* 790–801.

Choi, Y., & Luo, Y. (2011). *Three-month-old infants attribute preferences to others.* Unpublished manuscript.

Choi, Y., Luo, Y., & Baillargeon, R. (2011). *Infants' reasoning about agents' goals and perceptions.* Unpublished manuscript.

Csibra, G. (2008). Goal attribution to inanimate agents by 6.5-month-old infants. *Cognition, 107,* 705–717.

Csibra, G., Bíró, S., Koós, O., & Gergely, G. (2003). One-year-old infants use teleological representations of actions productively. *Cognitive Science, 27,* 111–133.

Csibra, G., & Gergely, G. (2009). Natural pedagogy. *Trends in Cognitive Sciences, 13,* 148–153.

Csibra, G., Gergely, G., Bíró, S., Koós, O., & Brockbank, M. (1999). Goal attribution without agency cues: The perception of "pure reason" in infancy. *Cognition, 72,* 237–267.

Farroni, T., Csibra, G., Simion, F., & Johnson, M. H. (2002). Eye contact detection in humans from birth. *Proceedings of the National Academy of Sciences of the United States of America, 99,* 9602–9605.

Farroni, T., Massaccesi, S., Pividori, D., Simion, F., & Johnson, M. H. (2004). Gaze-following in newborns. *Infancy, 5,* 39–60.

Flombaum, J. I., & Santos, L. R. (2005). Rhesus monkeys attribute perceptions to others. *Current Biology, 15,* 447–452.

Gergely, G., Bekkering, H., & Király, I. (2002). Rational imitation in preverbal infants. *Nature, 415,* 6873.

Gergely, G., & Csibra, G. (2003). Teleological reasoning in infancy: The naïve theory of rational action. *Trends in Cognitive Sciences, 7,* 287–292.

Gergely, G., Nádasdy, Z., Csibra, G., & Bíró, S. (1995). Taking the intentional stance at 12 months of age. *Cognition, 56,* 165–193.

Gopnik, A., Glymour, C., Sobel, D. M., Schulz, L. E., Kushnir, T., & Danks, D. (2004). A theory of causal learning in children: Causal maps and Bayes nets. *Psychological Review, 111,* 1–30.

Goswami, U. (2001). Analogical reasoning in children. In D. Gentner, K. Holyoak, & B. Kokinov (Eds.), *Analogy: Interdisciplinary perspectives*. Cambridge, MA: MIT Press.

Guajardo, J. J., & Woodward, A. L. (2004). Is agency skin-deep? Surface attributes influence infants' sensitivity to goal-directed action. *Infancy, 6,* 361–384.

Hamlin, J. K., Wynn, K., & Bloom, P. (2007). Social evaluation by preverbal infants. *Nature, 450,* 557–559.

Hamlin, J. K., Wynn, K., & Bloom, P. (2010). Three-month-olds show a negativity bias in their social evaluations. *Developmental Science, 13,* 923–929.

Hare, B., Call, J., & Tomasello, M. (2001). Do chimpanzees know what conspecifics know? *Animal Behaviour, 61,* 139–151.

Heider, F., & Simmel, M. (1944). An experimental study of apparent behavior. *American Journal of Psychology, 57,* 243–259.

Johnson, S. C. (2003). Detecting agents. *Philosophical Transactions of the Royal Society of London, 358,* 549–559.

Johnson, S. C. (2005). Reasoning about intentionality in preverbal infants. In P. Carruthers, S. Laurence, & S. Stich (Eds.), *The innate mind: Structure and contents* (pp. 254–271). New York: Oxford University Press.

Johnson, S. C., Bolz, M., Carter, E., Mandsanger, J., Teichner, A., & Zettler, P. (2008). Calculating the attentional orientation of an unfamiliar agent in infancy. *Cognitive Development, 23*, 24–37.

Johnson, S. C., Ok, S., & Luo, Y. (2007). The attribution of attention: Nine-month-olds' interpretation of gaze as goal-directed action. *Developmental Science, 10*, 530–537.

Johnson, S. C., Shimizu, Y., & Ok, S. (2007). Actors and actions: The role of agent behavior in infants' attribution of goals. *Cognitive Development, 22*, 310–322.

Johnson, S. C., Slaughter, V., & Carey, S. (1998). Whose gaze will infants follow? The elicitation of gaze-following in 12-month-olds. *Developmental Science, 1*, 233–238.

Kovacs, A. M., Teglas, E., & Endress, A. D. (2010). The social sense: Susceptibility to others' beliefs in human infants and adults. *Science, 330*, 1830–1834.

Kuhlmeier, V. A., & Wynn, K. (2003). 9-month-olds' understanding of intentionality. Paper presented at the Biennial meeting of the Society for Research in Child Development.

Kuhlmeier, V. A., Wynn, K., & Bloom, P. (2003). Attribution of dispositional states by 12-month-olds. *Psychological Science, 14*, 402–408.

Leslie, A. M. (1994). ToMM, ToBY, and agency: Core architecture and domain specificity. In L. A. Hirschfeld & S. A. Gelman (Eds.), *Mapping the mind: Domain specificity in cognition and culture* (pp. 119–148). New York: Cambridge University Press.

Leslie, A. M. (1995). A theory of agency. In D. Sperber, D. Premack, & A. J. Premack (Eds.), *Causal cognition: A multidisciplinary debate* (pp. 121–141). Oxford: Clarendon Press.

Luo, Y. (2010). Do 8-month-old infants consider situational constraints when interpreting others' gaze as goal-directed action? *Infancy, 15*, 392–419.

Luo, Y. (2011a). Do 10-month-old infants understand others' false beliefs? *Cognition, 121*, 289–298.

Luo, Y. (2011b). Three-month-old infants attribute goals to a non-human agent. *Developmental Science, 14*, 453–460.

Luo, Y., & Baillargeon, R. (2005). Can a self-propelled box have a goal? Psychological reasoning in 5-month-old infants. *Psychological Science, 16*, 601–608.

Luo, Y., & Baillargeon, R. (2007). Do 12.5-month-old infants consider what objects others can see when interpreting their actions? *Cognition, 105*, 489–512.

Luo, Y., & Baillargeon, R. (2010). Toward a mentalistic account of early psychological reasoning. *Current Directions in Psychological Science, 19*, 301–307.

Luo, Y., & Beck, W. (2010). Do you see what I see? Infants' reasoning about others' incomplete perceptions. *Developmental Science, 13*, 134–142.

Luo, Y., & Johnson, S. C. (2009). Recognizing the role of perception in action at 6 months. *Developmental Science, 12*, 142–149.

Luo, Y., Markson, L., Fawcett, C., Mou, Y., & vanMarle, K. (2010). Infants' inferences about agents' preferences. Paper presented at the Biennial Meeting of the International Conference on Infant Studies.

Mascalzoni, E., Regolin, L., & Vallortigara, G. (2010). Innate sensitivity for self-propelled causal agency in newly hatched chicks. *Proceedings of the National Academy of Sciences of the United States of America, 107*, 4483–4485.

Maye, J., Werker, J. F., & Gerken, L. (2002). Infant sensitivity to distributional information can affect phonetic discrimination. *Cognition, 82*, B101–B111.

Medin, D. L., & Ortony, A. (1989). Psychological essentialism. In S. Vosniadou & A. Ortony (Eds.), *Similarity and analogical reasoning* (pp. 179–195). Cambridge: Cambridge University Press.

Meltzoff, A. N. (1995). Understanding the intentions of others: Re-enactments of intended acts by 18-month-old children. *Developmental Psychology, 31*, 838–850.

Meltzoff, A. N. (2005). Imitation and other minds: The "like me" hypothesis. In S. Hurley & N. Chater (Eds.), *Perspectives on imitation: From neuroscience to social science* (Vol. 2, pp. 55–77). Cambridge, MA: MIT Press.

Meltzoff, A. N., & Moore, M. K. (1977). Imitation of facial and manual gestures by human neonates. *Science, 198*, 75–78.

Needham, A., Barrett, T., & Peterman, K. (2002). A pick-me up for infants' exploratory skills: Early simulated experiences reaching for objects using "sticky mittens" enhances young infants' exploration skills. *Infant Behavior and Development, 25*, 279–295.

Onishi, K. H., & Baillargeon, R. (2005). Do 15-month-old infants understand false beliefs? *Science, 308*, 255–258.

Perner, J. (2010). Who took the cog out of cognitive science?—Mentalism in an era of anti-cognitivism. In P. A. Frensch & R. Schwarzer (Eds.), *Cognition and neuropsychology: International perspectives on psychological science* (Vol. 1, pp. 241–262). New York: Psychology Press.

Phillips, A., & Wellman, H. M. (2005). Infants' understanding of object-directed action. *Cognition, 98*, 137–155.

Premack, D. (1990). The infant's theory of self-propelled objects. *Cognition, 36*, 1–16.

Premack, D., & Premack, A. J. (1995). Origins of human social competence. In M. S. Gazzaniga (Ed.), *The cognitive neurosciences* (pp. 205–218). Cambridge, MA: MIT Press.

Rakison, D. H., & Lupyan, G. (2008). Developing object concepts in infancy: An associative learning approach. *Monographs of the Society for Research in Child Development, 73,* 1–110.

Repacholi, B., & Gopnik, A. (1997). Early understanding of desires: Evidence from 14- and 18-month-olds. *Developmental Psychology, 33,* 12–21.

Saffran, J. R., Aslin, R. N., & Newport, E. L. (1996). Statistical learning by 8-month-old infants. *Science, 274,* 1926–1928.

Shimizu, Y. A., & Johnson, S. C. (2004). Infants' attribution of a goal to a morphologically unfamiliar agent. *Developmental Science, 7,* 425–430.

Simion, F., Regolin, L., & Bulf, H. (2008). A predisposition for biological motion in the newborn baby. *Proceedings of the National Academy of Sciences of the United States of America, 105,* 809–813.

Sobel, D. M., & Kirkham, N. Z. (2007). Bayes nets and babies: Infants' developing statistical reasoning abilities and their representation of causal knowledge. *Developmental Science, 10,* 298–306.

Sodian, B., Thoermer, C., & Metz, U. (2007). Now I see it but you don't: 14-month-olds can represent another person's visual perspective. *Developmental Science, 10,* 199–204.

Sommerville, J. A., Woodward, A. L., & Needham, A. (2005). Action experience alters 3-month-old infants' perception of others' actions. *Cognition, 96,* B1–B11.

Song, H., & Baillargeon, R. (2007). Can 9.5-month-old infants attribute to an agent a disposition to perform an action on objects? *Acta Psychologica, 124,* 79–105.

Song, H., & Baillargeon, R. (2008). Infants' reasoning about others' false perceptions. *Developmental Psychology, 44,* 1789–1795.

Song, H., Baillargeon, R., & Fisher, C. (2005). Can infants attribute to an agent a disposition to perform a particular action? *Cognition, 98,* B45–B55.

Spelke, E. S., & Kinzler, K. D. (2007). Core knowledge. *Developmental Science, 10,* 89–96.

Surian, L., Caldi, S., & Sperber, D. (2007). Attribution of beliefs by 13-month-old infants. *Psychological Science, 18,* 580–586.

Tomasello, M. (1999). Having intentions, understanding intentions, and understanding communicative intentions. In P. D. Zelazo, J. W. Astington, & D. R. Olson (Eds.), *Developing theories of intention: Social understanding of self control* (pp. 63–75). Mahwah, NJ: Erlbaum.

Tomasello, M., Carpenter, M., Call, J., Behne, T., & Moll, H. (2005). Understanding and sharing intentions: The origins of cultural cognition. *Behavioral and Brain Sciences, 28,* 675–735.

Tomasello, M., & Haberl, K. (2003). Understanding attention: 12- and 18-month-olds know what is new for other persons. *Developmental Psychology, 39,* 906–912.

Uller, C. (2004). Disposition to recognize goals in infant chimpanzees. *Animal Cognition, 7,* 154–161.

Vallortigara, G., Regolin, L., & Marconato, F. (2005). Visually inexperienced chicks exhibit a spontaneous preference for biological motion patterns. *PLoS Biology, 3,* 1312–1316.

Wang, S., & Baillargeon, R. (2008). Detecting impossible changes in infancy: A three-system account. *Trends in Cognitive Sciences, 12,* 17–23.

Wellman, H. M. (1991). From desires to beliefs: Acquisition of a theory of mind. In A. Whiten (Ed.), *Natural theories of mind: Evolution, development, and simulation of everyday mindreading* (pp. 19–38). Cambridge, MA: Blackwell.

Wellman, H. M., Cross, D., & Watson, J. (2001). Meta-analysis of theory of mind development: The truth about false belief. *Child Development, 72,* 655–684.

Woodward, A. L. (1998). Infants selectively encode the goal object of an actor's reach. *Cognition, 69,* 1–34.

Woodward, A. L. (2003). Infants' developing understanding of the link between looker and object. *Developmental Science, 6,* 297–311.

Woodward, A. L. (2009). Infants' grasp of others' intentions. *Current Directions in Psychological Science, 18,* 53–57.

Woodward, A. L., & Guajardo, J. J. (2002). Infants' understanding of the point gesture as an object-directed action. *Cognitive Development, 17,* 1060–1084.

Woodward, A. L., Sommerville, J. A., & Guajardo, J. J. (2001). How infants make sense of intentional action. In B. Malle, L. Moses, & D. Baldwin (Eds.), *Intentionality: A key to human understanding* (pp. 146–169). Cambridge, MA: MIT Press.

Xu, F. (2007). Rational statistical inference and cognitive development. In P. Carruthers, S. Laurence, & S. Stich (Eds.), *The innate mind: Foundations and the future* (Vol. 3, pp. 199–215). London: Oxford University Press.

Xu, F., & Denison, S. (2009). Statistical inference and sensitivity to sampling in 11-month-old infants. *Cognition, 112,* 97–104.

Xu, F., & Garcia, V. (2008). Intuitive statistics by 8-month-old infants. *Proceedings of the National Academy of Sciences of the United States of America, 105,* 5012–5015.

12 The Social Perception of Helping and Hindering

Valerie A. Kuhlmeier

Abstract

This chapter presents a series of studies describing the social perception of helping and hindering—an ability that likely is integral to the maintenance of cooperative interactions. Two aspects of this perception are considered both in infants and adults: the detection of the actions (e.g., the interaction of motion patterns and goal attribution systems that ultimately provides the percept of helping or hindering) and the interpretation of the action (e.g., as positive or negative, as justifiable or praiseworthy, or as indicative of the actors' underlying dispositions). It will be suggested that the human brain appears to be ready, at an early age, to create a meaningful understanding of even sparse, stylized representations of helping and hindering.

There is a common thread that connects events as seemingly varied as a person opening the door for someone carrying a large box, a child picking up and returning a dropped puzzle piece to someone who is trying to complete a puzzle, and an action hero grabbing the hand of the damsel just as she loses her grasp on the cliff's edge. We perceive them all as instances of helping behavior in which the protagonist provides instrumental aid to someone in need. They are also the psychological opposite of hindering behaviors in which instrumental goals are impeded: The door is held shut, the puzzle piece is moved even farther way, or the rock the damsel is tenuously holding onto is kicked away.

This chapter describes the social perception of helping and hindering, including both the detection of the actions (e.g., the interaction of motion patterns and the goal attribution systems that provide the percept of helping or hindering) and the interpretation of the action (e.g., as positive or negative, as justifiable or praiseworthy, or as indicative of the actors' underlying dispositions). The detection of the actions must arguably include elements discussed in other chapters in this volume (e.g., animacy detection, goal attribution); like chasing (Frankenhuis & Barrett, Scholl, this volume), helping and hindering are social interactions between at least two agents. Perceiving the behavior of others as an instance of helping or hindering potentially poses challenges over and above that of simple, individual goal attribution such as reaching for an object or climbing up a hillside. At least for adults, helping and hindering necessarily involve the interplay between multiple agents' goals; one agent is acting to complete a goal (e.g., reaching for an object) and another agent has the goal to enable (or impede) completion of the first agent's goal (e.g., picking up and delivering the desired object).

The interpretation of the actions is perhaps even more complex. Each example of hindering behavior presented at the beginning of this chapter likely is construed as negative ("How rude to hold the door shut when that poor woman wants to get out!"); yet interpretation can also vary based on biases, preexisting models of typical social interactions, and additional information regarding the agents. The same situation and actions seem much more positive when we know that the person holding the door shut is a police officer who is keeping a burglar from escaping out the door with your new television.

Helping and hindering are likely particularly important behaviors to detect and interpret. Detecting instances of helping and interpreting agents as helpers may point us to potential partners who can be trusted to engage in reciprocal exchanges, and noting hinderers allows for strategic avoidance. The latter part of this argument is not unlike the basis of social contract theory (SCT; e.g., Cosmides & Tooby, 1992). SCT suggests that human cognitive processing is designed to detect cheaters—those who "defect" by taking a benefit without fulfilling their end of a social contract. That is, cheaters cheat by not reciprocating. In fact, it is argued that this cheater detection system is necessary for cooperation to be an evolutionarily stable strategy because it allows for conditional helping (e.g., Cosmides & Tooby, 1989). Here, I simply add that people may do best to detect *both* helpers and hinderers, as the former would also support preferential interaction with those who may provide benefits (Brown & Moore, 2000; Camilleri, Kuhlmeier, & Chu, 2010).

This chapter will first consider the early development of the ability to detect and interpret helping and hindering during infancy, and then look at adult perception of these behaviors. Most of the findings presented here are based on studies using nonhuman agents (e.g., animated or puppeteered shapes such as triangles and squares), which may seem strange given the importance of these behaviors in real-world, *human* interaction. However, this work has grown from the findings of Heider and Simmel (1944, and described in chapter 1 of this volume), which demonstrated that we readily attribute goals to these types of agents, and we now know that infants do as well (e.g., Gergely et al., 1995; Luo & Baillargeon, 2005; Luo & Choi, this volume; Schlottmann & Ray, 2010).

There are other reasons for the use of these stimuli as well. For example, the looking-time methodologies used with infants demand control for contact, event length, and agent appearance within stimuli, as does work with adults specifically focused on the examination of component action elements and the manipulation of intentionality. Further, the understanding of these stimuli in infancy would indicate an early developing representation of these actions that is abstracted away from specific observed human actions.

Early Interpretation of Helping and Hindering Action

The importance of helping and hindering would suggest that detection and basic interpretation might begin early in development, creating a foundation for effective social interaction within and outside of the family unit. Some likely requisite abilities are already present during the first year-and-a-half of life. For example, infants detect individual goals (e.g., Woodward, 1998) even when goal actions fail (e.g., Meltzoff, 1995), and they recognize that goals are specific to individuals (Buresh & Woodward, 2007). Indeed, by fourteen- to eighteen-months, toddlers are already engaging in their own helping actions (e.g., Dunfield, O'Connell, Kuhlmeier, & Kelley, 2011; Warneken & Tomasello, 2006; Zahn-Waxler, Radke-Yarrow, Wagner, & Chapman, 1992).

Research with infants is limited by the difficulty in examining the detection of helping and hindering; infants cannot be asked to directly report which movements convey the percept of either action. Instead, the suggestion that these actions are detected by infants in a manner that has some similarity to the manner in which adults detect them comes from examinations of how infants interpret the actions. Interpretation that is consistent with adult interpretation of the same actions and agents presupposes the detection of the helping and hindering actions.

In the first study if its kind, Premack and Premack (1997) reported that twelve-month-old infants recognize the underlying valence of helping and hindering behavior as positive and negative, respectively. Infants were habituated[1] to one of four interactions displayed by animated circles:

helping (one ball lifted and pushed the second, enabling it to exit a door), hindering (one ball prevented the other from exiting the door), caressing, and hitting. To the adult observer, these events can be categorized at different levels. At one level, the helping and hindering events both consist of action showing intention to exit a door (and, for that matter, the presence of a door), and the hitting and caressing events both depict the approach of one agent toward another in an otherwise empty scene. At another level, the events could also be categorized by valence, such that helping and caressing share a sense of positivity, and hindering and hitting share negativity. The authors proposed that infants categorized the events by valence since the infants showed dishabituation to the hitting event if habituated to either helping or caressing, but not if habituated to hindering or hitting.

Premack and Premack's seminal paper raises some intriguing questions. What constitutes a valence attribution? Do the infants experience a positive emotional state when viewing helping and caressing actions? Existing data do not speak to this. Do the infants in some way prefer the helping and caressing actions to hindering and hitting? There was no evidence for a visual preference (i.e., longer looking duration) for helping and caressing when these events were first presented in the habituation phase; yet a general preference might not necessarily lead to this particular type of looking behavior. What do infants think about the agents involved in the helping and hindering actions? Here, recent research, described in the following section, provides some answers.

Disposition Attribution

At least for adults, the interpretation of goal-directed actions is often elaborated by attributing the dispositions from which the actions originate. A goal is, thus, a "particular state of affairs that an agent wants to achieve" (Luo & Baillargeon, 2007, p. 490), whereas a disposition refers to the "tendency or state that helps to explain *why* [italics added] an agent may choose to pursue a particular goal or engage in a particular activity" (p. 491). We may observe that an agent has a goal of reaching for a cookie, and we might also posit that this goal is held because of the agent's positive disposition toward (i.e., liking) cookies.

To examine infants' attribution of dispositions to the recipients of helping and hindering, we created computer-animated events depicting, for example, a square helping a circle and a triangle hindering that same circle. We reasoned that after observing such events, infants, like adults, may recognize that the circle should have a positive disposition toward the helper and a negative disposition toward the hinderer. In turn, we examined whether infants would now have an expectation regarding the circle's behavior in a novel situation based on the past experiences with the helper and the hinderer (see Rosati et al., 2001, for a review of tasks and scenarios used to test trait attribution in older children).

In one study following this design (Kuhlmeier, Wynn, & Bloom, 2004), we showed infants events depicting a red circle with eyes and a nose attempting to climb a hill. On some attempts, a yellow square went behind the circle and helped it up, and in others, a green triangle moved to the front of the circle and pushed (hindered) it down. Infants then watched two types of test events in which, with the hillside no longer present, the circle approached either the square or the triangle (figure 12.1). Since the test events differed only in terms of whom the circle approached, distinguishing between them would necessitate transfer of dispositional states derived from the helping and hindering scenes to these new approaching events. In other words, to differentiate between the two approaches would require an attribution of the "why" underlying the goal-directed approach behavior. Nine- and twelve-month-old infants, but not five- and six-month-olds, differentiate between the two events; the older infants looked longer at the inconsistent scene in which the circle approached the agent who hindered it rather than helped it (Kuhlmeier, Wynn, & Bloom, 2004; Hamlin, Wynn, & Bloom, 2007, using live-action puppets; also see

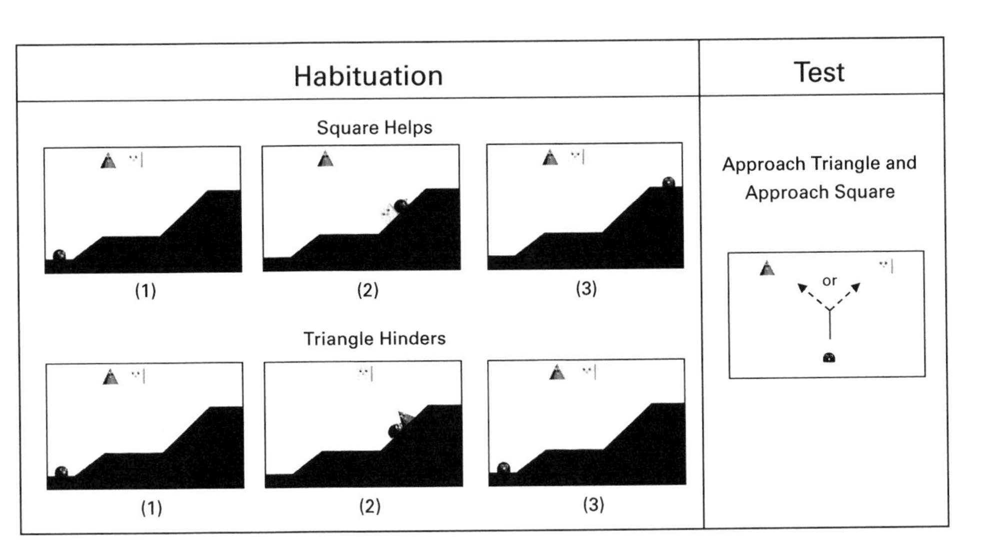

Figure 12.1
Helping and hindering actions, and test events, from Kuhlmeier et al. (2004).

Kuhlmeier, Wynn, & Bloom, 2003, for results with simple, faceless stimuli). Thus, by at least nine months, infants recognize that the circle should be more positively predisposed toward the helper than the hinderer and be more likely to approach the former than the latter.

As adults, we recognize that a disposition resides within an individual—it is specific to that individual. If Jane helps Bob, for instance, Bob might develop the disposition of liking Jane, and he might seek further social interactions with her. Further, Bob might like Jane even if, say, John does not. We have recently examined whether infants also recognize the individual-specificity of dispositions toward helpers and hinderers (Kuhlmeier et al., in preparation). Infants were provided with information about the interactions of two agents (circle and triangle) with a third agent (square): One agent was helped by the square, and one was hindered. Do infants track the different experiences of the circle and triangle, attribute specific dispositions toward the square for each, and consider that their subsequent actions toward the square should, in fact, differ?

In two different test trials, the circle and triangle each approach the square. If infants recognize that the circle and triangle should hold different dispositions (e.g., positive and negative) toward the square because of being either helped or hindered by the square, they should find the scene in which an agent approached its helper to be consistent, and scenes in which an agent approached its hinderer to be inconsistent, and looking time would differ between the two trials. Indeed, fourteen-month-olds looked longer at events in which the agent who was hindered approached the square.

My colleagues and I suggest that infants are attributing distinct dispositions to each of the climbers when observing the helping and hindering interactions (e.g., "The circle does not like the square" and "The triangle likes the square"). The subsequent actions in the second environment are then interpreted by tracking these individual dispositions and applying a preexisting rule such as "agents tend to approach things they like and avoid things they don't like." Indeed, infants themselves appear to prefer agents whom they see helping others, which will be discussed in the next section.

Preference for Helping Agents

After witnessing (1) a wooden square (puppeteered from behind a stage) enabling a circle to reach the top of a hill and (2) a triangle hindering the circle's climb, infants as young as nine months reach for the square more often than the triangle (Hamlin et al., 2007). Interestingly, in a control condition, no preference is found for squares that sit passively to the side of the hill while the circle completes or fails to complete its climb. (Active, causal connection with the achievement of the circle's goal is apparently relevant to subsequent choice behavior.) Similar results are found even when using multiple types of helping and hindering events (Hamlin & Wynn, 2011). Thus, these researchers suggest that infants engage in social evaluation when viewing basic helping and hindering events. That is, some agents, like the helper, are preferred and chosen over those that hinder. Again, as in the work on disposition attribution discussed in the previous section, infants' behavior is consistent with how adults might consider each of the characters involved in a helping and hindering action.

Is It Truly a Helper?

Claiming that infant interpretation is consistent with adult interpretation of the same actions and agents presupposes the *detection* of the helping and hindering actions, and in the studies presented in the previous section, infant looking time and reaching behavior seem to fit basic adult construals. It seems likely that infants are detecting an interaction between two agents in which the second agent either enables the goal completion of the first or hinders achievement of that goal. In this way, it can be said that infants "see" helping and hindering events as such. However, even though infants seem to (1) apply a sense of valence to the actions, (2) make assumptions about the likely subsequent behavior of the agent who was helped/hindered toward the helper and hinderer, and (3) show a preference for helpers (i.e., behave appropriately given basic evaluations that seem logical to the adult observer), much is still not known about how sophisticated infant interpretation of helping and hindering behavior really is.

It may be tempting to assume, for example, that infants also attribute to the *helper* a positive disposition toward the *helped* (circle), but this question remains untested. One thing that is known thus far is that infants do not react differently to test movies in which the helper or hinderer subsequently approaches the circle in another context. Kuhlmeier et al. (2003) report a control condition in which infants look equally when the triangle or square approach the circle; however, this condition offers limited evidence for the current question not only because it is a null result, but also because even adults may find it unsurprising for a hinderer to approach his former victim (perhaps to hinder again). Thus, we simply do not know whether infants attribute positive dispositions (e.g., preferences), or even traits such as helpfulness or generosity, to helpers toward the targets of the helping behavior.

Although we know that adults typically conclude that even the stylized, animated helpers in these scenarios are "good guys" and the hinderers "bad guys" (McAndrew, 2013), we do not yet know if infants socially perceive them in this way. It is possible that the agents are seen as more akin to tools that enable goal completion. That is, infants may prefer goal enablers and assume that those who are enabled prefer them as well. This is not just a trivial semantic quibble. It has implications for whether infants would, say, still prefer those who try but fail to help (or even accidental helpers who happen to enable goal completion), or whether they have expectations that a previously observed helper will continue to be helpful in other contexts, both of which are important interpretations that would likely affect evaluation of effective cooperators in one's social world.

Recent research suggests that it is possible that infants perceive helpers as something more than tools, though this work has been done with human actors instead of animated agents, and the participants were older (21 months; Dunfield & Kuhlmeier, 2010). Toddlers showed elements of reciprocity in their own helping behavior toward previously helpful actors, and, importantly, it seems that the intention behind an actor's actions was what mattered to toddlers, and not just the enabling of goal attainment.

The study described by Dunfield and Kuhlmeier (2010) was founded on the widely accepted proposal that one way to maintain helping between unrelated individuals is by monitoring and remembering past interactions and selectively providing aid accordingly (reciprocal altruism; e.g., Trivers, 1971). By providing aid to those who have helped you, the costliness of providing help to an individual who may not help back can be avoided. Dunfield and Kuhlmeier (2010) tested whether intentions or outcomes of previous interactions with two unfamiliar adults might mediate toddlers' subsequent choice of one adult as the recipient of their aid (i.e., retrieving an out-of-reach object). Toddlers selectively helped an individual who, in a previous interaction, intended to provide a desired toy over one who did not; furthermore, infants considered this positive intention even without a positive outcome (e.g., if the actor tried but failed to deliver the toy).

Although this study was primarily designed to examine early selectivity in helping behavior, it also begins to address how toddlers construe individuals who are willing to provide. It is possible that toddlers attributed dispositions to the actors who displayed the intention to provide that served to define them as good partners—those with whom it might be beneficial to enter into a reciprocal, helping relationship. However, the nature of the concept of "helper" for younger infants still requires further study.

On the Detection of Helping and Hindering in Infancy

If we are to accept that infants detect interactions in which one agent enables or hinders the goal completion of another agent, then a reasonable question is how the actions are processed to lead to this percept. As mentioned in the previous section, there is no direct research addressing this question, as most of the research with infants has focused on the interpretation of events vis-à-vis the agents themselves (e.g., attributing dispositions to the helped/hindered agent, infant preferences for the helping agent). However, existing models of infant goal attribution provide at least two possibilities.

One possibility, derived from work by Sommerville and colleagues (Sommerville, Hildebrand, & Crane, 2008; Sommerville & Woodward, 2005; Sommerville, Woodward, & Needham, 2005), is that infants' experiences with helping and hindering directed toward them (e.g., helping to crawl up an incline, or stopping them from crawling toward a particular location) provide action-production information that leads to action understanding and the development of a representational model of these actions. The action representation may, for example, be informed by the production of goal-directed action on the part of the infant (e.g., trying to crawl in a certain direction) coupled with the dispositional response to any agent that instrumentally acts to enable or hinder the goal. This possibility, however, gets muddied when one tries to incorporate an understanding of the goal of the helping agent; this is because infants themselves are not readily producing helping actions at this age, and thus the helping goal representation would not have self-action production to support it. Yet, as raised earlier, it is possible that the understanding of the helper's goal is limited, such that infants only see the helper as a goal enabler and not as an agent with the goal of enabling another's goal; experience with producing helping actions may subsequently provide a more mature understanding of the goals of both agents (actor and recipient) in a helping or hindering event.

The second possibility regarding the early detection of helping and hindering stems from the teleological stance, as detailed by Gergely, Csibra, and colleagues (e.g., Csibra & Gergely, 1998; Csibra et al., 1999; Gergely et al., 1995). In this account, the ability to recognize a goal-directed action is not supported primarily by one's own experience with action production but, instead, by a cognitive system that works with a rationality principle. That is, actions are seen as goal directed if the motion path and end state fit the environmental constraints in an efficient, rational manner (e.g., an agent jumps over a barrier to get to another agent, but does not perform the jumping motion if no barrier is present). The teleological stance even has predictive power; if only motion and environmental information are present, a likely end state can be posited (e.g., Csibra, Bíró, Koós, & Gergely, 2003).

In the hillside events of Kuhlmeier et al. (e.g., 2004) and Hamlin et al. (2007) described above, the action of the circle is rational for an end state of getting up the hill (i.e., repeated forward motion upward), and the actions of the helper and hinderer are rational for the end state of pushing the circle up and down, respectively. Thus, behavioral representations of, say, helping could be created if one posits that the infant can simultaneously consider the rationality information regarding both the circle and square agents. The teleological stance may thus form the *foundation* of subsequent disposition attribution—though as currently defined, would not itself provide it. (The reader is also directed to the chapter by Frankenhuis and Barrett in this volume, in which they present an alternative to the teleological stance as originally defined by Gergely, Csibra, and colleagues, instead considering the existence of multiple, specific action schemas that each hold teleological explanatory systems.)

Adult Detection and Interpretation of Helping and Hindering Action

The fields of developmental psychology, social psychology, experimental philosophy, and social cognitive neuroscience have recently provided detailed analysis and theory regarding moral judgment, including how adults and older children interpret harmful and helpful actions (e.g., Haidt, 2001; Haidt & Kesebir, 2010; Knobe, 2005; Spinrad, Eisenberg, & Bernt, 2007; Young, Cushman, Hauser, & Saxe, 2007; Young & Saxe, 2009). While acknowledging the importance and productive interdisciplinary approach of that work, I focus here on the adult detection and interpretation of stripped-down, stylized instances of helping and hindering similar to those used in the infant research described above. The purpose for this is twofold: (1) Research findings with adults may then be more precisely compared with those from infants and young children, in turn aiding developmental studies; and (2) the abstract, representational nature of the interactions of computer-animated stimuli allow for control over the observer's previous experience with the agents, the agents' general appearance, the presence of emotion expressions, and the duration of events, among other features.

What are the aspects of the agents' behavior that give rise to the perception of helping and hindering in stimuli such as these? At this time, there is little research examining either the mechanisms underlying the detection of helping and hindering by adults or the interpretation of the events. On the one hand, the lack of work on the interpretation of these stylized events may be because there is no real need to study it. For example, the movie events described earlier were designed so as to deliver a certain percept and cause a likely disposition attribution (e.g., the helped object will like the helper and dislike the hinderer). Pilot tests in my own lab with adults verify that this is the case; almost everyone agrees with the animator. On the other hand, the fairly universal interpretation of the actions does allow for the examination of subtle manipulations that may alter interpretations and, in the case of differing interpretations, even suggest important individual differences. Thus, both detection and interpretation are discussed in the next sections.

Did You See Helping (Hindering)?

As discussed in the first section of this volume, much work has been done on how adults detect biological motion. Research has focused, for example, on local and global processing of point-light biological motion and the specifics of the ballistic velocity signature of this movement (e.g., Johansson, 1973; Pavlova & Sokolov, 2000; Troje, 2002). The (limited) work examining the detection of helping and hindering has taken a different form, however, with researchers focusing on the fact that helping and hindering are goal-directed social interactions between two agents. Here, I consider two approaches in the existent literature on the detection of social goals—one that is often characterized as "bottom-up, motion cue based," and the other as "theory-based" (Baker, Goodman, & Tenenbaum, 2008).

Blythe and colleagues (e.g., Blythe, Todd, & Miller, 1999; Barrett, Todd, Miller, & Blythe, 2005) have posited that some types of basic social interactions are important and frequent enough that humans (and possibly other animals) have adaptive, domain-specific motion perception that is tuned to each interactions' corresponding action pattern. The prototypical movement patterns (e.g., changes in direction or velocity) of an action such as chasing, for example, provide visual cues that are computed through bottom-up perceptual processes; the movements can be then be quickly and frugally categorized as chasing. In this model, intention is provided directly from motion trajectories that fit a stereotypical category, and there is no need for higher cognitive processing. These researchers have focused primarily on actions such as chasing, fleeing, playing, courting, and fighting, but they predict that this general model may apply to helping and hindering behaviors as well (see, e.g., Frankenhuis and Barrett, this volume).

An alternative approach is that taken by Tenenbaum and colleagues, who consider a more cognitive, top-down process for goal attribution, including the social-goal attribution that accompanies the detection of helping and hindering in adults (Baker et al., 2008; Baker, Saxe, & Tenenbaum, 2009; Ullman et al., 2010). Under this account, detection is not based solely on perceptual motion cues, but comes from the workings of a cognitive system that considers the causal mental states that likely underlie the motion that is observed. Like Dennett (1987) and Gergely, Csibra, and colleagues (Csibra & Gergely, 1998; Csibra et al., 1999; described above), this group posits that humans have an intuitive theory of psychology that includes the principle of rationality, such that agents are assumed to engage in action that is efficient given their beliefs about the world.[2] One difference from the previous models, however, is that this is not posited to be a deductive system. It is instead a probabilistic system, which means that detection should become more confident with increased observation.

Ullman et al. (2010) formalized the social goals of helping and hindering into a computational framework for inference that was based on this proposed cognitive system. The model effectively predicted goals in computer-animated helping and hindering events in a maze—even when environments changed slightly. Further, it correlated well with adults' judgments, and, in some cases, it outperformed a computational model based on motion cues alone, like that of Blythe and colleagues (e.g., Blythe et al., 1999). These models, however, are only in their initial stages, and further elaboration and augmentation will likely occur.

Indeed, is possible that the detection of some social actions, such as chasing, may require a fundamentally different process from the one used for detecting the social goals of helping/hindering. The general action of chasing is an interaction consisting of following and attempting to avoid. The exact motion patterns and subgoals (e.g., turn right, hide behind barrier) may differ given the environmental constraints, but the basic approach-and-avoid motion structure stays the same. Helping, we have argued (e.g., Dunfield & Kuhlmeier, forthcoming), is an enabling response to an instrumental need in which a recipient is initially unable to complete a goal-directed behavior. Thus, the helper's goal behaviors can range widely from those involving contact motion (e.g., pushing up an incline), to those altering the physical environment to enable motion (e.g., removing an obstacle or fixing a flat tire), to those involving only communicative behavior (e.g., sharing information). Detecting and categorizing helping and hindering would likely require flexibility of a sort that a system based on motion cues alone might not provide. Also, it is also possible that a detection system used in infancy may differ, either quantitatively or qualitatively, from one used by adults.

Remembering Helpers and Hinderers

Detecting helping and hindering events allows for detection of particular agents who are helpers and those who are hinderers. As mentioned earlier (see p. 290), remembering those who have shown a propensity to help allows for selective provision of aid (e.g., Trivers, 1971), and noting hinderers allows for strategic avoidance. Recently, we examined adults' ability to recognize helpers and hinderers over a time delay (Camilleri et al., 2010). The project was inspired by previous work on detecting and remembering cheaters—that is, those who break a defined social contract (e.g., Barclay & Lalumière, 2006; Cosmides & Tooby, 1989; Mealey, Daood, & Krage, 1996). In the exploration of memory for helpers and hinderers, we opted to utilize the hillside events previously used in infant studies (e.g., Kuhlmeier et al., 2003) instead of verbal vignettes such as those used in cheater-detection studies. This allowed for matching of helping and hindering events on many dimensions, and we were able to avoid concerns that had been raised in cheater-detection work regarding attempts to equate the severity of cheating and cooperating (see Barclay & Lalumière, 2006).

Adult participants watched eight movies. Four of the movies depicted triangles of varying colors pushing (helping) a circle up the hillside, and the other four movies depicted triangles hindering the circle. Importantly, half of the participants saw the depicted helping and hindering behavior as intentional, and half saw it as unintentional (i.e., the triangles fell from a ledge in such a manner as to push the circle down or up). We tested the participants' memory for the helpers and hinderers by showing them sixteen triangles (eight previously-viewed colors and eight new colors) one at a time. Each was followed by the question, "Do you remember this triangle from the previous videos?"

We examined (1) whether memory for helpers differed from memory for hinderers and (2) whether the intentional nature of the helping/hindering action made a difference. We found no evidence for a difference in memory for agents who previously helped compared to agents who previously hindered the circle; this is consistent with a growing body of research suggesting that enhanced memory for recalling either cheaters or hinderers over altruists or helpers (or vice versa), if existent, is subtle (e .g., Barclay & Lalumière, 2006; Mehl & Buchner, 2008).

We did find that the intentionality of helping and hindering actions affected memory of the triangles. In both the intentional and unintentional events, the goal of the circle was ultimately aided or thwarted; yet memory for the helpers and hinderers was enhanced when their actions were intentional. An important consideration is that this effect is not confounded with other factors (e.g., the goal of the circle and its final end states remain consistent, all physical features of the characters remain identical, the duration of contact between characters is equal); the only difference between the movies in the two conditions was the intention of the agents to help or hinder. It is possible that participants encoded and remembered these agents more because they were intentional helpers and hinderers, and thus their actions were more predictive of future helpful or unhelpful actions. Alternatively, it is possible that enhanced memory was observed because, in general, intentional objects may simply engender more interest. Yet participants in the unintentional condition did pay enough attention to the unintentional agents to remember them at above-chance levels. Also, recent research has found no difference in memory for human actors involved in intentional vs. accidental actions unrelated to helping and hindering (Fausey & Boroditsky, 2011); this suggests that helping and hindering may be particularly important actions.

Our work also pointed to the importance of considering individual differences in the encoding of helping and hindering agents. We included the Levenson Psychopathy Scale (LPS; Levenson, Kiehl, & Fitzpatrick, 1995), which is a self-report psychopathy measure often used in nonclinical settings. Interestingly, we found that psychopathy was related to memory for helpers, but not for hinderers. Those who scored higher on the self-report measure showed increased memory for helpers. Psychopaths are considered social predators who manipulate and exploit others without regret for harm; they may therefore be predisposed to attend to and remember people who are particularly exploitable, such as those who help (e.g., Barclay & Lalumière, 2006). Our sample likely did not contain any true psychopaths, however, and can only speak to possible subtle individual differences in encoding. If the hypothesis of helper and/or hinderer detection among psychopaths is true, we expect larger and more robust effects in studies that include clinical psychopaths. Until such studies are conducted, we remain cautious, yet intrigued, about the effects of psychopathy on the encoding of helpers.

Adult Interpretations of Helping and Hindering: Some Considerations

Not only may there be individual differences in remembering helpers and hinderers, but there may be individual differences in the interpretation of the actions—even in stylized computer-animated representations of the actions. In informal questioning, most people describe a hindering event, such as that depicted in the hillside movies, as an instance of one agent stopping the progress of another and being mean or "a bad guy." However, occasionally a response stands out, like one from a teenage participant who described the hinderer as a "teacher" who thought that the circle ("a student") wasn't ready to get to the top of the hill. So, yes, the shape hindered, but it did not have a malicious motive in this observer's construal.

There is little research on the adult interpretation of animated helping and hindering events such as these, as the majority of work has focused solely on the detection of the actions (e.g., pilot data presented in Kuhlmeier et al., 2003; Ullman et al., 2010). Heider and Simmel (1944) documented elaborate interpretations of the social interactions in their animated events, but these depicted much more complicated stories that did not contain instances of instrumental aid or hindrance. Yet this and subsequent work on social inference in the field of social psychology (typically using live action and story vignettes) might point to important directions for researching the social perception of helping and hindering in terms of action interpretation.

Research has focused on the shared theories that give rise to inferences about others' actions (e.g., Heider, 1958; Kelley, 1973) as well as the differences in theories that may affect interpretations (e.g., Choi, Nisbett, & Norenzayan, 1999; Molden, Plaks, & Dweck, 2006). There are a number of ways that individual's theories of the social world might differ (Morris, Ames, & Knowles, 2001), which in turn may lead to differences in the interpretation of helping and hindering actions. Some candidate influences include differences in beliefs regarding whether a person's attributes are stable or dynamic (e.g., entity vs. incremental theorists, Molden et al., 2006) or differences in the belief in a just world (e.g., Lipkus, 1991). The former, for example, might lead to interpretations based more on the agents' traits (entity theorist) or the situation (incremental theorist). For the latter, those with a strong belief in a just world might possibly posit reasons to justify hindering (e.g., the victim is in some way deserving of hindrance). Future work on the interpretation of helping and hindering that takes into account both similarities and differences across observers will thus further our knowledge of the social perception of these actions and, in turn, may extend our knowledge of the social theories that people hold.

Conclusions

For humans, a species with a complex and dynamic social structure, the detection and interpretation of social goals are likely important cognitive processes. Of these social goals, the perception of helping and hindering is particularly integral to the maintenance of our species-unique cooperative interactions. Basic, foundational elements of this perception are present within the latter part of the first year of life, even before physical development allows for the enactment of the behaviors. Yet, some exemplars of helping and hindering may be detected and understood earlier than others, as the recognition of a need—and the appropriate intervention—requires an understanding of the relevant physical or mental constraints in any given situation. For now, one can say that in the first year of life, infants appear to recognize the enabling or blocking of goal-directed motion on an inclined plane (e.g., Kuhlmeier et al., 2004) and perhaps the blocking of access (e.g., Hamlin & Wynn, 2011; Premack & Premack, 1997); the recognition of other instances of helping and hindering may occur later in development. Indeed, what is learned from studies of early development may influence models of social-goal perception in adults, and vice versa.

A final thought is in order regarding the study of helping and hindering perception with stylized, evolutionarily novel stimuli. Any of the models presented here would propose that the mechanisms involved in the detection of the actions would consider certain cues, but not all available features; the rationality of the motions within the environmental constraints might be informative for detecting, say, helping, but the detection of helping does not likely rely on identification of the agents' hair color. Using stylized stimuli that can specifically control motion cues is thus integral to examining detection. Such stimuli also appear to provide information for the interpretation of the actions. Even though the agents are simple, infant and adult observers readily remember the agents' specific roles and attribute dispositions. Of course, interpretations may become richer and more nuanced—and show more individual differences—when displays are set within more naturalistic scenes (e.g., with emotional cues or indicators of agents' race, in-group status, age). That said, the human brain appears to be ready, at an early age, to create a meaningful understanding of even sparse representations of helping and hindering.

Notes

1. Habituation looking-time methodology presents infants with repeated exposure to a stimulus until looking time decreases to a predetermined criterion. Stimuli that are either similar or different on the dimension of interest are then presented, and looking-time increases to the dissimilar stimulus are examined.

2. Gergely, Csibra, and colleagues talk about this slightly differently in their account of infant goal attribution, as they are not positing that infants represent agents' belief states (e.g., Csibra & Gergely, 1998). Instead, their model is based on the rational relation among action, situational constraints, and end states. Additionally, they typically use the phrase "teleological stance" instead of "intentional stance," which is used by Dennett (1987) and Baker et al. (2009).

References

Baker, C. L., Goodman, N. D., & Tenenbaum, J. B. (2008). Theory-based social goal inference. In *Proceedings of the Thirtieth Annual Conference of the Cognitive Science Society*, 1447–1452. Austin, TX: Cognitive Science Society.

Baker, C. L., Saxe, R., & Tenenbaum, J. B. (2009). Action understanding as inverse planning. *Cognition, 113*, 329–349.

Barclay, P., & Lalumière, M. L. (2006). Do people differentially remember cheaters? *Human Nature, 17*, 98–113.

Barrett, H. C., Todd, P. M., Miller, G. F., & Blythe, P. W. (2005). Accurate judgments of intention from motion cues alone: A cross-cultural study. *Evolution and Human Behavior, 26*, 313–331.

Blythe, P. W., Todd, P. M., & Miller, G. F. (1999). How motion reveals intention: Categorizing social interactions. In G. Gigerenzer, P. M. Todd, & the ABC Research Group (Eds.), *Simple heuristics that make us smart* (pp. 257–286). New York: Oxford University Press.

Brown, W. M., & Moore, C. (2000). Is prospective altruist-detection an evolved solution to the adaptive problem of subtle cheating in cooperative ventures? Supportive evidence using the Wason selection task. *Evolution and Human Behavior, 21*, 25–37.

Buresh, J. S., & Woodward, A. (2007). Infants track action goals within and across agents. *Cognition, 104*, 287–314.

Camilleri, J. A., Kuhlmeier, V. A., & Chu, J. Y. Y. (2010). Recognizing helpers and hinderers depends on behavioral intentions of the character and psychopathic characteristics of the observer. *Evolutionary Psychology, 8*, 303–316.

Choi, I., Nisbett, R. E., & Norenzayan, A. (1999). Causal attribution across cultures: Variation and universality. *Psychological Bulletin, 125*, 47–63.

Cosmides, L., & Tooby, J. (1989). Evolutionary psychology and the generation of culture, part II: Case study: A computational theory of social exchange. *Ethology and Sociobiology, 10*, 51–97.

Cosmides, L., & Tooby, J. (1992). Cognitive adaptations for social exchange. In J. Barkow, L. Cosmides, & J. Tooby (Eds.), *The adapted mind: Evolutionary psychology and the generation of culture*. New York: Oxford University Press.

Csibra, G., Bíró, S., Koós, O., & Gergely, G. (2003). One-year-old infants use teleological representations of actions productively. *Cognitive Science, 27*, 111–133.

Csibra, G., & Gergely, G. (1998). The teleological origins of mentalistic action explanations: A developmental hypothesis. *Developmental Science, 1*, 255–259.

Csibra, G., Gergely, G., Bíró, S., Koós, O., & Brockman, O. (1999). Goal attribution without agency cues: The perception of "pure reason" in infancy. *Cognition, 72*, 237–267.

Dennett, D. C. (1987). *The intentional stance*. Cambridge, MA: MIT Press.

Dunfield, K. A., & Kuhlmeier, V. A. (2010). Intention-mediated selective helping in infancy. *Psychological Science, 21*, 523–527.

Dunfield, K. A., & Kuhlmeier, V. A. (forthcoming). Classifying prosocial behavior: Children's responses to instrumental need, emotional distress, and material desire. *Child Development*.

Fausey, C. M., & Boroditsky, L. (2011). Who dunnit? Cross-linguistic differences in eye-witness memory. *Psychonomic Bulletin & Review, 18*, 150–157.

Gergely, G., Nadasdy, Z., Csibra, G., & Bíró, S. (1995). Taking the intentional stance at 12 months of age. *Cognition, 56*, 165–193.

Haidt, J. (2001). The emotional dog and its rational tail: A social intuitionist approach to moral judgment. *Psychological Review, 108*, 814–834.

Haidt, J., & Kesebir, S. (2010). Morality. In S. Fiske, D. Gilbert, & G. Lindzey (Eds.), *Handbook of social psychology* (5th ed.). Hoboken, NJ: John Wiley.

Hamlin, J. K., & Wynn, K. (2011). Young infants prefer prosocial to antisocial others. *Cognitive Development, 26*, 30–39.

Hamlin, J. K., Wynn, K., & Bloom, P. (2007). Social evaluation by preverbal infants. *Nature, 450*, 557–559.

Heider, F. (1958). *The psychology of interpersonal relations*. New York: Wiley.

Heider, F., & Simmel, M. (1944). An experimental study of apparent behavior. *American Journal of Psychology, 57*, 243–259.

Johansson, G. (1973). Visual perception of biological motion and model for its analysis. *Perception & Psychophysics, 14,* 201–211.

Kelley, H. H. (1973). The process of causal attribution. *American Psychologist, 28,* 107–128.

Knobe, J. (2005). Theory of mind and moral cognition: Exploring the connections. *Trends in Cognitive Sciences, 9,* 357–359.

Kuhlmeier, V.A., Dunfield, K.A., Stewart, J., Wynn, K., and Bloom, P. (in preparation). Infants interpret dispositions as individual attributes.

Kuhlmeier, V. A., Wynn, K., & Bloom, P. (2003). Attribution of dispositional states by 12-month-olds. *Psychological Science, 14,* 402–408.

Kuhlmeier, V. A., Wynn, K., & Bloom, P. (2004). Reasoning about present dispositions based on past interactions. Paper presented at the International Conference on Infant Studies, Chicago, Illinois.

Levenson, M. R., Kiehl, K. A., & Fitzpatrick, C. M. (1995). Assessing psychopathic attributes in a noninstitutionalized population. *Journal of Personality and Social Psychology, 68,* 151–158.

Lipkus, I. (1991). The construction and preliminary validation of a global belief in a just world scale and the exploratory analysis of the multidimensional belief in a just world scale. *Personality and Individual Differences, 12,* 1171–1178.

Luo, Y., & Baillargeon, R. (2005). Can a self-propelled box have a goal? Psychological reasoning in 5-month-old infants. *Psychological Science, 16,* 601–608.

Luo, Y., & Baillargeon, R. (2007). Do 12.5-month-old infants consider what objects others can see when interpreting their actions? *Cognition, 105,* 489–512.

McAndrew, A. 2013. *The social perception of helping and hindering: The role of team bias.* Unpublished houour's thesis, Queen's University, Kingston, Ontario, Canada.

Mealey, L., Daood, C., & Krage, M. (1996). Enhanced memory for faces of cheaters. *Ethology and Sociobiology, 17,* 119–128.

Mehl, B., & Buchner, A. (2008). No enhanced memory for faces of cheaters. *Evolution and Human Behavior, 29,* 35–41.

Meltzoff, A. M. (1995). Understanding the intentions of others: Re-enactments of intended acts by 18-month-old children. *Developmental Psychology, 31,* 838–850.

Molden, D. C., Plaks, J. E., & Dweck, C. S. (2006). "Meaningful" social inferences: Effects of implicit theories on inferential processes. *Journal of Experimental Social Psychology, 42,* 738–752.

Morris, M. W., Ames, D. R., & Knowles, E. D. (2001). What we theorize when we theorize that we theorize: Examining the "implicit theory" construct from a cross disciplinary perspective. In G. B. Moskowitz (Ed.), *Cognitive social psychology: The Princeton Symposium on the legacy and future of social cognition* (pp. 143–161). Mahwah, NJ: Lawrence Erlbaum.

Pavlova, M., & Sokolov, A. (2000). Orientation specificity in biological motion perception. *Perception & Psychophysics, 62*, 889–899.

Premack, D., & Premack, A. J. (1997). Infants attribute value +/– to the goal-directed actions of self-propelled objects. *Journal of Cognitive Neuroscience, 9*, 848–856.

Rosati, A. D., Knowles, E. D., Kalish, C. W., Gopnik, A., Ames, D. R., & Morris, M. W. (2001). The rocky road from acts to dispositions: Insights for attribution theory from developmental research in theories of mind. In B. Malle, L. Moses, & D. Baldwin (Eds.), *Intentions and intentionality: Foundations of social cognition* (pp. 287–303). Cambridge, MA: MIT Press.

Schlottmann, A., & Ray, E. (2010). Goal attribution to schematic animals: Do 6-month-olds perceive biological motion as animate? *Developmental Science, 13*, 1–10.

Sommerville, J. A., Hildebrand, E. A., & Crane, C. C. (2008). Experience matters: The impact of doing versus watching on infants' subsequent perception of tool use events. *Developmental Psychology, 4*, 1249–1256.

Sommerville, J. A., & Woodward, A. L. (2005). Pulling out the intentional structure of human action: The relation between action production and processing in infancy. *Cognition, 5*, 1–30.

Sommerville, J. A., Woodward, A. L., & Needham, A. (2005). Action experience alters 3-month-old infants' perception of others' actions. *Cognition, 6*, B1–B11.

Spinrad, T. L., Eisenberg, N., & Bernt, F. (2007). Introduction of the special issues on moral development: Part 1. *Journal of Genetic Psychology: Research and Theory on Human Development, 168*, 101–104.

Trivers, R. L. (1971). The evolution of reciprocal altruism. *Quarterly Review of Biology, 46*, 189–226.

Troje, N. F. (2002). Decomposing biological motion: A framework for analysis and synthesis of human gait patterns. *Journal of Vision, 2*, 371–387.

Ullman, T. D., Baker, C. L., Macindoe, O., Evans, O., Goodman, N. D., & Tenenbaum, J. B. (2010). Help or hinder: Bayesian models of social goal inference. *Advances in Neural Information Processing Systems, 22*, 1874–1882.

Warneken, F., & Tomasello, M. (2006). Altruistic helping in human infants and young chimpanzees. *Science, 311*, 1301–1303.

Woodward, A. L. (1998). Infants selectively encode the goal object of an actor's reach. *Cognition, 69*, 1–34.

Young, L., Cushman, F., Hauser, M., & Saxe, R. (2007). The neural basis of the interaction between theory of mind and moral judgment. *Proceedings of the National Academy of Sciences of the United States of America, 104*, 8235–8240.

Young, L., & Saxe, R. (2009). An fMRI investigation of spontaneous mental state inference for moral judgment. *Journal of Cognitive Neuroscience, 21*, 1396–1405.

Zahn-Waxler, C., Radke-Yarrow, M., Wagner, E., & Chapman, M. (1992). Development of concerns for others. *Developmental Psychology, 28*, 1038–1047.

III RECOGNIZING AND INTERPRETING GOAL-DIRECTED BEHAVIOR IN HUMAN ACTORS

Section Introduction: Recognizing and Interpreting Goal-Directed Behavior in Human Actors

Jeff Loucks

Without a doubt, human beings in action—running, writing, cooking, dancing—offer some of the most important social stimuli that infants must process. From birth, infants are immediately immersed in a world of dynamic human action; it is one of the most ubiquitous stimuli infants are exposed to, on par with human language. Infants must become efficient processors of human action early on, as their analysis of others' actions underlies their ability to learn from others about human activities and the world more broadly.

However, human action is not a simple stimulus to process. Actions are often carried out rapidly and without pauses; multiple body parts and objects are often involved, and an action may only be visible once for a brief window of time. The ease with which most of us process action therefore seems to imply the use of relatively sophisticated perceptual and cognitive processes. It is only in the last few decades that researchers have begun to explore the nature of these processes in adults and chart the development of this system in infancy.

The following chapters all recognize the importance of studying infants' processing of human action events as it relates to broader aspects of the development of social perception. Baldwin and Sage's chapter highlights the complex, hierarchical nature of human action and indicates that infants possess an early emerging sensitivity to this structure in action. Hauf's chapter examines how changes in self-produced movements influence infants' attention to global properties of human actions. Loucks and Sommerville's chapter investigates changes in infants' attention during action perception and how those attentional changes relate to infants' ability to predict future actions and action outcomes. And finally, Woodward and Cannon's chapter highlights the early development of efficiency in action perception, as manifested in infants' ability to anticipate the goals of other people's action.

All of the research described in these chapters relies on paradigms that capture infants' looking behavior as the dependent measure. The research in Loucks and Sommerville, and some of the research in Woodward and Cannon, relies on classic habituation/dishabituation looking-time paradigms. Baldwin and Sage describe a novel technique in which infants' looking time is measured as they progress through a slideshow of human action sequences. Finally, Hauf, as well as Woodward and Cannon, take advantage of emerging technologies in infant eye-tracking to measure important subtleties about infants' observation of human actions that standard looking-time paradigms might miss.

In addition, many of the chapters also measure some aspect of infants' developing motor capabilities and emphasize the role of motor development in infants' perception of other people's action. There is now a considerable body of research indicating that (1) perception and action are fundamentally intertwined in the human mind and (2) action observation capitalizes on the observer's own motor representations to aid in perception. Importantly, the research presented in these chapters highlights the potential role of motor development as a *mechanism* of change in infants' developing social perception.

13 Dwelling on Action

Dare Baldwin and Kara D. Sage

Abstract

We present a series of studies using a new methodology—the dwell-time paradigm—to investigate observers' sensitivity to structure within unfolding action sequences. Viewers advance at their own pace through slides extracted from digitized videos; dwell times for each slide are recorded. Findings reveal that adults, preschoolers, and even infants display sensitivity to both segmental and hierarchical structure within events. Other work investigates whether dwell times may also reflect infants' and adults' sensitivity to violations of causal structure. In addition, work underway investigates whether social context (e.g., pedagogy) influences both infants' dwell-time patterns and their later successful enactment of a sequence. Although questions remain about the precise way to characterize dwell-time effects, the paradigm shows promise for providing new information about action processing and how it changes with development.

Dwell in possibilities.
—Emily Dickinson

Have you ever leaned expectantly in for a kiss being offered, only to realize that your erstwhile kisser was actually leaning forward just by way of standing up, and not in lip-puckery anticipation? Or have you ever slammed on the brakes while driving to avoid crushing a little live thing, only to realize that your hazardous act of care was made on behalf of a windblown plastic bag? Blushworthy mistakes of intentional and causal inference like these might seem frustratingly common, at least on a bad day, but in fact they are probably vanishingly rare in the context of the multitude of seemingly error-free inferences we make every day. Part of what is striking about their rarity is that the flow of information we witness—that is, the actual movements actors undertake or the relations we actually observe between things in the world across time—would seem to be characterized by a high degree of local ambiguity vis-à-vis causal and intentional inference. For example, someone's act of leaning forward in a chair is consistent with a sizable range of possible intentions (e.g., offering a kiss, standing up, preparing to take a sip from an overfull glass, easing a sore bottom, unsticking one's legs from the chair on a humid day, dodging a fly, general hyperactivity, balancing on a wobbly chair, etc.). This ambiguity raises the question of how we actually execute relatively error-free causal and intentional inferences as we witness events unfold, as well as how infants and children come to be able to do this from an early age (e.g., Baldwin, 1991; Bíró & Leslie, 2007; Carpenter, Akhtar, & Tomasello, 1998; Csibra, 2010; Gergely, Nadasdy, Csibra, & Bíró, 1995; Kuhlmeier, Wynn, & Bloom, 2003; Meltzoff, 1995; Olofson & Baldwin, 2011; Woodward, 1998, 1999; Woodward & Sommerville, 2000). In this chapter we will not try to answer these questions. What we will do instead is introduce a methodology that we have been developing—the dwell-time paradigm—that we hope ultimately will play a helpful role in providing such answers.

To date, methodologies that make it possible to measure how observers process actions as they unfold across time are limited and often challenging to execute. For example, a handful of behavioral techniques have been developed to illuminate how adults (e.g., Newtson, 1973; Newtson & Enquist, 1976; Zacks, Tversky, & Iyer, 2001) and infants (e.g., Baldwin, Baird, Saylor, & Clark, 2001; Hespos, Saylor, & Grossman, 2009, 2011) extract segmental and hierarchical structure, but the information these techniques provide tends to be restrictive. For example, most demonstrations of infants' sensitivity to segmental structure hinge on the use of classic looking-time paradigms, such as the habituation and violation of expectation paradigms. While looking-time data have been extremely useful in documenting the fundamental fact that infants are sensitive to segmental structure within at least some kinds of everyday intentional action, they clarify relatively little about how infants' processing of such structure actually unfolds across time. Regarding research with adults, existing behavioral techniques provide some information about the nature of the segmental structure adults recover, but provide relatively little information about how such structure is detected within unfolding events. Moreover, technologies such as eye-tracking—potentially useful for clarifying how adults allocate attention during action processing and how they detect relevant structure—are prohibitively challenging to mount with streaming video. Finally, behavioral techniques used with adults have not been amenable to research with preschoolers or infants, and vice versa, limiting the ability to trace developmental change in action processing.

Our hope is that the dwell-time paradigm, a methodology innovated by Hard and colleagues (Hard, 2006; Hard, Recchia, & Tversky, 2011), will offer a new window into how observers—infant or otherwise—deploy their attention to guide processing as action unfolds across time. We are particularly hopeful that dwell time will be useful across a broad developmental span, facilitating investigation of how action processing changes with development and increasing expertise. In what follows we describe findings from a series of recent studies that speak to the promise of the dwell-time paradigm, and we outline ways in which this method might fruitfully be used in future research.

Dwell Times Index Sensitivity to Segmental and Hierarchical Structure

Offering adult observers the opportunity to view generally familiar intentional action sequences in a self-paced slide-show format, Hard and colleagues recently discovered that the observers' looking times ("dwell-times") reveal sensitivity to both the segmental and hierarchical structure of the unfolding event sequence. In this research, digitized videos were employed depicting four different everyday intentional action sequences (organizing a bedroom, eating at a cafeteria, building a bookshelf, and applying makeup). Still frames were extracted from each video at a constant increment, such as two frames per second. Observers clicked through the frames at their own pace, and their dwell times for each slide were measured. Subsequently, viewers were asked to (1) watch the videos from which the frames had been extracted and (2) nominate meaningful junctures within the action—termed breakpoints—following Newtson's (1973) classic segmentation method. Viewers performed this segmentation task with the video several times to provide segment-boundary judgments at several levels of analysis (i.e., fine-, intermediate-, and coarse-grained levels). Because dwell times tended to decrease across time in accordance with a power function, each individual's dwell times were fit to a power function, and dwell time scores were calculated from the resultant residuals. Analyses then examined relations between these dwell-time scores and the segmentation judgments viewers themselves had provided. Hard and colleagues found that adult observers dwelt longer on slides coinciding with segment boundaries (e.g., the region within action at which one goal is completed and the next is initiated), and this effect increased as segmentation level increased. That is, coarse-grained segment boundaries (e.g., boundaries between large-scale activities, such as transitioning from washing dishes to folding laundry in a housecleaning scenario) received longer dwell times than intermediate (e.g., transitions between activities such as loading the top level of a dishwasher and loading the bottom level of the dishwasher) or fine-grained boundaries (e.g., transitions between activities such as grasping a dish and setting a dish down). One interpretation of these findings, proposed by Hard and colleagues, is that transitions from one segment to the next serve as bridges to conceptual understanding during event processing, and adults thus allocate increased attention to them. Coarse-grained boundaries are especially information rich in these terms and thus receive particular attention.

Hard and colleagues also tested an alternative account that considered the possibility that breakpoints receive enhanced attention simply because they are inherently more attention attracting in some way. If this is correct, dwell times should increase at segment boundaries even when frames from an event are presented in scrambled order. As it happens, Hard et al. found that this is indeed the case. According to their data, however, that is not the full story. That is, dwell times increase at breakpoints to a greater degree when breakpoint frames occur in the context of an unfolding event relative to when they occur in scrambled order. This finding helps to confirm that enhanced breakpoint dwell times indeed reflect event processing that goes beyond a simple response to salient portions of the event stream.

A related but slightly different question is whether there might be a low-level, perceptually based correlate of breakpoints; such a correlate might provide a direct explanation for the dwell-time surges that tend to occur at breakpoints. Hard et al. tested this possibility by developing an objective measure of one important source of perceptual variation between slides—the degree of position change from one slide to the next. To derive this position-change index, edge-extraction (a convolution filter) was applied to each slide to isolate edges of the actor, and the average degree of pixelation change from each slide to the immediately subsequent slide was then computed relative to these edges. These average pixilation differences thus largely reflected pixelation change resulting from position change arising from the actor's movements. Of particular interest was whether average position change would be greater as slides preceding breakpoints transitioned to breakpoint slides themselves, relative to slides occurring midstream transitioning to other slides midstream within the same action segment. As it turned out, position change was in fact enhanced at breakpoints in this analysis. Moreover, position change was found to be greatest for breakpoints at a coarse-grained level relative to finer-grained levels.

This finding raises the possibility that dwell-time increases at breakpoints (seemingly indicative of segmentation) and at coarse-grained, relative to finer-grained, breakpoints (seemingly indicative of hierarchical organization) might instead be a straightforward response to magnitude of position change. In other words, perhaps dwell times increased at these junctures merely because observers were more attentive whenever position change was sizable. Against this, the fact that breakpoints garnered longer dwell times when these frames were presented in the context of unfolding events relative to when they were presented in scrambled order might, on the face of it, seem to suggest that position change could not fully account for dwell-time surges. However, it turned out to be the case in the Hard et al. stimuli that *position change* was also greater when slides were presented in context relative to being presented in scrambled order. Putting this all together, the Hard et al. findings raise the intriguing possibility that dwell times index viewers' extraction of segmental and hierarchical structure, but the findings are also subject to an alternative account in which dwell times simply index viewers' response to position change as events unfold.

A recent study of ours (Baldwin, Hard, Meyer, & Sage, in preparation) speaks directly to the position-change issue and shifts the balance away from position change as a complete account of dwell-time patterns. In particular, this study confirmed that dwell-time effects can be observed for segmental and hierarchical structure when this structure definitely is not grounded in position change. To address these issues, our research utilized a unique unfolding event sequence created for prior research regarding statistical learning of event segments (Baldwin, Andersson, Saffran, & Meyer, 2008; Meyer & Baldwin, 2011). Details of the methodology and findings from the prior research were important to our conceptualization of the new work to be described, and thus we will describe the earlier work in some detail.

Baldwin and colleagues' (2008) research aimed to investigate whether adults can discover segmental structure within a novel event sequence based solely on statistical regularities among the motion elements comprising that sequence. Prior work had demonstrated just such a role for statistical learning in bootstrapping adults' and infants' extraction of segmental structure in the language domain (e.g., Saffran, Aslin, & Newport, 1996) and with static visual arrays (e.g., Fiser & Aslin, 2002, 2005); thus, Baldwin and colleagues were curious whether statistical learning might likewise support detection of segmental structure within dynamic human action. To make this idea concrete, imagine an event scenario in which a woman

(1) grasps a dog by the collar, (2) grasps a leash by the fastener, (3) links the leash fastener to the collar, then (4) grasps a hat, (5) places it on her head, and (6) tucks in her hair. In this naturally flowing action sequence, some of the motion elements predict subsequent ones at high rates (e.g., grasp leash, fasten leash) whereas others predict subsequent motion elements to a lesser degree (e.g., fasten leash, grasp hat). The idea here is that high transitional probabilities tend to link motion elements that fall within an action segment (leashing a dog), whereas low transitional probabilities potentially signal where boundaries might occur between distinct action segments (e.g., leashing a dog and donning a hat).

To test this idea, Baldwin and colleagues created a novel event stream specifically designed such that statistical regularities were the only source of information available vis-à-vis segment boundaries; thus, prior knowledge about action, as well as other possible factors (e.g., sensitivity to position change), could not account for discovery of segmental structure within the flow of behavior. This was achieved by creating digital videos of a small inventory of motion elements, all involving a bottle and several other objects, each of which began and ended with the actor in the same position. The twelve motion elements used were readily recognizable small acts, such as taking a drink from the bottle, inspecting the bottom of the bottle, shaking the bottle near the ear, and so forth. Because action began and ended with the actor in the same neutral position, it was possible to string motion elements together in any possible order; the flow of action would appear causally possible and relatively natural, but also the sequence would seem unfamiliar and nonsensical. In this sense, viewers of the action sequence would have no *a priori* understanding of what intentions the actor might be undertaking within the *sequence* of action; they would have prior understanding for the intentional content only at the level of each individual motion element. Thus top-down knowledge of intention/goal structure could not assist observers in discovering the higher-level segmental structure within the action sequence; again, the only available route to segment discovery at the higher level was the statistical structure of the sequence.

Four triads were randomly selected from the inventory of twelve motion elements to serve as actions, and an exposure corpus—a roughly twenty-minute-long event sequence—was constructed by randomly intermixing the four actions for a total of twenty-eight presentations of each of the four actions. Research participants watched the entire exposure corpus and then provided forced-choice responses to sixteen test trials. In the test trials, they were shown two triads in sequence: one was an "action" as defined statistically in their previous viewing of the exposure corpus, and the other was a foil triad—a set of motion elements not previously presented together. In one study, the foil was a "non-action" (a random combination, which they had previously never seen, of three of the twelve motion elements), and in other studies, the foil was a "part-action" (a triad that combined motion elements spanning two actions, e.g., a motion element from one of the actions with two motion elements from a different action). Participants were asked to nominate the triad within a test pair (e.g., an action versus a non-action in one study, or an action versus a part-action in another study) that struck them as most familiar based on their previous viewing. Across four studies, participants systematically identified the actions as most familiar relative to foils; this indicates that they had discovered the action segments based solely on statistical regularities governing relationships among motion elements. A control study confirmed that viewers indeed detected statistical regularities within action, per se, to discover segmental structure, and they did so not just via recoding into language and detecting statistical regularities across linguistic elements. In particular, in this control study participants succeeded at discriminating actions from part-actions even when their linguistic resources were siphoned off by a secondary and demanding linguistic task.

For present purposes, the detail of interest to us in the Baldwin et al. (2008) research is that it provided a set of stimuli that could be used to investigate whether dwell-time patterns indicative of sensitivity to segmental structure occur even when position change is unavailable as a source of information about segment boundaries or breakpoints. We put the Baldwin et al. (2008) stimuli to work this way in a recent study from which we present preliminary analyses here (Baldwin, Hard, Meyer, & Sage, in prep.). In this research, one group of viewers watched the exposure corpus employed in the Baldwin et al. (2008) study; immediately thereafter, they paged through a self-paced slide show that included slides extracted from the exposure corpus at one-second intervals. Finally, they provided forced-choice nominations (i.e., selecting which among a pair of triads were familiar based on the previous viewing), with part-action triads serving as the foils to "actions" in each pair. Dwell-time patterns during the slide show were then analyzed with respect to whether slides depicted portions of the first, second, or third motion element within an action triad. If dwell-time patterns index segmentation in this task, then they should surge at the first motion element within a triad and decline from there for slides depicting portions of the second and third elements. That is, if segmentation via statistical learning occurred, and dwell times reflected this, then dwell times should increase as a segment boundary occurs (during the first motion element) and decline thereafter as the segment proceeds—only to increase again as the next segment is registered. This dwell-time prediction was confirmed. Moreover, dwell-time patterns bore a clear relation to viewers' ability to successfully perform the forced-choice task that diagnosed their detection of the segmental structure in a different, more standard way. A median split produced a group who performed well at the forced-choice discrimination task, and these participants displayed an especially strong linear trend indicating a tendency to dwell longest on slides depicting a portion of the first motion element, with decreasing dwell times for slides extracted from the second and third motion elements, respectively. Those who performed poorly on the forced-choice discrimination task, in contrast, displayed no such linear trend in their dwell times. Thus dwell-time patterns coincided with performance on the forced-choice discrimination task in terms of indexing whether viewers had successfully extracted the segmental structure from the exposure corpus based on statistical regularities embedded within it.

A second group participated in an important control: They viewed the slide show depicting the exposure corpus—but without any prior viewing of that exposure corpus. Thus, they had no prior opportunity to discover the segmental structure of action segments within the exposure corpus at the time they began the dwell-time procedure. As predicted, this group of participants did not display the linear-trend pattern in their dwell times that we observed with the participants who had first viewed the exposure corpus prior to experiencing the slide show. These findings provide an important confirmation: The linear trend in dwell times, which was observed in the group of participants who first viewed the exposure corpus, did indeed index the segmental structure that participants had discovered as a result of their ability to track the statistical regularities within the exposure corpus.

These findings have several implications. First, they replicate prior findings (e.g., Baldwin et al., 2008) that adult viewers can extract segmental structure via statistical learning. Second, they confirm, with an entirely new set of action sequences, that dwell times index viewers' unfolding processing of segmental structure. Third, given that statistical structure was the only available source of information regarding segmental structure in the action stimuli, the findings point to dwell-time patterns indexing segmental structure without position change playing any role in the detection of segment boundaries. Thus it is possible to safely conclude that dwell times go beyond simply reflecting registration of superficial aspects of the shifting visual display, but instead genuinely illuminate sensitivity to structure—segmental and hierarchical structure in particular—that viewers have discovered within the unfolding event stream. And finally, our findings show that dwell times can index not only the segmental structure of highly familiar action sequences like those used in prior research, but also newly acquired segmental structure (e.g., Hard, Recchia, & Tversky, 2011). These findings thus help to clarify the general usefulness of the dwell-time paradigm to elucidate important aspects of human action processing.

The Dwell-Time Paradigm across the Developmental Gamut

Given the inherent methodological simplicity of the self-paced slide show that enables measurement of dwell time, we wondered whether the same technique might be suitable for studying event processing in young children and possibly even infants. Three studies now speak to considerable promise here. In one study (Meyer, Baldwin, & Sage, 2011), we asked preschoolers to pace themselves, much as we had instructed adults to do in prior studies, through a slide show; the primary difference was that we showed children an event sequence designed to be of interest to preschoolers. In the event, a woman (1) waved, (2) stacked a series of rings, (3) nested a graduated set of cups, (4) placed two stuffed toys in a box, and (5) waved again. We extracted slides from the digitized video of this event at the rate of one per second. After pacing themselves through the slide show by clicking on a computer mouse, children answered a set of memory questions regarding what objects and actions they had seen during their slide show viewing. Analyses revealed findings paralleling those that Hard and colleagues (2011) reported for adults: preschoolers displayed longer dwell times for frames coinciding with breakpoints within the action (as nominated by expert coders) relative to slides depicting points within action segments, and coarse-grained breakpoints received longer dwell times than fine-grained breakpoints. Especially striking was that children's memory for the event sequence made systematic contact with their dwell-time patterns. In particular, children displaying consistently high levels of performance on the memory probe displayed strong dwell-time patterns indicative of segmental and hierarchical structure, whereas dwell-time patterns in children who made errors on the memory test were unsystematic. These latter findings help to clarify that segment- and hierarchy-related dwell-time patterns are related to other aspects of children's cognitive processing, such as their memory for aspects of the event stream, thereby providing additional validation for the psychological reality of these dwell-time effects.

Two studies with infants (e.g., Sage, Ross, & Baldwin, 2012) recently showcased the promise of the dwell-time paradigm for illuminating infant event-processing capabilities as well. Strikingly, infants in these studies are entirely capable of undertaking the self-paced slide show when one simple but important modification was introduced: Rather than clicking a mouse to advance the slides, infants tap a touch screen mounted at an incline just below the computer monitor displaying the slide. A first incarnation of the methodology, in which infants touched the computer monitor displaying the slides, was abandoned because infants often placed their hand in a way that occluded the slide in view. Some important adaptations to the methodology were also needed. Adults and preschoolers tend to watch the monitor attentively throughout the self-paced slide show, meaning that time between mouse clicks serves as a straightforward measure of dwell time. Infants, in contrast, frequently look away from the monitor during the self-paced slide show. This raises some important methodological issues. First, tapping the touch screen when they are looking away from the monitor likely would undercut infants' understanding of the contingency that taps cause a change in slides. This possibility is particularly concerning given that across several studies we have found that, on average, infants look at the slides when their tap advances to the next slide only about 50 percent of the time. As a result, among the modifications we are testing is to modify the MatLab-based software to advance a slide only when both of two conditions are met: infants (1) tap the touch screen and (2) are looking at the monitor. Of course, the fact that infants often look away from a slide during the slide show also means that the latency between infants' taps is not in and of itself an accurate index of dwell time. For infants, frame-by-frame coding of the videotaped procedure is thus needed to ascertain when infants are in fact looking at individual slides, and dwell times are derived from this after-the-fact coding. Even with these improvements, the touch-screen technology is imperfect, leading to error variance that we hope to reduce with future versions of the technology.

In our first study with infants, we gave ten- to thirteen-month-olds the opportunity to advance through slides depicting a simple event designed to be of interest to their age range: a woman stacking a series of child cups. We chose the ten- to thirteen-month age window given prior research indicating that infants this age are sensitive to the segmental structure of at least some everyday action sequences (e.g., Baldwin, Baird, Saylor, & Clark, 2001; Saylor, Baldwin, Baird, & LaBounty, 2007; Hespos, Saylor, & Grossman, 2009; Hespos, Grossman, & Saylor, 2011). Infants received a brief training phase in which they were encouraged by both the experimenter and their parent to tap the touch screen and to notice that this tapping caused a slide to advance. These training slides depicted random attractive objects, such as a puppy and a baby. After ten training slides, infants began the slide show proper, which depicted still frames extracted from the action sequence of interest. This slide show was constructed in a manner that departed in one important way from the method utilized previously with adults and preschoolers: Rather than extracting slides from the digitized video of the event sequence at a regular interval (e.g., one per second), for the infants' slide show we specifically selected still frames that coincided with segment boundaries and alternated those with still frames depicting a point in the midst of a segment. This meant that all breakpoints within the sequence were depicted in the slide show (which was not true of slide shows used with adults and preschoolers, because the regular extraction interval meant that breakpoints sometimes were missed). As well, in the infants' slide show, one, and only one, within-segment point was ever depicted for a given segment. This also was not necessarily true of adults' and preschoolers' slide shows; again, the regular extraction interval used in these prior studies sometimes delivered multiple within-segment slides for a given segment or, sometimes, missed a within-segment slide altogether when action was rapid. Our rationale for making methodological changes in the infants' slide show centered on the fact that the infants' slide show was necessarily brief, involving only eighteen slides. It was important to maximize observations for both breakpoint and within-segment slides, which this alternation method achieved. One consequence of the alternation approach was that the infants' slide show contained an equivalent number of breakpoint and within-segment slides; this was in contrast to prior studies with adults and preschoolers in which the regular extraction rate rendered a slide show with considerably more midsegment than breakpoint slides.

Despite the methodological differences just described, it turned out that infants, like preschoolers and adults, looked longer when slides depicted breakpoints than when they captured a point in the midst of an unfolding action segment. This finding simultaneously provided the first evidence that (1) infants can participate effectively in the dwell-time procedure, and (2) dwell times systematically index their sensitivity to segmental structure within the unfolding event. It is worth noting here that infants' systematic segment-related dwell times tell us only that they upregulate attention to breakpoints relative to within-unit slides. In particular, it is important to recognize that these findings do not provide information regarding the extent to which infants are able to gain a conceptual understanding of the action stream they are observing in the slide show in terms of goals, intentions, and the like. A question of considerable interest for future research is the extent to which infants' sensitivity to the segmental structure of an event is typically a precursor, and perhaps even causally facilitative, of their ability to interpret the event on a conceptual level. The dwell-time paradigm has the potential to assist in investigating these questions. Another important point regarding the findings of this first study is that, unfortunately, it was not possible to examine whether infants' dwell times displayed signs of indexing hierarchical organization of the event structure as indicated by longer dwell times for slides depicting coarse-grained relative to fine-grained breakpoints. The relatively simple event depicted in the slide show effectively included only one internal coarse-grained breakpoint; thus, far too few relevant observations regarding dwell times for coarse-grained breakpoints were produced to undertake an analysis concerning hierarchical organization. One of the goals of a second study was to present infants with a slide show depicting an event with a more complex structure, as this would provide greater opportunity to investigate the issue of infants' possible dwell-time sensitivity to hierarchical event structure. Another important goal was to replicate the basic segmental structure finding with different action sequences.

In the second study, infants advanced through slides depicting a woman, seated at a table, (1) progressively unpacking a toy block from a backpack that was contained in a larger box and (2) briefly playing with two felt bugs that she extracted from the block. This time, we opted to construct the slide show using a regular extraction rate (one slide per second) in order to observe infants' dwell times under the same viewing conditions that preschoolers and adults had experienced. Again, despite the methodological change, infants dwelt significantly longer on breakpoint slides than they did on midsegment slides. These findings thus replicate dwell-time patterns indexing infants' sensitivity to segmental structure with a quite different action sequence and using an unbiased, regular extraction-rate approach to construction of the slide show. Most strikingly, infants' dwell times in the second study displayed a significant linear trend of progressively longer looking from within-segment slides, to fine-grained level breakpoints, and, longest, to coarse-grained level breakpoints. These findings are among the first evidence to date indicating that infants as young as ten-to-thirteen months of age are sensitive to hierarchical relations among event segments. This suggests a level of complexity to infants' event processing—at least for relatively familiar kinds of events depicting everyday intentional activity—beyond what has been previously known.

Among other things, this finding helps to explain how infants can rapidly acquire new knowledge as they observe others' actions, in that infants can achieve an organized processing of an event structure reflecting partonomic relations among subevents. Recognition of such relations is known to support understanding, encoding, and retrieval of events in adults' processing, and it likely serves a similar function for infants' event processing.

Beyond Segmental Structure

The systematicity apparent in infants' dwell times in our first two studies suggests that the self-paced slide show procedure can provide a valuable window into infants' extraction of segmental structure and, seemingly, their hierarchical organization of extracted segments as well. Several immediate questions arise out of the success of these first two studies. The first such question is whether dwell times might be sensitive to infants' apprehension of other kinds of structure, beyond segmental or hierarchical structure, within unfolding events. One candidate for investigation is causal structure. Causal understanding changes markedly across many domains as infants develop and gain knowledge (e.g., Gopnik et al., 2001; 2004, Schulz & Gopnik, 2004; Klahr, 2000; Kuhn, 1989); at the same time, even children in early infancy are known to possess certain fundamental kinds of causal understanding (e.g., naïve physics; Spelke, 1994; Baillargeon, Spelke, & Wasserman, 1985; Bullock, Gelman, & Baillargeon, 1982; Johnson & Aslin, 1996). It is exciting to consider the possibility that the dwell-time paradigm might provide a nuanced window into the particular junctures within an unfolding event that infants attend to in relation to the causal structure of that event. If this were to pan out, dwell times would offer a new window into developmental changes in processing that emerge as causal understanding progresses.

A second question arising out of our first two studies is whether the dwell-time paradigm might be sensitive to the influence of contextual factors on infants' processing of unfolding events. For example, prior research points to infants being sensitive to parental pedagogical cues—cues indicating that a parent is intending to help them learn something. In particular, the presence of parental pedagogy seems to significantly alter the way infants and preschoolers (as well as adults, for that matter) process an event stream (e.g., Bonawitz et al., 2011; Csibra & Gergely, 2006; 2009; Gergely & Csibra, 2005; 2006; Gergely, Egyed, & Kiraly, 2007; Sage & Baldwin, 2011; Topal et al., 2008). Conceivably, dwell times might reveal specific kinds of changes that pedagogy induces in infants' event processing. In what follows, we describe a first effort that is underway to probe these two possibilities.

Dwell Times for a Tool-Use Event. Tool use is a domain in which causal understanding and action processing intersect. An understanding of how a tool can be effectively used as a means to an end is reflected not only in our ability to produce successful action with that tool, but also in our ability to readily interpret others' actions involving the tool. Thus, infants' dwell times for tool-use events might reflect their understanding of the causal structure underlying such events. In addition, however, infants might be more inclined to process the causal structure of tool-use events when the actor indicates an intention to teach them about how to engage the tool. To test these possibilities, we (Sage & Baldwin, in prep) are examining 10- to 13-month-old infants' dwell times for a simple tool-use event—pulling a hook to bring a toy within reach—after they have had the opportunity to witness the tool being used either with accompanying social/pedagogical cues (e.g., the experimenter speaks and emotes to infants as she demonstrates the use of the tool) or without (e.g., the experimenter is silent and infants see only the experimenter's hands manipulating the tool. The actions themselves are identical to those in the social/pedagogical condition). Subsequently, dwell times are measured as infants advance at their own pace through slides depicting the hook-pulling sequence in which the experimenter (1) reaches out from behind a curtain to grasp the stalk of a hook, (2) pulls the hook, and (3) grasps the toy. Infants advance through two versions of the slide show. One version is the possible event: it involves twelve slides depicting a toy positioned within the crook of the hook and the actor's pulling action bringing the toy within reach so that the actor can grasp it. In the other version—the impossible event—the hook is oriented such that the toy is not within its crook; yet, impossibly, the actor's pull on the stalk of the hook nevertheless brings the toy within reach so that the actor can grasp it. All infants view both slide show versions twice each.

The slide shows depicting the possible and impossible sequences are carefully designed such that they display identical hand and arm motions; they differ only in the spatial relations between the toy and the hook. Moreover, the two slide show versions are frame-locked, in that the positioning of the hand and arm are identical in each of the twelve slides in the two sequences.

Finally, after completing the self-paced slide show, infants have the opportunity to attempt hook-pulling a toy for themselves. Their success at pulling the hook to retrieve the toy can thus be compared to their ability to do so in an identical baseline task that is undertaken prior to everything else.

Two additional control conditions are incorporated into the study. In one control—the "dwell-time baseline control"—a different group of infants participates in the dwell-time slide-show without any prior viewing of a hook-pulling demonstration. This control will enable us to determine whether viewing a demonstration of hook-pulling benefits infants' ability to detect segmental structure and/or causal violation, as indexed by dwell-time patterns, relative to no such viewing opportunity. A second control condition—the "scrambled slide-show control"—involves yet another group of infants participating in all parts of the social/pedagogical and non-social/pedagogical conditions described earlier, with the only change that the slides within both the possible and impossible self-paced slide-shows are randomly scrambled in their sequence. This control will enable us to ascertain whether dwell-time patterns indicative of sensitivity to segmental structure and/or causal violation reflect enhanced attention to a particular set of slides (e.g., breakpoint slides and/or causal violation slides), or whether these slides garner interest only when embedded within the context of a comprehensibly sequenced hook-pulling event. In other words, the scrambled slide-show control findings will illuminate whether dwell-time patterns of interest that may emerge are indicative of infants' event processing, per se.

A first question to ask regarding infants' dwell times in this study centers on whether they display sensitivity to the segmental structure of the hook-pulling sequence such as they have shown in our prior research with rather different action sequences. In particular, the prediction is that infants will dwell longer on slides coinciding with breakpoints (e.g., when the staff of the hook is grasped; when the pulling motion is completed) than on slides depicting points midstream within action segments. Also of interest is whether social/pedagogical cues in any way affect infants' sensitivity to segmental structure within the action sequence, as indexed by their dwell times. One especially intriguing possibility is that social/pedagogical cues accompanying the hook-pulling demonstration infants viewed might help infants to sharpen or organize their processing of that event such that they are more sensitive to the segmental structure of the event than infants who viewed a nonsocial demonstration, or no demonstration at all (i.e., in the dwell-time baseline control).

To address whether or not infants' dwell times reveal sensitivity to the causal structure of the tool-use action, we will compare dwell times for the first causal-violation slide in the impossible event (e.g., the point at which, impossibly, the toy begins moving despite being outside the crook of the hook) to the time-locked matched slide in the possible event (e.g., the point at which the toy begins moving within the crook of the hook). Again, an interesting possibility is that infants who experience a social/pedagogical demonstration of hook-pulling might show an especially pronounced sensitivity to the causal violation that occurs in the impossible event (i.e., when the hook-pulling results in the toy magically moving forward despite not being in the crook of the hook).

Lastly, the design of this study will enable us to test for replication of our previous finding (Sage & Baldwin, 2011) that infants who view a demonstration accompanied by social/pedagogical cues improve significantly more in their own attempts at hook-pulling, relative to their success at baseline, than infants who view a nonsocial demonstration. Additionally, it will be possible to examine whether infants' dwell-time patterns—both those related to extracting segmental structure and those reflecting sensitivity to causal violation of hook-pulling in the impossible event—relate in any systematic way to their subsequent ability to enact hook-pulling themselves. If so, this will provide additional confirmation that infants' dwell-time patterns reflect aspects of event processing with downstream cognitive consequences.

In sum, we are hopeful that findings from this new study will (a) replicate prior findings, while also shining further light on (b) how social/pedagogical cues may alter the manner in which infants visually examine and process events, and (c) whether such processing differences may, in turn, influence infants' own ability to later perform the action sequence themselves.

The predicted pattern of findings from this study, while groundbreaking in their own right, would also raise a number of questions. One among these would concern how to interpret the dwell-time surge that infants might display to the causal violation of the impossible tool-use slide show. Are they genuinely recognizing a violation of causal structure? That is, is it appropriate to interpret their dwell-time patterns as indicative of causal understanding? An alternative possibility is that they respond with increased dwell time to an unexpected occurrence rather than to a causal violation, per se. In other words, they might display a surge in dwell time simply because they have seldom, if ever, previously witnessed a toy moving in the configuration depicted in the impossible event and not because they recognize the scenario as causally impossible. As outlined thus far, our findings could not distinguish between these alternative accounts. However, it is possible to do so. One way to approach this would be to examine possible developmental change in dwell-time patterns vis-à-vis the causal violation. Presumably, the impossible version of the hook-pulling event is unfamiliar to infants regardless of age. If unfamiliarity is key to the dwell-time effects, then, given the unfamiliarity of the impossible event, the dwell-time surge should emerge across all ages. Alternatively, we might find age-related change in a dwell-time surge to the causal violation inherent in the impossible event. If such developmental change is in some way related to infants' ability to enact the tool-use sequence themselves, this would suggest that the dwell-time effects index an understanding of the causal structure of the hook-pulling event. Our next step will be to investigate this interesting developmental question.

Putting the Dwell-Time Paradigm to Work

We hope to have provided a general sense of the ways in which the dwell-time paradigm, while currently under development, displays promise as a useful tool for investigating the structure and meaning that viewers detect within unfolding events. Dwell-time patterns reveal nuances in event processing that illuminate how attention is allocated, and these patterns appear to be sensitive to changes resulting from increasing knowledge and understanding. Moreover, the fact that this method can be used across a broad age span contributes further to its value for observing developmental and expertise-driven changes in event processing. The dwell-time paradigm potentially can be used in a variety of other ways as well. For example, dwell times might index processing nuances that are important in face, gesture, and emotion processing, as well as the processing of social interactions. In addition, because the dwell-time paradigm provides a behavioral anchor for important event-processing phenomena (e.g., segmental structure, causal structure, perhaps goal structure, and even event-similarity structure), it could be used to ground neuroimaging approaches to investigating event processing. In other words, the dwell-time paradigm can provide a behavioral index indicative of attentional modulations during event processing to anchor analyses of neuroimaging data. With luck, altogether new information may thus be attainable regarding the neurophysiological concomitants of event processing.

Conclusion

We began with the recognition that event processing involves the challenge of constantly resolving ambiguity in the unfolding behaviors that are actually witnessed as events occur. Much remains mysterious about how we actually resolve such ambiguity and achieve an integrated analysis of event structure to guide encoding and recall. The dwell-time paradigm we present here has the potential to provide a new window on important aspects of processing that ensue as viewers—from infancy through adulthood—observe events unfold. It thus seems to offer real potential for assisting in ultimately illuminating the processes that subserve event processing—including ambiguity resolution. We are cautiously encouraged by findings showing that dwell time indexes systematic aspects of attentional allocation during event processing, and we are eager to see how this paradigm can promote further understanding of new facets of event processing and the systems that support it.

Acknowledgments

Our gratitude to the many collaborators who have contributed centrally to the research we describe in this chapter, including Bridgette Martin Hard, Meredith Meyer, Phillip DeCamp, Deb Roy, Jason Dooley Garrison, and Jessica Kosie. We are also grateful to numerous research assistants for their help, and to the Office of Naval Research and the University of Oregon for helping to fund this research.

References

Baillargeon, R., Spelke, E., & Wasserman, S. (1985). Object permanence in 5-month-old infants. *Cognition, 20,* 191–208.

Baldwin, D. (1991). Infants' contribution to the achievement of joint reference. *Child Development, 62,* 875–890.

Baldwin, D., Andersson, A., Saffran, J., & Meyer, M. (2008). Segmenting dynamic human action via statistical structure. *Cognition, 10,* 1382–1407.

Baldwin, D. A., Baird, J. A., Saylor, M. M., & Clark, M. (2001). Infants parse dynamic action. *Child Development, 72,* 708–717.

Baldwin, D., Hard, B., Meyer, M. and Sage, K. (in preparation). Dwell times reorganize as segmental structure is acquired.

Bíró, S., & Leslie, A. (2007). Infants' perception of goal-directed actions. Development through cues-based bootstrapping. *Developmental Science, 10,* 379–398.

Bonawitz, E., Shafto, P., Gweon, H., Goodman, N., Spelke, E., & Schulz, L. (2011). The double-edged sword of pedagogy: Instruction limits spontaneous exploration and discovery. *Cognition, 120,* 322–330.

Bullock, M., Gelman, R., & Baillargeon, R. (1982). The development of causal reasoning. In W. Friedman (Ed.), *Development of time concepts*. New York: Academic Press.

Carpenter, M., Akhtar, N., & Tomasello, M. (1998). Fourteen- through 18-month-old infants differentially imitate intentional and accidental actions. *Infant Behavior and Development, 21*, 315–330.

Csibra, G. (2010). Recognizing communicative intentions in infancy. *Mind & Language, 25*, 141–168.

Csibra, G., & Gergely, G. (2006). Social learning and social cognition: The case for pedagogy. In Y. Munakata & M. H. Johnson (Eds.), *Processes of change in the brain and cognitive development: Attention and performance XXI*. Oxford: Oxford University Press.

Csibra, G., & Gergely, G. (2009). Natural pedagogy. *Trends in Cognitive Sciences, 13*, 148–153.

Fiser, J., & Aslin, R. (2002). Statistical learning of new visual feature combinations by infants. *Proceedings of the National Academy of Sciences of the United States of America, 99*, 15822–15826.

Fiser, J., & Aslin, R. (2005). Encoding multielement scenes: Statistical learning of visual feature hierarchies. *Journal of Experimental Psychology. General, 132*, 521–537.

Gergely, G., & Csibra, G. (2005). The social construction of the cultural mind: Imitative learning as a mechanism of human pedagogy. *Interaction Studies: Social Behaviour and Communication in Biological and Artificial Systems, 6*, 463–481.

Gergely, G., & Csibra, G. (2006). Sylvia's recipe: Human culture, imitation, and pedagogy. In S. Levenson & N. Enfield (Eds.), *Roots of human sociality: Culture, cognition, and human interaction*. Oxford: Berg Publishers.

Gergely, G., Egyed, K., & Kiraly, I. (2007). On pedagogy. *Developmental Science, 10*, 139–146.

Gergely, G., Nadasdy, Z., Csibra, G., & Bíró, S. (1995). Taking the intentional stance at 12 months of age. *Cognition, 56*, 165–193.

Gopnik, A., Glymour, C., Sobel, D., Schulz, L., Kushnir, T., & Danks, D. (2004). A theory of causal learning in children: Causal maps and Bayes nets. *Psychological Review, 111*, 1–31.

Gopnik, A., Sobel, D., Schulz, L., & Glymour, C. (2001). Causal learning mechanisms in very young children: Two, three, and four-year-olds infer causal relations from patterns of variation and covariation. *Developmental Psychology, 37*, 620–629.

Hard, B. 2006. *Reading the language of action: Hierarchical encoding of observed behavior*. PhD Dissertation. Palo Alto, CA: Stanford University.

Hard, B., Recchia, G., & Tversky, B. (2011). The shape of action. *Journal of Experimental Psychology: General, 140*, 586–604.

Hespos, S., Grossman, S., & Saylor, M. (2011). Infants' ability to parse continuous action series: further evidence. *Neural Networks, 23*, 1026–1032.

Hespos, S., Saylor, M., & Grossman, S. (2009). Infants' ability to parse continuous actions series. *Developmental Psychology, 45*, 575–585.

Johnson, S., & Aslin, R. (1996). Perception of object unity in young infants: The roles of motion, depth, and orientation. *Cognitive Development, 11*, 161–180.

Klahr, D. (2000). *Exploring science: The cognition and development of discovery processes*. Cambridge, MA: MIT Press.

Kuhlmeier, V., Wynn, K., & Bloom, P. (2003). Attribution of dispositional states by 12-month-olds. *Psychological Science, 14*, 402–408.

Kuhn, D. (1989). Children and adults as intuitive scientists. *Psychological Review, 96*, 674–689.

Meltzoff, A. (1995). Understanding the intentions of others: Re-enactment of intended acts by 18-month-old children. *Developmental Psychology, 31*, 838–850.

Meyer, M., & Baldwin, D. (2011). Statistical learning of action: The role of conditional probability. *Learning & Behavior, 39*, 383–398.

Meyer, M., Baldwin, D., & Sage, K. (2011). Assessing young children's hierarchical action segmentation. In L. Carlson, C. Hölscher, & T. Shipley (Eds.), *Proceedings of the 33rd annual conference of the Cognitive Science Society*. Austin, TX: Cognitive Science Society.

Newtson, D. (1973). Attribution and the unit of perception of ongoing behavior. *Journal of Personality and Social Psychology, 28*, 28–38.

Newtson, D., & Enquist, G. (1976). The perceptual organization of ongoing behavior. *Journal of Experimental Social Psychology, 12*, 436–450.

Olofson, E., & Baldwin, D. (2011). Infants recognize similar goals across dissimilar actions involving object manipulation. *Cognition, 118*, 258–264.

Saffran, J., Aslin, R., & Newport. E. (1996). Statistical learning by 8-month-old infants. *Science, 274*, 1926–1928.

Sage, K., & Baldwin, D. (2011). Disentangling the social and the pedagogical in infants' learning about tool use. *Social Development, 20*, 825–844.

Sage, K. D., Ross, R. A., & Baldwin, D. (2012). Assessing infants' hierarchical action segmentation. Poster presented at the International Conference on Infant Studies, Minneapolis, Minnesota.

Saylor, M., Baldwin, D. A., Baird, J. A., & LaBounty, J. (2007). Infants' on-line segmentation of dynamic human action. *Journal of Cognition and Development, 8,* 113–128.

Schulz, L., & Gopnik, A. (2004). Causal learning across domains. *Developmental Psychology, 40,* 162–176.

Spelke, E. (1994). Initial knowledge: Six suggestions. *Cognition, 50,* 431–445.

Topal, J., Gergely, G., Miklosi, A., Erdohegyi, A., & Csibra, G. (2008). Infants' perseverative search errors are induced by pragmatic misinterpretation. *Science, 321,* 1831–1834.

Woodward, A. (1998). Infants selectively encode the goal object of an actor's reach. *Cognition, 69,* 1–34.

Woodward, A. (1999). Infants' ability to distinguish between purposeful and nonpurposeful behaviors. *Infant Behavior and Development, 22,* 145–160.

Woodward, A., & Sommerville, J. (2000). Twelve-month-old infants interpret action in context. *Psychological Science, 11,* 73–77.

Zacks, J. M., Tversky, B., & Iyer, G. (2001). Perceiving, remembering, and communicating structure in events. *Journal of Experimental Psychology: General, 130,* 29–58.

14 The Role of Self-Produced Movement in the Perception of Biological Motion and Intentional Actions in Infants

Petra Hauf

Abstract

Recent research has focused on the question of how infants come to understand others' actions and intentions and how this understanding might be related to their agentive experience. Framed by the idea of a bidirectional link between action perception and action production, this chapter (1) emphasizes the role of infants' self-produced movements in their perception of biological motion and intentional actions in the context of reaching and grasping and (2) discusses possible implications for motor development in particular and social-cognitive development in general.

As human beings, we act and interact in a social environment from early on in our lives. We explore our world, watch others acting, and process a variety of different impressions. We represent and reflect on what we perceive, and we plan our own actions on this basis. Most of these actions take place in a social context. In social interaction, we (1) act in the environment and (2) relate our actions to the actions of other people. Our own actions affect the behavior of the person toward whom they are directed, and the actions of others affect our own behavior (e.g., Hauf & Försterling, 2007). For that reason, it is essential to recognize that the understanding of other individuals' actions and intentions is tightly linked to one's own actions and intentions. Furthermore, this relatedness is not only of major importance for infants' growing understanding of others' actions (see Hauf, 2007 for an overview), but it is also significant for the development of social-cognitive skills such as social referencing (Moses, Baldwin, Rosicky, & Tidball, 2001; Repacholi & Gopnik, 1997), language learning (Baldwin & Moses, 2001; Tomasello, 1999), and imitative learning (Carpenter, Akthar, & Tomasello, 1998; Meltzoff, 1995).

This chapter will approach the question of how infants come to understand others' actions and intentions and how this growing understanding might be related to their agentive experience. Following a brief introduction to the idea of a bidirectional influence of action perception and action production, the influence of infants' self-produced movements on their perception of biological motion will be discussed. A brief overview will outline how infants perceive and understand goal-directed reaching and grasping movements performed by others and how they perceive and understand others' crawling and walking movements. Following this, the main section will relate infants' own motor capacities to their motion perception. Finally, the role of self-produced movement on motion perception will be discussed with respect to possible implications for motor development in particular and social-cognitive development in general.

Bidirectionality in Action Control

Actions are defined as movements toward a goal associated with a desired outcome or effect. In our everyday lives, we perform and perceive actions without paying much attention to the components involved. For example, we reach for and grasp a cup to get a sip of coffee, which is the desired action outcome. Thereby, we almost automatically adjust our movements to the situational context. We enlarge the opening of the hand if we intend to grasp the whole mug, or we prepare our thumb and index finger if we plan to grasp the handle only. In doing so, we draw on an impressive, yet unnoticed, interplay of sensory system (e.g., what we see) and motor system (e.g., how we move). But how is this interplay of action perception and action production actually functioning? And how does this interplay develop?

The *common coding* approach originated by Wolfgang Prinz (1990, 1997) suggests that our actions are controlled by means of information gathered through action perception and action production and stored in common representations. These common representations include visuospatial representations based on perception experience as well as sensorimotor representations based on motor experience. Importantly, these representations are used for both action perception and action production; thus, they emphasize a bidirectional exchange from perception to production and vice versa (Hauf & Prinz, 2005; Meltzoff, 2007). Research with adult populations has demonstrated a privileged connection between action production and action perception in cognitive psychology (for an overview, see Hommel, Müsseler, Aschersleben, & Prinz, 2001; Prinz, 2002). Furthermore, research in neuroscience has shown that the same brain regions are activated both when actions are observed and when actions are produced (see Brass & Heyes, 2005 for a recent review).

Alongside this research on action control in adults, recent developmental research has started to focus on the question of how action perception and action production are related to each other and how these two aspects of action control develop during early childhood (see Hauf, 2007 for an overview). For example, Meltzoff's "like me" developmental framework (Meltzoff, 2002, 2007) postulates that infants' fundamental interpersonal relations are based on their representation of action. Thereby, infants monitor their bodily acts and detect cross-modal equivalents between their own acts-as-felt and the acts-as-seen in others (Meltzoff, 2007).

Current research with infants supports the notion that such a relation is already established during the first year of life, as indicated by findings highlighting the impact of action production on action perception and vice versa (Hauf, 2007; Hauf & Prinz, 2005; Longo & Bertenthal, 2006; Sommerville & Woodward, 2005). Furthermore, recent research examining when and how infants start to understand goal-directed actions performed by others has emphasized the role of infants' agentive experience on action perception (Barresi & Moore, 1996; Daum, Prinz, & Aschersleben, 2011; Hauf, Aschersleben, & Prinz, 2007; Meltzoff, 2005; Tomasello, 1999).

Taken together, the findings on the bidirectionality in action control link action perception to action production and emphasize an interdependence of sensory and motor systems. But this also raises the question of whether such interdependence might be restricted to actions or might also be involved in motor movements per se.

Infants' Perception of Goal-Directed Reaching and Grasping Actions

Long before infants are actually able to perform intentional actions on their own, they are attentive observers of motion in general and movements performed by others in particular. Consequently, recent research in the area of early cognitive development has focused on the question of when and how infants come to understand actions performed by others; such research particularly concentrates on reaching and grasping actions.

Recent research on action control emphasizes the interplay of action perception and action production. Thereby, actions are defined as movements toward a goal associated with a desired effect. Accordingly, typical research paradigms (1) show infants' reaching or grasping movements toward interesting objects and (2) examine how infants perceive and understand the goals of the presented actions. Using such paradigms, recent studies in the area of action understanding show that infants understand reaching and grasping movements as being goal directed at around six months of age. In her seminal studies, Woodward (1998, 1999) used a habituation paradigm and presented infants with an actor reaching for and grasping one of two toys sitting side-by-side on a stage for several habituation trials. In subsequent test trials, the position of the two toys was reversed, and the actor reached for and grasped either the same toy at a new location or the new toy at the old location. Infants looked longer at the event in which the actor changed the goal and reached for the new toy; this suggests that they had represented the reaching movement as directed toward a goal—in this case a certain toy (Woodward, 1998). Interestingly, these results were not found if the reaching and grasping action was performed by a mechanical claw (Woodward, 1998) or if an unfamiliar movement—like a touch with the back of the hand—was presented (Woodward, 1999). This indicates that infants' early action understanding is restricted to (1) familiar grasping movements and (2) experience with human actions. Even though recent studies have demonstrated that six-month-old infants can understand the goal of unfamiliar human actions if the presented action leads to a salient effect (e.g., Hofer, Hauf, & Aschersleben, 2007; Jovanovic et al., 2007; Király et al., 2003), it is still questionable whether these findings demonstrate infants' understanding of the underlying action goal or only reflect an early understanding of object-related movements. To investigate this question further, recent research focused on purposeful and accidental actions in relation to the action outcome. For example, Behne and colleagues (2005) interacted with nine- to eighteen-month-old infants. During this interaction, an adult attempted to hand toys to the infants. Even though the transaction always failed—thus the infant never received the desired toy—infants displayed more impatience when the adult was unwilling to give the toy (e.g., teasing) compared to when the adult was unable to give it (e.g., accidentally dropping the toy). The fact that infants' reactions varied with respect to the actor's intention provides evidence for an early understanding of goal-directedness in reaching actions.

Jovanovic and Schwarzer (2007) addressed the same question by means of a looking paradigm. Seven- and nine-month-olds watched a hand entering a stage and moving to the contralateral side toward an object. On the way to that object, the hand touched another object standing in the motion path and overthrew it. Nevertheless, the hand continued moving along the path until it reached the other object and lifted it. During subsequent test trials, a new object replaced either the one that had been accidentally overthrown or the one that had been intentionally lifted. Infants looked longer when the lifted object was replaced than when the overthrown object had been replaced. This differential looking behavior indicates that seven- and nine-month-old infants understood the underlying intention of a manual reaching action when watching a grasp-and-lift event compared to a touch-and-overthrow event. In both cases, the movement itself elicited a salient effect (object motion); however, this effect was intentional in the first case and accidental in the latter. Therefore, these findings suggest that infants preferentially encode action goals rather than action outcomes.

Taken together, the reported findings provide evidence that infants begin to understand goal-directed reaching and grasping actions at around six-to-nine months of age. This early action understanding takes into consideration the fact that infants have to be able to understand the movement and the goal. But do infants of this age also understand the biological motion involved in a reaching movement, or do they solely focus on the action goal?

The importance of biological motion for infants' action understanding has only recently gained attention and been addressed in studies looking at grasping comprehension in infants. For example, Daum and Gredebäck (2011) presented three-, five-, and seven-month-old infants with a static picture of a grasping hand. The hand was opened as if grasping for an unseen object, and the hand opening was directed either to the right or the left. Following this cueing trial, an object was presented either to the same side as the previous hand opening (congruent trial) or to the opposite side (incongruent trial). The latency of infants' gaze shift from the hand to the object was recorded with an eye-tracker. Shorter saccadic reaction times for congruent trials would indicate that infants encode a hand opening as the starting point of a grasping action and thus expect an object to appear on the same side as the hand opening rather than on the opposite side. Indeed, results revealed such a congruency effect by five month of age—but not younger—indicating an early understanding of grasping actions. Interestingly, this difference in gaze shift was not found when presenting a mechanical claw instead of the static hand (Daum & Gredebäck, 2011), which suggests that infants' early grasping comprehension might be restricted to biological motion and/or the human body form.

To further explore this question, Wronski and Daum (2013) presented seven-month-old infants with movies that depicted a reaching and grasping hand in motion (biological motion condition). In line with the depicted motion, the hand opening was directed to either the right or left side of the screen (e.g., hand moving from right to left with the hand opening to the left). Following this demonstration, an object appeared either in line with the depicted motion and hand opening (congruent trial) or in the opposite direction of such (incongruent trial). In the mechanical-motion condition, infants saw a block linearly moving across the screen followed by the test trials just described. Shorter saccadic eye movements in congruent trials occurred only in the biological motion condition (Wronski & Daum, 2013), which indicates that biological motion—but not motion per se—is an important aspect of early grasping comprehension.

But do infants also understand the biological motion involved in a reaching movement when portrayed by point-light displays (PLDs) only? In this case, instead of using a full body-part display, movement is depicted with points of lights that are attached to the arm, hand, fingers, and major joints involved in the movement of an otherwise unseen human body form. A recent study by Elsner, Falck-Ytter, & Gredebäck (2011) examined whether sixteen-month-old infants anticipate another person's reaching action based solely on biological motion information depicted in a PLD of a reaching hand. In the biological motion condition a point-light hand entered the scene, individual fingers of the hand tapped on a table, and the hand then waved toward the infants. Following this, the hand reached for an object partly covered by an occluder. Note that the point-light hand then disappeared behind the occluder, so infants could not see the actual grasp of the object. In the mechanical-motion condition, the point-light display was manipulated to create a linear—and therefore nonbiological—motion with trajectories similar to the point-light hand (Elsner et al., 2011).

Analysis of infants' gaze shifts from the initial position of the hand toward the goal-object revealed that infants shifted their gaze more often during the biological reaching action as compared to the nonbiological version of the action. These findings indicate that infants, who are experienced in producing reaching and grasping movements, are able to anticipate others' action goals based on biological motion.

Taken together, the findings on infants' understanding of reaching and grasping actions outlined in this section indicate that children in early infancy have an advanced understanding of action goals only when biological motion is involved. If this is indeed the case, then infants' perception of biological motion is crucial for action understanding and thus has an important impact on social cognitive development.

Infants' Perception of Biological Motion

From the first hours following birth, infants begin to observe the world around them. Some researchers have demonstrated that humans have an innate mechanism for perceiving biological motion (Bardi, Regolin, & Simion, 2011; Simion, Regolin, & Bulf, 2008). This sensitivity to biological motion, combined with infants' early abilities to recognize human body forms (Bertenthal, Proffitt, & Kramer, 1987; Moore et al., 2007), has inspired recent research on how infants' perception of biological motion develops as they gain perceptual experience by watching the movements of others and agentive experience as they themselves begin to explore the world and move around.

Infants' early ability to distinguish biological motion from other types of motion is evident from research using point-light displays that depict locomotion only through points of light attached to the head and major joints of an otherwise unseen actor. Bertenthal and colleagues (1987) demonstrated that three-month-old infants preferentially look at the biological motion depicted in the normal human walking PLD compared to randomly moving dots. Furthermore, three-month-olds differentiate between biologically possible and impossible human walking PLDs (Bertenthal et al., 1987; Proffitt & Bertenthal, 1990), and at 4 months of age, infants are aware of differences in the orientation of PLDs as evidenced by their ability to distinguish between upright and inverted human walkers (Bertenthal, Proffitt, & Cutting, 1984). In addition to this research demonstrating infants' ability to both perceive biological motion and detect violations of human locomotion, recent research has shown that three-month-old infants are able to distinguish walking from running (Booth, Pinto, & Bertenthal, 2002), which indicates their ability to identify information about the action being performed. Kuhlmeier, Troje, and Lee (2010) further investigated whether six-month-old infants could detect directionality in displays of point-light walkers. Infants were habituated to a PLD walking on either a leftward or rightward trajectory. In subsequent test trials, infants were able to detect changes in the direction of motion only in upright human PLD walkers, but not when the PLD walkers were inverted. Kuhlmeier et al. (2010) suggest that infants do appear to be extracting information about the action depicted in the normal walking motion, but the infants had difficulty extracting a global human form in the inverted motion.

If infants are able to extract action-related information, such as speed and direction, from biological motion, do they also perceive features of the actor that might help to inform their developing body schema? Reid, Hoehl, Landt, and Striano (2008) presented eight-month-old infants with a biomechanically possible PLD of a kicking human and a biomechanically impossible PLD of same; the latter was created through altering the rotation of knee, ankle, and toe. Electroencephalographic (EEG) recordings and calculated event-related potentials (ERPs) revealed that infants distinguished between both movements. Furthermore, infants were able to discriminate between a walking PLD movement based on a normal human body schema and one based on an altered body schema where joints circumflexed backward during the walking motion. These findings indicate that infants are able to detect violations of their developing representations of human body forms.

Indeed, recent research suggests that (1) PLDs activate a whole-body schema in infants, and (2) infants apply this body schema to their perception of biological motion. To test this notion, Moore et al. (2007) habituated six- and nine-month-old infants to a walking human PLD moving behind a table. In subsequent test trials, some of the point-lights were partially occluded by the tabletop, and others were passing in front of the tabletop. Accordingly, the PLD appeared to walk through the tabletop—an impossible event for a solid human body. Both age groups looked longer at this event, which indicates that the PLD triggered a body schema with a solid vertical form. Interestingly, additional conditions with inverted PLDs and scrambled dots did not have the same effect on infants; this provides further evidence that infants' early understanding of biological motion seems to be restricted to the upright human form.

Research on motion perception indicates that, from early on, infants are able to discriminate between normal human walking PLDs and randomly moving dots (e.g., Bertenthal et al., 1984, 1987). Shortly after this, infants are able to extract information about directions, actions, and solidity from moving PLDs, which indicates the development of an increasingly elaborate body representation throughout infancy (e.g., Booth et al., 2002; Kuhlmeier et al., 2010). However, further research is needed to clarify how infants develop a body representation and how this body schema might impact their perception and production of biological movements (Hauf & Power, 2011).

The Interrelation of Infants' Action Production and Action Perception

It is assumed that infants reach a level of functional equivalence between perceived and produced actions at around nine months of age (Hauf, 2007). Having achieved functional equivalence means that representations of knowledge about one's own actions are similar to representations of knowledge about others' actions. This knowledge can be used by the perceptual system to perceive and understand actions performed by others and by the motor system to initiate and produce one's own actions. Different theoretical approaches have emphasized different underlying mechanisms, such as intermodal mapping (Meltzoff, 2002, 2007) and action control (Prinz, 1997, 2002), but they all share the idea of a direct interrelation between action perception and action production.

According to this notion, infants' action understanding might be based on knowledge gained through both (1) experience in action perception and (2) action production. As a consequence, it could be assumed that extensive experience in action perception influences how infants produce actions and vice versa.

For example, infants start to understand others' actions during their first year of life. At the same time, infants are able to produce actions to bring about desired action effects. To demonstrate, two- to five-month-old infants understand the relationship between leg kicks and the contingent movement of a mobile (e.g., Rovee & Rovee, 1969; Rovee-Collier & Hayne, 2000; Rovee-Collier & Shyi, 1993). However, in order to produce more complex actions, infants have to be able to differentiate between the movements and the goals involved in their own behavior. Infants acquire this capacity at around nine months of age as indicated by their ability to, for instance, use a cloth as a support to obtain a desired, out-of-reach object (Goubet et al., 2006) or remove an obstacle to reach an object (Uzgiris & Hunt, 1975). Taken together, these findings suggest that both aspects of action control—namely action perception and action production—develop at the same time. Based on this assumption, further research focuses on the question of how action perception and action production are interrelated.

To address this issue, Sommerville and Woodward (2005) tested ten-month-olds' ability to (1) retrieve an out-of-reach toy by pulling the supporting cloth (production task) and (2) understand such a sequence presented only visually (perception task). Interestingly, only those infants who were able to produce the action sequence on their own understood the same sequence performed by another person—thus indicating a link between production and perception. Furthermore, Longo and Bertenthal (2006) investigated the same question using a task eliciting the Piagetian A-not-B error. To do so, nine-month-old infants watched an experimenter (1) waving a small toy with one hand, (2) removing the lid of a hiding location with the other hand, (3) placing the toy in the location, and (4) closing the lid again. Importantly, ipsilateral reaching was performed as well as contralateral reaching. In the reaching condition, the infants themselves recovered the toy in location A three times; in the observation condition, they watched the experimenter recovering the toy three times. Following these A-trials, the toy was hidden in location B and the infants were given the chance to look for the hidden toy. Infants displayed only perseverated reaching behavior after they watch an ipsilateral reaching action, which they were able to produce, but not after watching a contralateral reaching, which infants at this age are not able to produce.

Recent studies provide further evidence for a close link between action production and action perception by emphasizing the impact of agentive experience, and thus action production, on infants' ability to understand actions performed by others. A study by Hauf, Aschersleben, and Prinz (2007), for example, demonstrated a profound influence of action production on action perception in nine- and eleven-month-old infants. First, infants gained action experience by actively exploring a toy. Following this production phase, infants participated in a perception phase by watching two adults acting with either the same toy the infants had played with or a novel toy. Both age groups preferred to watch the same-toy action. These and control results indicated that infants' active action experience influenced their interest in watching actions performed by others.

To investigate the impact of agentive experience, Hauf and colleagues (2007) used movements, such as sliding or waving, that already existed in infants' repertoire. Another approach to address the impact of agentive experience would be to (1) provide infants with motor experiences at an age earlier than such experiences are usually available to them and (2) investigate how these early exposures alter infants' behavior subsequently. According to this approach, Sommerville, Woodward, and Needham (2005) tested three-month-old infants' understanding of a goal-directed reaching action in a production and perception phase. At this early age, infants are not yet able to perform a sophisticated reaching action. However, the infants in the study were given the opportunity to produce reaching and grasping actions by wearing special mittens (so-called sticky mittens with Velcro attached) allowing them to pick up a toy. Following this production phase, infants participated in a perception task, where they watched an actor reach for and grasp one of two toys arranged next to each other on a stage several times. In subsequent test trials, the position of the two toys was reversed and the actor reached for and grasped either the same toy at the new location or the new toy at the old location. Infants looked longer at the scenario in which the actor reached for the new toy than at the scenario in which the actor reached for the old toy, which indicates that they detected the goal change. Interestingly, a second group of three-month-olds participated in the perception task only; those infants looked equally long at both scenarios. These findings provide further evidence for the role of agentive experience in infants' developing action understanding.

Whereas the study by Sommerville and colleagues (2005) implemented reaching experience earlier than usually available by means of special mittens, Daum, Prinz, and Aschersleben (2011) assessed infants' natural grasping skills individually. During the production task, six-month-old infants were presented with three objects, and grasping skills were categorized as either (1) palmar, where the infant grasps objects by pressing the fingers against the palm, or (2) thumb-opposite, where thumb and fingers are positioned in opposition to hold the object. During the perception task, all participating infants watched a video of an actor's grasping movement toward an occluded object. Importantly, the actors' grasping movement always depicted a thumb-opposite grasp; however, in one condition the hand opening was small (grasping for the handle of a cup) and in the other condition the hand opening was large (grasping for the whole cup). During subsequent test trials, the occluder had been removed and the grasping movement was shown with both an expected outcome and unexpected outcome, simultaneously. Looking time for both events indicated that only those infants who were able to perform a thumb-opposite grasp differentiated between the two events; infants who previously had demonstrated only the palmar grasp did not show evidence of this discrimination (Daum et al., 2011).

To further explore the impact of reaching and grasping skills on the perception of such movements, Melzer, Prinz, and Daum (2012) investigated the production and perception of contralateral reaching in six- and twelve-month-old infants by introducing anticipatory looking behavior as a measure of infants' goal understanding in a reaching action. In a production task, infants' contralateral reaching capacities were assessed; in a perception task, a contralateral reaching movement was presented. Anticipatory gaze shifts revealed that in twelve-month-old infants, but not in six-month-olds, the production and perception of contralateral reaching movements were positively correlated. Thus, twelve-month-old infants with advanced reaching and grasping skills displayed faster anticipatory gaze shifts while watching contralateral reaching movements performed by another person (Melzer et al., 2012).

Regarding infants' ability to produce movements based on their perception and comprehension of such movements when performed by others, recent studies offer further evidence for this capacity by adding a training component to provide infants with additional production or perception experience. For example, ten-month-old infants benefited from active training in how to use a special tool to retrieve an out-of-reach toy; only infants with this training experience were able to identify the goal of such a tool-use event performed by another person (Sommerville, Hildebrand, & Crane, 2008).

Libertus and Needham (2010) investigated how reaching experience might influence the development of reaching and grasping skills, as well as the development of understanding reaching and grasping movements performed by others, in more detail. For this study, three-month-old infants participated in daily training sessions for two weeks. During these sessions, infants were either allowed to explore objects manually by means of special mittens that allowed them to touch and lift the objects (active training) or to watch an adult perform the same actions (passive training). Results revealed that active and passive training altered infants' manual and visual exploration skills differently. Infants who received manual training, and who thus had agentive experience in reaching and grasping for objects, displayed advanced reaching skills and also showed changes in their visual exploration of the objects. Passive observation had no impact on reaching production or reaching perception (Libertus & Needham, 2010).

Interestingly, infants in the active training condition also displayed higher levels of visual engagement with the agent performing the actions in the perception tasks. This finding emphasizes the role of self-produced reaching actions not only for the movement per se, but also for the actor performing the movement. Furthermore, increased active reaching experience also boosted infants' interest in faces (Libertus & Needham, 2011); this emphasizes the role of self-produced actions for social-cognitive development.

Taken together, the reported findings indicate a major impact of infants' capacities to produce actions on how they perceive and understand actions performed by others. Nevertheless, the presented studies investigate action understanding without separating the execution of an action (the movement) and the action outcome or action goal. Therefore, further research needs to address whether an interrelation can also be found in movement production and motion perception per se.

The Interrelation between Infants' Movement Production and Motion Perception

Infants are able to discriminate between randomly moving dots and walking PLDs (Bertenthal et al., 1987; Booth, Pinto, & Bertenthal, 2002), as well as other movements, at around four months of age. Shortly after this, infants can also detect directionality in human walking movements (Kuhlmeier et al., 2010). This is exactly the same age at which infants start to move on their own—at first by rolling, but soon by crawling and walking (Adolph, 2008; Bly, 1994). Could it be that infants' developing motor skills have an impact on how they come to understand movements performed by others?

Along with other researchers in this field, Reid, Belsky, and Johnson (2005) investigated how infants perceive and process possible and impossible biological motion. After eight-month-old infants had been familiarized with a human actor reaching for, grasping, and exploring several objects, both a biologically possible and a biologically impossible reaching and grasping movement toward an object were displayed. Longer looking time for impossible compared to possible body movements, as well as gamma-band frequency analysis of EEG recordings, indicated that only infants with high fine motor skills (e.g., pressing buttons, holding objects, twisting toys) processed the stimuli differently (Reid et al., 2005).

To support the notion that the production of a certain movement is connected to motion perception in infancy, Sanefuji, Ohgami, and Hashiya (2008) used PLDs of crawling and walking adults to investigate whether crawling and walking infants display a preference for watching one or the other motion. The findings demonstrated a clear preference related to the level of motor skills: crawling infants preferred to watch crawling PLDs, whereas walking infants preferred to watch walking PLDs; this indicates a close relation between motor skills and motion perception early in life.

In line with this assumption, Hauf, MacDonald, and Giese (2013) used PLDs to investigate the relationship between motor experience and motion perception in noncrawling, crawling, and walking infants. The aim of this study was to clarify (1) whether infants would identify moving dots as crawling and walking humans and (2) whether different levels of motor skills would influence the perception of these movements.

Noncrawling and nonwalking infants (six-and-a-half months of age), crawling but nonwalking infants (nine-and-a-half months of age), and crawling and walking infants (thirteen-and-a-half months of age) watched videos that presented adult crawling and walking PLDs. Each of the movements was presented in (1) a normal version, (2) a phase-shifted version, and (3) a scrambled version. Note that all stimuli were presented as biological motion; however, only the normal movements were bodily possible movements, whereas the phase-shifted version violated body constraints and the scrambled version depicted no body at all. Infants' looking behavior was recorded with an eye-tracker and analyzed with respect to looking time for (1) the overall display and (2) the lower body area, which encompassed the body parts performing the actual movement (arms, hands, knees, and feet for crawling; feet, knees, and legs for walking).

There were no differences in overall looking time. Also, no differences were found with respect to the upper body area, which encompassed the body parts not directly involved in the production of the movement. However, differences in looking time for the lower body area indicate a relation between motor skills and motion perception. To demonstrate, noncrawling infants looked longer at the lower body area while watching the normal crawling as compared to phase-shifted and scrambled crawling. In contrast, both of the crawling-age groups looked longer at this area while watching the phase-shifted crawling as compared to the normal and scrambled crawling. Similar results were found for the perception of walking movement. Both of the nonwalking age groups looked longer at the lower body area while watching the normal walking as compared to the phase-shifted and scrambled walking, whereas walking infants spent more time looking at the lower body area while watching the phase-shifted motion as compared to the normal and scrambled version of the walking movement (Hauf et al., 2012).

These findings support the idea that infants' level of motor experience influences how they perceive crawling and walking movements. In addition to Sanefuji et al.'s (2008) findings, which demonstrate that crawling infants prefer to look at crawling movements whereas walking infants prefer walking displays, these findings show that infants detect violation of body constraints, as indicated by longer looking time, only once they are able to produce this movement themselves.

Conclusions

Understanding others' actions is fundamental to human experience. From early childhood on, we see the motions of others as they perform actions as being structured by goals and intentions. This basic aspect of social cognition provides a critical foundation for early social-cognitive skills. Thus, it is not surprising that an increasing amount of research focuses on the question of how infants come to understand actions performed by others. While previous research focused on infants' action perception, recent research investigates how infants' own agentive experience influences their action perception. In the meantime, there is a growing body of evidence emphasizing the impact of self-produced reaching and grasping actions on the perception of these actions performed by others (e.g.,

Daum et al., 2011; Longo & Bertenthal, 2006; Melzer et al., 2012; Sommerville & Woodward, 2005; Sommerville et al., 2005) and on the interest in actions performed by others (Hauf et al., 2007). Furthermore, recent studies highlight the impact of early experience in reaching skills on infants' motor development and visual exploration behavior (Libertus & Needham, 2010). However, it is important to note that these studies focus on infants' understanding of the goals underlying reaching and grasping actions, but they do not look into the movement execution per se. As outlined previously, an action is constituted of two equally important components: the movement (e.g., stretching the arm, rotating the wrist, positioning the fingers) and the anticipated goal or action outcome (e.g., holding the object). Further research has to take both components into consideration. Whereas most of the research on infant action understanding focuses on the action outcome and the anticipated action goal, research investigating infants' understanding of biological motion focuses on the role of the movement execution in infants' action understanding. Interestingly, recent research findings provide evidence for a similar relation between production and perception of biological motion in reaching and grasping (Reid et al., 2005) as well as crawling and walking (Hauf et al., 2012; Sanefuji et al., 2008).

Taken together, there is converging evidence emphasizing the role of self-produced actions on the perception of actions performed by others. In addition, the presented findings indicate a similar influence of self-produced movements on the perception of movements executed by others. Accordingly, theoretical approaches emphasizing the bidirectionality of perception and production in actions have to be viewed more generally to include the perception and production of movements as well. However, further research needs to address how the influence of self-performed movements and actions on infants' perception and understanding is established. Equally important, it needs to be addressed whether infants' indeed have to be able to perform a movement and action in order to understand a movement or action performed by others or whether their agentive experience is helpful, but not necessary, to developing a sophisticated action understanding.

It is also important to note that, as the link between the onset of self-locomotion and developmental changes in perception and cognition has been suggested for decades, contributions to the question of how and when infants develop a sophisticated understanding of movements and actions might have further implications for infants' social-cognitive development in general (Gibson, 1988; Piaget, 1953; for a review, see Campos et al.,

2000). For example, the onset of crawling and walking seems to correspond with (1) major advances in social development as indicated by changes in social referencing in relation to motor skills (Clearfield, 2011) and (2) cognitive development as indicated by an increased flexibility in infants' memory retrieval in relation to infants' crawling skills (Herbert, Gross, & Hayne, 2007). Beyond this, recent research suggests that providing infants with early reaching and grasping experience will further enhance (1) the development of their motor skills, (2) the way they visually explore objects, and (3) their interest in social partners (Libertus & Needham, 2011).

Taken together, the findings reported in this chapter highlight the close relationship between self-produced movements and the perception of biological motion in particular, but also emphasize the relationship between motor development and social-cognitive development in general.

Acknowledgments

Special thanks to Mel Rutherford and Valerie Kuhlmeier for the effort they provided on this project and for their tremendous support all the way along. The Canada Research Chairs (CRC) program, the Canada Foundation for Innovation (CFI), and a grant from the Natural Sciences and Engineering Research Council of Canada (NSERC) supported this research.

References

Adolph, K. E. (2008). Learning to move. *Current Directions in Psychological Science, 17*, 213–218.

Baldwin, D. A., & Moses, L. (2001). Links between social understanding and early word learning: Challenges to current accounts. *Social Development, 10*, 309–329.

Bardi, L., Regolin, L., & Simion, F. (2011). Biological motion preference in humans at birth: Role of dynamic and configural properties. *Developmental Science, 14*, 353–359.

Barresi, J., & Moore, C. (1996). Intentional relations and social understanding. *Behavioral and Brain Sciences, 19*(1), 107–122.

Behne, T., Carpenter, M., Call, J., & Tomasello, M. (2005). Unwilling versus unable: Infants' understanding of intentional actions. *Developmental Psychology, 41*, 328–337.

Bertenthal, B. I., Proffitt, D. R., & Cutting, J. (1984). Infant sensitivity to figural coherence in biomechanical motions. *Journal of Experimental Child Psychology, 3*, 213–230.

Bertenthal, B. I., Proffitt, D. R., & Kramer, S. J. (1987). The perception of biomechanical motions by infants: Implementation of various processing constraints. *Journal of Experimental Psychology: Human Perception and Performance, 13,* 577–585.

Bly, L. (1994). *Motor skills acquisition in the first year.* San Antonio, TX: Therapy Skill Builders.

Booth, A. E., Pinto, J., & Bertenthal, B. I. (2002). Perception of the symmetrical patterning of human gait by infants. *Developmental Psychology, 38*(4), 554–563.

Brass, M., & Heyes, C. (2005). Imitation: Is cognitive neuroscience solving the corresponding problem? *Trends in Cognitive Sciences, 9,* 489–495.

Campos, J. J., Anderson, D. I., Barbu-Roth, M. A., Hubbard, E. M., Hertenstein, M. J., & Witherington, D. (2000). Travel broadens the mind. *Infancy, 1,* 149–219.

Carpenter, M., Akthar, N., & Tomasello, M. (1998). Fourteen- and 18-month-old infants differentially imitate intentional and accidental actions. *Infant Behavior and Development, 21,* 315–330.

Clearfield, M. W. (2011). Learning to walk changes infants' social interaction. *Infant Behavior and Development, 34,* 15–25.

Daum, M. M., & Gredebäck, G. (2011). The development of grasping comprehension in infancy: Covert shifts of attention caused by referential actions. *Experimental Brain Research, 208,* 297–307.

Daum, M. M., Prinz, W., & Aschersleben, G. (2011). Perception and production of object-related grasping in 6-month-olds. *Journal of Experimental Child Psychology, 108,* 810–818.

Elsner, C., Falck-Ytter, T., & Gredebäck, G. 2011. Do 3.5-year-old children anticipate others' actions based on biological motion information? Poster presented at the 2011 Biennial Meeting of the Society for Research in Child Development, Montreal, Quebec, Canada, March 31–April 2.

Gibson, E. J. (1988). Exploratory behaviour in the development of perceiving, acting, and the acquiring of knowledge. *Annual Review of Psychology, 39,* 1–41.

Goubet, N., Rochat, P., Maire-Leblond, C., & Poss, S. (2006). Learning from others in 9- to 18-month-old infants. *Infant and Child Development, 15*(2), 161–177.

Hauf, P. (2007). Infants' perception and production of intentional actions. In C. von Hofsten & K. Rosander (Eds.), *Progress in brain research: From action to cognition* (pp. 285–301). Amsterdam: Elsevier Science.

Hauf, P., Aschersleben, G., & Prinz, W. (2007). Baby do—baby see! How action production influences action perception in infants. *Cognitive Development, 22,* 16–32.

Hauf, P., & Försterling, F. (2007). *Making minds: The shaping of human minds through social context.* Amsterdam: John Benjamins.

Hauf, P., MacDonald, K. M., & Giese, M. (2013). Are you moving? How motor experience influences infants' perception of crawling and walking point-light-displays. Manuscript in preparation.

Hauf, P., & Power, M. (2011). Infants' perception and production of crawling and walking movements. In V. Slaughter & C. A. Brownell (Eds.), *Early development of body representations* (pp. 227–246). Cambridge: Cambridge University Press.

Hauf, P., & Prinz, W. (2005). The understanding of own and others' actions during infancy: "You-like-me" or "Me-like-you"? *Interaction Studies: Social Behaviour and Communication in Biological and Artificial Systems, 6*(3), 429–445.

Herbert, J., Gross, J., & Hayne, H. (2007). Crawling is associated with more flexible memory retrieval by 9-month-old infants. *Developmental Science, 10*(2), 183–189.

Hofer, T., Hauf, P., & Aschersleben, G. (2007). Infants' perception of goal-directed actions on video. *British Journal of Developmental Psychology, 25*, 485–498.

Hommel, B., Müsseler, J., Aschersleben, G., & Prinz, W. (2001). The theory of event coding: A framework for perception and action planning. *Behavioral and Brain Sciences, 24*, 849–937.

Jovanovic, B., Király, I., Elsner, B., Gergely, G., Prinz, W., & Aschersleben, G. (2007). The role of effects for infant's perception of action goals. *Psychologia, 50*(4), 273–290.

Jovanovic, B., & Schwarzer, G. (2007). Infant perception of the relative relevance of different manual actions. *European Journal of Developmental Psychology, 4*(1), 111–125.

Király, I., Jovanovic, B., Prinz, W., Aschersleben, G., & Gergely, G. (2003). The early origins of goal attribution in infancy. *Consciousness and Cognition, 12*(4), 752–769.

Kuhlmeier, V. A., Troje, N. F., & Lee, V. (2010). Young infants detect the direction of biological motion in point-light displays. *Infancy, 15*, 83–93.

Libertus, K., & Needham, A. (2010). Teach to reach: The effects of active vs. passive reaching experience on action perception. *Vision Research, 50*, 2750–2757.

Libertus, K., & Needham, A. (2011). Reaching experience increases face preferences in 3-month-old infants. *Developmental Science, 14*(6), 1355–1364.

Longo, M. R., & Bertenthal, B. I. (2006). Common coding of observation and execution of action in 9-month-old infants. *Infancy, 10*(1), 43–59.

Meltzoff, A. N. (1995). Understanding the intentions of others: Re-enactment of intended acts by 18-month-old children. *Developmental Psychology, 31*, 1–16.

Meltzoff, A. N. (2002). Elements of a developmental theory of imitation. In A. N. Meltzoff & W. Prinz (Eds.), *The imitative mind: Development, evolution, and brain bases* (pp. 19–41). Cambridge: Cambridge University Press.

Meltzoff, A. N. (2005). Imitation and other minds: The "Like Me" hypothesis. In S. Hurley & N. Charter (Eds.), *Perspectives on imitation: From neuroscience to social science* (Vol. 2, pp. 55–77). Cambridge, MA: MIT Press.

Meltzoff, A. N. (2007). The "like me" framework for recognizing and becoming an intentional agent. *Acta Psychologica, 124*(1), 26–43.

Melzer, A., Prinz, W., and Daum, M. M. (2012). Production and observation of contralateral reaching: A close link by 12 months of age. Manuscript submitted for publication.

Moore, D. G., Goodwin, J. E., George, R., Axelsson, E. L., & Braddick, F. M. B. (2007). Infants perceive human point-light displays as solid forms. *Cognition, 104*, 377–396.

Moses, L. J., Baldwin, D. A., Rosicky, J. G., & Tidball, G. (2001). Evidence for referential understanding in the emotions domain at twelve and eighteen months. *Child Development, 72*, 718–735.

Piaget, J. (1953). *The origins of intelligence in the child.* London: Routledge & Kegan Paul.

Prinz, W. (1990). A common coding approach to perception and action. In O. Neumann & W. Prinz (Eds.), *Relationships between perception and action: Current approaches* (pp. 167–201). Berlin: Springer.

Prinz, W. (1997). Perception and action planning. *European Journal of Cognitive Psychology, 9*(2), 129–154.

Prinz, W. (2002). Experimental approaches to imitation. In A. N. Meltzoff & W. Prinz (Eds.), *The imitative mind: Development, evolution, and brain bases* (pp. 143–162). Cambridge: Cambridge University Press.

Proffitt, D. R., & Bertenthal, B. I. (1990). Converging operations revisited: Assessing what infants perceive using discrimination measures. *Perception & Psychophysics, 47*, 1–11.

Reid, V. M., Belsky, J., & Johnson, M. H. (2005). Infant perception of human action: Toward a developmental cognitive neuroscience of individual differences. *Cognition, Brain, and Behavior, 9*(2), 35–52.

Reid, V. M., Hoehl, S., Landt, J., & Striano, T. (2008). Human infants dissociate structural and dynamic information in biological motion: Evidence from neural systems. *Social Cognitive and Affective Neuroscience, 3*, 161–167.

Repacholi, B. M., & Gopnik, A. (1997). Early reasoning about desires: Evidence from 14- and 18-month-olds. *Developmental Psychology, 33*, 12–21.

Rovee, C., & Rovee, D. T. (1969). Conjugate reinforcement of infants' exploratory behaviour. *Journal of Experimental Child Psychology, 8*, 33–39.

Rovee-Collier, C., & Hayne, H. (2000). Memory in infancy and early childhood. In E. Tulving & F. Craik (Eds.), *The Oxford handbook of memory* (pp. 267–282). New York: Oxford University Press.

Rovee-Collier, C., & Shyi, G. (1993). A functional and cognitive analysis of infant long-term retention. In M. L. Howe, C. J. Brainerd, & V. F. Reyna (Eds.), *Development of long-term retention*. New York: Springer.

Sanefuji, W., Ohgami, H., & Hashiya, K. (2008). Detection of the relevant type of locomotion in infancy: Crawlers versus walkers. *Infant Behavior and Development, 31*, 624–628.

Simion, F., Regolin, L., & Bulf, H. (2008). A predisposition for biological motion in the newborn baby. *Proceedings of the National Academy of Sciences of the United States of America, 102*(2), 809–813.

Sommerville, J. A., Hildebrand, E., & Crane, C. C. (2008). Experience matters: The impact of doing versus watching on infants' subsequent perception of tool use events. *Developmental Psychology, 44*, 1249–1256.

Sommerville, J. A., & Woodward, A. L. (2005). Pulling out the intentional structure of action: The relation between action processing and action production in infancy. *Cognition, 95*, 1–30.

Sommerville, J. A., Woodward, A. L., & Needham, A. (2005). Action experience alters 3-month-old infants' perception of others' actions. *Cognition, 9*, B1–B11.

Tomasello, M. (1999). *The cultural origins of human cognition*. Cambridge, MA: Harvard University Press.

Uzgiris, I. C., & Hunt, J. M. (1975). *Assessment in infancy: Ordinal scales of psychological development*. Chicago: University of Illinois Press.

Woodward, A. L. (1998). Infants selectively encode the goal of objects of an actor's reach. *Cognition, 69*, 1–34.

Woodward, A. L. (1999). Infants' ability to distinguish between purposeful and non-purposeful behaviors. *Infant Behavior and Development, 22*, 145–160.

Wronski, C., & Daum, M. M. (2013). Motion cues evoke anticipatory shifts of covert attention during action observation. Manuscript in preparation.

15 Human Action Perception across Development

Jeff Loucks and Jessica Sommerville

Abstract

What do we pay attention to when we view other people's actions? How does such attention change over the course of development? In this chapter, we will discuss developmental changes in infants' attention during action perception—changes that shift their attention from being initially undifferentiated to being selectively focused. This perceptual tuning process is tied to infants' development of motor abilities and also supports a functional analysis of other people's behavior. Importantly, attention during action perception following the tuning is not rigid and fixed, but is instead selective and flexible. These attributes aid in predicting other people's actions and action outcomes, thereby increasing the power and efficiency of action perception for infants' broader social-cognitive development.

As adults, we are especially keen observers of others' behavior. We readily identify simple actions—grasping an object, pushing a button—with ease. Part of our skill in this regard undoubtedly stems from real-world experience with scripts, or schemas, of action (Schank & Abelson, 1977). For example, imagine yourself in the checkout line at the grocery store. Your identification of the actions of the clerk is aided by top-down knowledge of what typically happens in this scenario—the clerk will grasp your items, scan them, sometimes punch in numbers on a keypad, and send them down to the bagger. While such top-down information renders our processing of others' action more efficient, there are certainly additional, bottom-up perceptual mechanisms at work. For instance, you would readily be able to identify actions of the clerk that are not part of a typical checkout-line script—for example, suddenly opening a drawer, grasping a notepad, and smacking a fly you hadn't noticed on the cash register. Although these actions would capture your attention as they unfold, and you might wonder *why* they are being executed, you would have no trouble identifying *what* is being executed.

How is this process achieved? In particular, what perceptual information is attended to and encoded that allows for such rapid interpretation of others' behavior? While an answer to this question would provide valuable information about how adults process action, it is even more relevant to the study of infants' action perception. Infants are disadvantaged in their processing of action in at least two ways. First, they possess relatively limited information-processing abilities, such as working memory, with which to attend to and encode these various dimensions of action. In addition, they lack a great deal of top-down knowledge about human activities, and thus must rely on bottom-up processes to a greater degree in order to perceive others' action.

In this chapter, we will explore how attention to different perceptual dimensions of action changes throughout development. We will discuss research indicating that even though adults can attend to multiple dimensions of action, they often selectively attend to one particular dimension over others. We provide evidence that this bias develops in infancy and is the result of a process of perceptual tuning that takes place over the course of the first year of life. We will explore the possibility that this perceptual tuning may be caused by infants' developing motor capabilities. We will also examine whether this bias might serve to help infants more efficiently predict the actions of others. Finally, we will demonstrate that this bias may not be fixed, in that attention to perceptual dimensions of action appears to be selective and flexible according to context.

Adults' Attention to Dimensions of Action

What do we attend to when we observe other people performing actions? Even a simple action contains a great deal of perceptual information that could be attended to and encoded by the visual system. Take, for example, a simple act of reaching for and grasping an object. One could attend to the object being grasped, the type of grasp the actor is using, the spatial trajectory of the reach, the speed with which the reach is carried out, and so forth.

Loucks and Baldwin (2009) explored this question by investigating the nature of adults' ability to discriminate actions—that is, to identify that one action is distinct from another action. Examining action discrimination can shed light on what perceptual dimensions adults encode during action observation. With relatively little research on action perception to serve as a guide, Loucks and Baldwin looked to another domain of perception in which the nature of identification and discrimination has been heavily researched: face perception. Since, in everyday experience, both faces and human action are ubiquitous stimuli that demand efficient processing for effective social interaction, Loucks and Baldwin wondered if there might be similarities in how adults perceive and encode faces and action.

Research has identified at least two perceptual dimensions that people attend to for identifying and discriminating faces: (1) Featural information, which is local detail regarding facial features, such as the eyes and mouth, and (2) configural information, which is more global spatial-relational information regarding the arrangement of features (Maurer, Le Grand, & Mondloch, 2002). Research indicates that these two dimensions are distinct from one another in three ways. First, inverting a face—turning it upside down—uniquely impairs the processing of configural information (Freire, Lee, & Symons, 2000; Leder & Carbon, 2006; Mondloch, Le Grand, & Maurer, 2002). Second, the processing of these dimensions relies on distinct spatial frequencies: Featural information relies on relatively high spatial frequencies, and configural information relies on relatively low spatial frequencies (Goffaux et al., 2005). Finally, featural information is selectively attended to over configural information, in that adults are more sensitive to changes in featural information (see Freire et al., 2000; Mondloch et al., 2002; Yovel & Kanwisher, 2004).

In order to examine whether adults encode similar dimensions during action observation, Loucks and Baldwin (2009) defined analogous perceptual dimensions in human action. Featural action information was defined as relatively local detail regarding fine body motion. For example, the perceptual information concerning the particular type of hand contact an actor uses on an object (e.g., whole-hand vs. precision grasp) would be featural action information. In contrast, configural action information was defined as relatively more global, spatial trajectory information. For example, the perceptual information concerning the trajectory an actor uses to move an object (e.g., direct versus arcing path of motion) would be configural action information.

Two experiments were conducted in order to explore the nature of featural and configural action information in adults' action perception. Stimuli consisted of eight scenarios of human action. For each scenario, there was (1) a standard video, (2) a featural change video, and (3) a configural change video. For example, the standard video for one scenario depicted an actor reaching in a direct path toward a touch lamp and pressing down on it with his or her fingers, thus turning the lamp on. In the featural change video for this scenario, the actor used a closed fist to turn on the lamp—changing the local details of the action while keeping the global motion unchanged. The configural change video for this scenario depicted the actor reaching for the lamp with an arcing spatial trajectory instead of a direct spatial trajectory—changing the global motion while leaving the fine-motor motion unchanged. In each trial, participants were shown serially presented videos and asked to judge whether the videos were the same or different. In half of the trials, the same video file was shown twice; in the other trials, the standard video was paired with either the featural or configural change video.

The first experiment examined whether processing of configural action information, as in its facial counterpart, is selectively impaired with inversion. Thus, for half of the trials, the videos were inverted. As is the case in face processing, accuracy at detecting configural changes was significantly lower when action was presented upside down versus upright. In contrast, there was no inversion effect for featural action information. The fact that configural action information is disproportionately affected by inversion suggests that it is processed distinctly from featural action information.

Data from the upright condition confirmed that adults attend to both action dimensions during perception: Accuracy at detecting both types of changes was well above chance. However, as in face processing, featural action information appears to be selectively attended to over configural action information, in that adults were more accurate at detecting featural changes over configural changes in the upright orientation. This was the case even though configural changes were *objectively larger* perceptual changes relative to featural changes (in terms of the amount of pixel change). Since arms occupy more space than hands, changing the path of motion is a larger visual alteration than changing the type of grasp used. However, even when configural changes have an objective advantage in detection, adults selectively focus on featural information.

A second experiment examined whether processing of these two perceptual dimensions relies on relatively distinct high and low spatial frequencies. Participants engaged in the same forced-choice discrimination task with the same stimuli, except this time in the stimuli were low-pass filtered half of the trials. Low-pass filtering removes high spatial frequencies, effectively blurring the videos. Detection of the featural changes was significantly lower in the low-pass filtered condition relative to the full frequency condition. This finding indicates that processing of featural action information relies on high spatial frequencies. Interestingly, there was also a small but significant reduction in accuracy for the configural changes, which suggests that processing of configural action information may rely somewhat on high spatial frequencies. However, the effect of low-pass filtering on featural change detection was much more drastic—detection actually dropped significantly below chance for featural changes, while detection remained above chance for configural changes.

The findings of Loucks and Baldwin (2009) outlined in this section indicate that adults attend to two relatively distinct perceptual dimensions during action observation—local, detail-oriented featural information, and more global, spatial-relational configural information—but that they selectively attend to featural over configural information. Although faces and action are quite different classes of stimuli, these results suggest that they may share deeper processing characteristics (see also Loucks, 2011). In both domains, perception of global configural information is orientation specific, and featural information is selectively attended to or encoded with greater fidelity.

Although the evidence presented in the previous paragraphs suggests commonalities in the processing of faces and action, one key difference between these stimuli is the fact that action is inherently dynamic in nature. As such, an additional, salient perceptual dimension is temporal action information. Temporal information is perceptual information concerning the speed at which actions are carried out.[1] In unpublished research, Loucks and Baldwin have found that adults also attend to temporal action information during action perception, but that they attend to this dimension less than they do to featural information.

Thus, when adults observe the action of others, they attend to at least three different dimensions of action—configural, featural, and temporal. All three of these perceptual dimensions surely aid in goal inference. Returning to the checkout line example used at the beginning of this chapter, recall that the clerk might suddenly open a drawer, grasp a notepad, and use it to vanquish a fly. Attending to configural information would aid in perceiving the initial reach to toward the drawer as opposed to a reach toward the next food item on the conveyor belt. Attending to featural information would aid in perceiving the particular grasp used on the handle of the drawer and suggest that the clerk is opening it instead of pushing the drawer closed. And finally, attending to temporal information—in this case, information based on the speed at which the clerk is bringing the notebook down—aids in perceiving that the clerk's goal is to crush something.

If all three of these dimensions are useful for action identification, then why is featural information selectively attended to? Although the data from these studies cannot directly address this issue, adult observers of action presumably find this perceptual dimension particularly useful for some aspect of action processing. One possibility is that, although configural and temporal information are useful for processing the currently unfolding action, perhaps featural information is more often useful for predicting and anticipating subsequent actions. We will specifically address this possibility later on in this chapter.

Infants' Attention to Action Dimensions

Given that adults selectively attend to featural information, when and how does this perceptual bias develop? Although there are many possible routes to a featural bias, one possibility that we considered was that infants might initially attend broadly to multiple action dimensions equally, but then begin to lose sensitivity to configural and temporal information through a process of perceptual narrowing.

Perceptual narrowing is a term used to describe developmental trajectories in which younger infants begin with a broader set of perceptual capabilities relative to older children. The term focuses on the narrowing or loss of abilities; older infants have seemingly the same ability as younger infants, but they use this ability on a more restricted set of stimuli than younger infants do. The idea of perceptual narrowing runs counter to folk intuitions of development in that it documents an apparent loss of skill with increasing age or experience. Perceptual narrowing has been identified in a variety of domains, including phoneme perception (Werker & Tees, 1984), face perception (Kelly et al., 2007; Pascalis, de Haan, & Nelson, 2002), music perception (Hannon & Trehub, 2005a, 2005b), multisensory perception (Lewkowicz & Ghazanfar, 2006), and early visual perception (Dobkins, 2009). Because of its ubiquity, some have proposed that it is a domain-general phenomenon of development (Lewkowicz & Ghazanfar, 2009; Scott, Pascalis, & Nelson, 2007).

In the case of action perception, we hypothesized that younger infants might initially attend broadly to featural, configural, and temporal information equally, but that older infants would selectively attend to featural information over the other two action dimensions. In the language domain, perceptual narrowing typically refers to a loss in phonetic discrimination abilities. However, in face perception and music perception, the term usually refers to a decrease in discrimination abilities, but not an outright loss. Similarly, we presume that infants do not lose the ability to perceive and encode configural and temporal information. In this way, perhaps a better term to describe this hypothesized developmental pattern would be perceptual *tuning*.

In order to explore this possibility, we examined infants' attention to featural, configural, and temporal changes at two ages: four months and ten months (Loucks & Sommerville, 2012a). Stimuli consisted of one action scenario depicted in four different videos: (1) a standard video, (2) a featural change video, (3) a configural change video, and (4) a temporal change video. In the standard video, an actor reached to his right side for a toy cup, grasped around the side of the cup with a whole-hand grasp, and moved it in a straight path to his left side. In the featural change video, the actor used a precision grasp on the inside of the cup instead of a whole-hand grasp. In the configural change video, the actor used an arcing trajectory of motion to move the cup. In the temporal change video, the actor moved the cup twice as fast across the table (and performed the reach and release actions slightly slower to equate overall duration).

A habituation paradigm was employed, in which infants' attention to each dimension was measured via their looking duration. Infants in each group habituated to the standard video and, following habituation, were shown each of the featural, configural, and temporal change videos (twice each, in alternating orders). Plotted in figure 15.1 are infants' recovery scores for the test events: Their average looking time for each test event over and above their average looking time during the last three trials of habituation. As is evident, infants' attention differed between four and ten months of age. The ten-month-old infants demonstrated the same selective attention bias as adults: Recovery was significant for the featural change, but not for the other two changes (though there was a marginal trend for recovery to the temporal change). Recovery to the featural change was also significantly greater in comparison to scores for the other two changes. The four-month-olds, on the other hand, demonstrated significant recovery to all three changes and did not differ in their recovery. Thus, four-month-olds have relatively broad, undifferentiated attention to perceptual dimensions of action, and ten-month-olds have much narrower attention that is focused primarily on featural information.

Importantly, we believe these results indicate only that development in infancy leads to lessened attention to configural and temporal information. Since adults attend to all three perceptual dimensions, we would not expect that infants lose the ability to process configural and temporal information only to gain it back again. Furthermore, both configural and temporal information are key to goal inference on a daily basis. Spatial

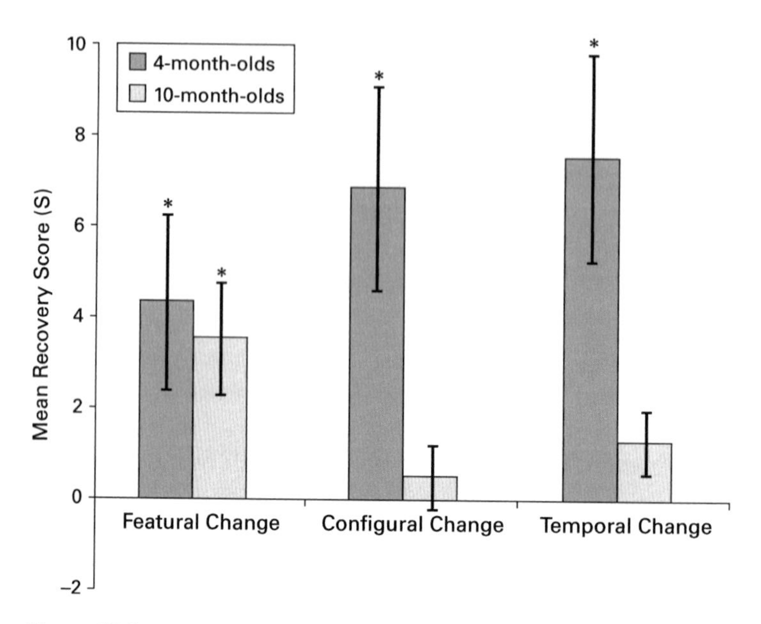

Figure 15.1
Mean recovery scores as a function of the type of action change for four-month-olds and ten-month-olds. The asterisk indicates significant recovery from habituation. Error bars represent standard error.

trajectories of heads and arms inform the observer as to the particular target of a person's reach, and perception of action speed can inform action category judgments (e.g., placing something gently vs. slamming it down). Thus, we believe our data indicate that infants' sensitivity to configural and temporal information declines with age, while sensitivity to featural information is maintained.

An exciting question for future research is whether perceptual tuning in action perception leads to an enhancement in the processing of featural information. In the speech domain, research indicates that the loss in discriminating nonnative contrasts is accompanied by an increase in the ability to discriminate native contrasts (Kuhl et al., 2006). In the same manner, perhaps infants' ability to discriminate changes in featural information is increased at ten months in comparison to four months. Such a finding would highlight a functional role for perceptual tuning in action perception.

As of yet, it is unknown whether there are any additional distinctions between infants' processing of featural and configural information. In particular, we know that adults' processing of configural information is uniquely sensitive to orientation, but we do not know whether this is true for infants' perception of this dimension. One possibility is that processing of configural information is not orientation specific at young ages (e.g., four months) but that this characteristic develops over time. If so, perhaps there is a qualitative change in the processing of configural information concurrent with the perceptual tuning we observed taking place between four and ten months of age. Further research is needed to elucidate this possibility.

Overall, these findings indicate that the featural bias in adults cannot be traced back to an innate or default bias in infancy. Instead, this bias develops out of a process of perceptual tuning in infancy in which infants initially attend broadly to multiple dimensions of action, and then decrease their attention to configural and temporal information over time. Featural information thus appears to be attended to at the same rate throughout the lifespan.

Motor Experience and Perceptual Tuning

Infants' perceptual experience plays a causal role in the phenomenon of perceptual tuning. In phoneme perception, it is clear that infants' auditory experience shapes their discrimination abilities, as infants only maintain sensitivity to phonetic contrasts that are audible in their everyday environment. Infants can also maintain sensitivity to nonnative contrasts if given nonnative language exposure prior to the onset of the tuning (Kuhl, Tsao, & Liu, 2003). Similarly, infants' experience with particular human faces in their everyday lives shapes perceptual tuning in the face domain, and experience with other-species faces and other-race faces can alter the typical course of tuning in face perception (Pascalis et al., 2005; Scott & Monesson, 2009). If perceptual tuning in the action domain is the result of experience, what is the relevant experience?

Research indicates that action production and observation use a shared computational system (Prinz, 1997; Rizzolatti & Craighero, 2004). It is thought that representations used for motor planning are used in the service of perceiving others' actions, and vice versa. Evidence to support this claim comes from behavioral research with adults (Brass, Bekkering, & Prinz, 2001; Knoblich & Flach, 2001) and neuroimaging studies with adults (Decety et al., 1997; Hari et al., 1998).

There is also considerable evidence that connections between action production and action perception are present in early infancy. The bulk of this work has focused on infants' understanding that actions are directed toward particular goals or targets (e.g., Woodward, 1998). Woodward and Guajardo (2002) demonstrated that infants' own ability to produce referential points correlated with their ability to understand pointing as a goal-directed action in others. Sommerville and Woodward (2005) also found a relation between infants' own ability to perform a means-ends action (i.e., pulling a cloth in order to obtain a toy that is resting on the cloth) and their understanding of the goal structure of that same means-ends action in others. Research also indicates that motor experience can play a causal role in this process: Active experience with a novel tool-use action facilitates goal understanding, but observation of those same actions does not (Sommerville, Hildebrand, & Crane, 2008), and providing nongrasping infants with grasp-like experience increases goal understanding relative to infants not given such experience (Sommerville, Woodward, & Needham, 2005).

Given the findings outlined in this section thus far, we wondered if the relevant experience that drives perceptual tuning in the action domain might be infants' first-person motor experience. Motor experience could contribute to perceptual tuning in two distinct ways. On the one hand, it could increase infants' observational experience with actions (visually observing their own behavior), and on the other hand, it could be the motor acts themselves that provide unique proprioceptive experience. We hypothesize that it is the actual motor experience that drives perceptual tuning, for two reasons. First, as mentioned previously, there is evidence that action production and perception share a common underlying neural architecture (Hari et al., 1998; Prinz, 1997). In addition, the link between infants' goal understanding and their active experience is based on motor experience, not observational experience (Sommerville et al., 2008).

In particular, we wondered if infants' experience with grasping might influence their attention to different action dimensions over the course of development. For infants below the age of five-to-six months, grasping is highly uncoordinated—they may be able to grasp something if it is placed in their hand, but reaching out and securely grasping an object is beyond their motor capabilities (White, Castle, & Held, 1964). Our findings indicate that infants younger than this age (four months) attend broadly to multiple dimensions of action. As the ability to reach in a directed fashion and securely grasp objects emerges, this may provide infants with unique proprioceptive information about the functional aspects of grasping. Infants may learn that the way in which their hand contacts an object has important consequences for whether a grasp is successful or not. Infants may come to learn that other dimensions of their own action—the spatial trajectory of their reaches or the speed of their reaches—are less important than the grasp itself. This motor information may then lead to paying less attention to configural and temporal information in others' behavior.

In order to explore this idea, we recently conducted an experiment with six-month-old infants (Loucks & Sommerville, in preparation). At this age, the process of perceptual tuning may already be underway, and since coordinated grasping is only a recent achievement at this age, we expected a relatively high degree of individual differences in this motor capacity. We hypothesized that individual differences in grasping skill would be correlated with individual differences in infants' attention to perceptual dimensions in others' action. In particular, we predicted that infants who were *better* at grasping would pay *less* attention to configural and temporal action information.

Infants participated in a habituation task and an action task. The habituation task was identical to the one described previously with four- and ten-month-olds. This task measured infants' attention to featural, configural, and temporal information in others' action. The action task was designed to measure infants' skill at grasping objects. Both tasks were administered, in counterbalanced order, to all infants.

The stimuli for the action task were four wooden balls that varied in size and color affixed to the ends of four wooden dowels. During the task, infants were presented with each of these objects and encouraged to grasp the ball end of the objects. During each trial, the ball end was placed midline in front of the infant, within the infant's reach, and the experimenter tapped the ball and verbally encouraged infants to grasp. Infants were given three blocks of trials, with each stick presented once in each block, for a total of twelve grasping trials.

Infants' raw grasping responses during the action task were coded offline by two naïve observers. Grasps were coded as being either *successful*, in which the infant securely grasped the ball end of the toy within their hand, or *unsuccessful*, in which the infant did not secure the ball in their hand (most often these were slips off the ball or grasps behind the ball, on the stick). We then calculated each infant's *efficiency score*: the proportion of successful grasps divided by the total number of grasps. Efficiency scores could range from zero (all grasps during the task unsuccessful) to one (all grasps during the task successful). Recovery scores for each infant were also calculated for each of the featural, configural, and temporal changes by subtracting the average looking time for the last two habituation trials (the standard video) from the average looking time for the average of the two test trials for a particular change.

As predicted, efficiency scores during the action task were significantly negatively correlated with recovery to the configural change and the temporal change. That is, the better infants were at securely grasping the ball, the less attention they paid to changes in configural and temporal information in another person's action. Interestingly, efficiency scores were not significantly correlated with recovery to the featural change. Thus, increased skill at grasping was not related to increased attention to featural action information.

We also performed categorical analyses of the data by categorizing infants, based on a median split, as either efficient or inefficient graspers. This analysis was consonant with the correlational analysis: Inefficient graspers recovered significantly more to the configural and temporal changes in comparison to the efficient graspers. In addition, efficient and inefficient graspers did not differ in their recovery to the featural change.

Perceptual tuning in action perception leads to a bias in attending to featural information relative to configural and temporal information. Thus, another way to analyze the data was to compute featural bias scores for each infant: The amount of recovery to the configural and temporal changes subtracted from the amount of recovery to the featural change. Featural bias scores were significantly positively correlated with efficiency scores: As grasping skill increased, so did the tendency to look longer at the featural change relative to the other two changes. In addition, categorical analyses confirmed that efficient graspers had significantly higher featural bias scores than did inefficient graspers. We also computed similar configural bias scores and temporal bias scores. Neither of these scores was significantly correlated with efficiency, nor did categorical analyses reveal any significant differences between the two grasping groups on these measures.

In sum, these findings indicate that the relevant experience that may be driving perceptual tuning in the domain of action may be infants' active experience with grasping. Prior to the onset of grasping, infants attend broadly to multiple perceptual dimensions in others' action. As they learn to grasp, they may begin to understand that the particular path one takes to get to an object (configural information) or the speed with which one reaches toward an object (temporal information) are often not as critical for the purposes of grasping as the particular way in which the digits of their hands contact an object (featural information). This change in their own behavior may then shape their perception of others' actions: They may attend less to configural and temporal information in others' action but maintain focus on featural information. In this way, attention to featural information remains steady across development. These findings also expand the role of motor experience in infants' perception of others' action: Motor experience plays a role not only in understanding the goals of action (e.g., Sommerville et al., 2005), but also in infants' online perception of action dimensions.

There are two important caveats for our study, however. First, these findings do not rule out the possibility that active experience merely provides infants with additional visual experience with grasping actions and that this increased experience drives perceptual tuning in action perception. Given previous findings, we believe this possibility is unlikely (Sommerville et al., 2008). However, if this were true, then giving pregrasping infants an opportunity to observe someone else grasping objects might result in the same reduction in attention to configural and temporal information. Second, our findings also do not specify the casual direction of the relation between motor experience and perceptual tuning. It is possible that perceptual tuning in the perception of others' action leads to infants' having insight into their own motor behavior. It is also possible that another factor, such as infants' goal understanding, is the causal factor. As stated earlier, infants' ability to understand the link between an actor's reach and that actor's goal emerges at between five and six months of age (Woodward, 1998). In addition, at as early as three months of age, infants' active experience with grasping—and more specifically, with obtaining objects—can causally influence their understanding that reaches are goal directed (Sommerville et al., 2005). Thus, there may be a complex three-way relationship between motor experience, goal understanding, and attention to action dimensions. Future studies may be able to uncover the underlying causal relationship among these factors.

Featural Information and Action Prediction

Up until now, we have focused in this chapter on (1) characterizing the nature and development of attention during action perception and (2) exploring a potential cause of the development of the featural bias. But what is the function of this bias? Why would development lead to this bias toward featural information at the potential cost to attending to configural and temporal information?

Perhaps selective attention to featural information is beneficial for *predicting* others' future actions. A wealth of research indicates that prediction is a fundamental aspect of action perception. Studies with adults (Flanagan & Johansson, 2003; Woodward & Cannon, this volume) and infants

(Gredebäck & Kochukhova, 2009; Kanakogi & Itakura, 2011; Kochukhova & Gredebäck, 2010; Paulus, Hunnius, & Bekkering, 2011; Paulus, Hunnius, van Wijngaarden, et al., 2011; Rosander & von Hofsten, 2011; Southgate & Csibra, 2009; Woodward & Cannon, this volume) indicate that observers predict or anticipate the goal of other people's hand movements with their eye gaze. This ability develops early on and seems to be tied to infants' motor development (Cannon et al., 2012; Kanakogi & Itakura, 2011).

Intuitively, we all know that particular types of grasps are used for particular purposes: A pen is grasped with a specific grip so that it may be used for writing, a fork with a different grip so that it may be used for eating, and so forth. Research with adults also indicates that people use particular grasps depending on how an object will ultimately be used. People are more likely to select a grasp that will result in a comfortable hand position at the end of an action (the "end state comfort effect," Rosenbaum, Vaughan, Barnes, & Jorgensen, 1992). Beginning around the age of twelve months of age, infants gradually become sensitive to such dimensions of grasps (McCarty, Clifton, & Collard, 1999; McCarty, Clifton, & Collard, 2001), and this sensitivity is ultimately crucial for learning about different tools in their culture (Lockman, 2000).

Along these same lines, using a particular grasp on an object constrains what can be done with that object in the future. For instance, imagine observing someone grasping a hairbrush. If the person grasps the hairbrush with a whole-hand grasp, you might predict that the next action the person will perform is to brush his or her hair. However, if the person grasps the hairbrush with a precision grasp, it would be difficult and awkward to use this grasp for hair brushing. In this way, using a particular grasp thus has *functional consequences*: It constrains the possible future actions one can take with an object.

Adults and ten-month-old infants may selectively attend to featural information because (1) they are sensitive to the functional consequences of actions, and (2) they use this information for the purposes of action prediction. Knowledge about what actions cannot be performed in the future constrains adults' and infants' predictions about possible future actions. Note that we are hypothesizing a more complex type of predictive process than action anticipation. However, the fact that even some six-month-old infants anticipate others' actions (Kanakogi & Itakura, 2011) supports the possibility that selective attention to featural information at ten months of age might be related to a predictive process.

We might also expect an understanding of the functional dimensions of grasps to be correlated with observers' motor experience with various grasps: Prior experience with particular grasps could provide grasp-specific proprioceptive information, which the action processing system may capitalize on for action prediction. For instance, experience using whole-hand grasps—with all fingers and thumb enclosing the object toward the palm—could result in knowledge that such grasps are useful for lifting heavy objects. When an observer with such experience views someone use a whole-hand grasp on a heavy object, they can use this information to predict that lifting is a possible subsequent action. Experience with a variety of grasps may be especially powerful in this respect. For example, contrasting how whole-hand grasps differ from precision grasps—those with one finger and thumb opposed—could result in additional knowledge about the functionality of each type of grasp.

Thus, observers who selectively attend to featural information may also be sensitive to the functional consequences of different grasps. In addition, sensitivity to functional information of this nature might also be related to an observers' active experience with various grasps. We conducted an experiment that explored both of these possibilities in ten-month-old infants (Loucks & Sommerville, 2012b). Recall that the featural bias is already present at this age, as perceptual tuning in action is already complete.

Infants in this experiment participated in (1) a habituation task designed to probe infants' understanding of the functional consequences of precision and whole-hand grasps and (2) a grasping task designed to measure infants' ability to perform precision grasps in their own behavior. The stimuli for the habituation task included a green bowl decorated in such a way that it was easy to tell if the bowl was upright (with the concave surface pointing upward) or inverted (see figure 15.2). For this particular bowl, when it was upright it was possible to grasp and move the bowl using a precision grip (with the index and middle fingers on the inside wall of the bowl and the thumb on the outside wall, in line with the fingers); a whole-hand grasp (with all of the fingers and thumb reaching across the top) was not functional for moving the bowl. However, when the bowl was inverted the functionality of these two grasps reversed.

Infants' understanding of the functional dimensions of precision and whole-hand grasps was assessed using two experimental conditions: the Upright-to-Inverted (UI) condition and the Inverted-to-Upright (IU) condition. Each infant participated in only one condition. The UI condition assessed infants' understanding of the functional consequences of using precision grasps, and the IU condition assessed infant's functional understanding of whole-hand grasps (figure 15.2).

Upright-to-Inverted Condition

Habituation

Test Events

Whole-hand grasp (New, Functional)

Precision Grasp (Same, Nonfunctional)

Figure 15.2
A schematic of the habituation task.

Inverted-to-Upright Condition

Habituation

Test Events

Precision Grasp (New, Functional)

Whole-hand Grasp (Same, Nonfunctional)

Figure 15.2
(continued)

In the UI condition, the bowl began in the upright configuration, and infants were habituated to an actor using a two-finger, precision grasp to move the bowl across a table. Following habituation, the bowl was turned upside down, and this change was shown to infants. Next, infants were shown two test trials, twice in alternating order: (1) The whole-hand event, in which the actor grasped the bowl with a new, but functional, whole-hand grasp, or (2) the precision event, in which the actor grasped the bowl with the old, but now nonfunctional, precision grasp. Importantly, the bowl was not moved across the table—the actor stopped her movement at the point of contact with the bowl.

In the IU condition, the bowl began in the inverted configuration, and infants were habituated to the actor using a whole-hand grasp to move the bowl across the table. Following habituation, the bowl was now turned upright, and infants were shown (1) the precision event, in which the actor used a new, but functional, precision grasp, and (2) the whole-hand event, in which the actor used the old, but now nonfunctional, whole-hand grasp.

The stimuli for the grasping task included five colorful "sticks" (stacked pieces of Lego). The sticks were presented inside of one of two transparent plastic containers: the wide container or the narrow container. While both containers were approximately the same height as the sticks, the wide container was approximately twice the width of the sticks, and the narrow container was only slightly wider than the sticks. Critically, when the sticks were in the narrow container, a precision grip was the easiest and most efficient way to remove the toy from the container. Each trial, an infant was presented with one of the sticks inside one of the containers and was encouraged to retrieve the stick. The first two trials were warm-up trials in which the sticks were placed in the wide containers. The remaining three trials were test trials in which the sticks were presented in the narrow container.

The grasp used to ultimately remove the stick from the narrow container was coded along two dimensions: (1) The type of grasp (i.e., precision or otherwise) and (2) the planfulness of the grasp (i.e., whether infants pre-configured their grasp). From this data, we categorized infants into groups of precision and nonprecision graspers. Precision graspers were those infants who produced at least one planful precision grasp, and nonprecision graspers were those infants who did not perform any planful precision grasps. Importantly, some nonprecision graspers may have produced some precision grasps, but they were never *planful* precision grasps.

As is clear from figure 15.3, infants' looking times were a function of both habituation condition and motor performance. In both conditions, precision graspers recovered to both the perceptual change (using a new grasp) and the functional change (using the old, nonfunctional grasp). Nonprecision graspers behaved differently: They only recovered to the perceptual change in each condition. We also found a significant correlation between infants' motor performance and their functional understanding: In both conditions, infants who produced more planful precision grasps tended to look longer at the nonfunctional grasp over the perceptually new grasp.

Thus, infants who are relatively skilled at precision grasping are able to evaluate the functional consequences of using both a precision grasp and a whole-hand grasp. Infants with less skill at precision grasping do not appear to understand the functional consequences of either type of grasp. These results indicate a connection between *prospective* motor behavior and *predictive* perceptual processing. Only infants who were able to plan a precision grasp in advance demonstrated an understanding of functional consequences, and the more prevalent this planning behavior was, the stronger the understanding.

As discussed earlier, these results suggest that contrastive experience with different grasps is critical for understanding functional consequences. If experience with a particular grasp was sufficient for understanding the functional dimensions of that grasp, then nonprecision graspers should have recognized that the whole-hand grasp on the upright bowl was nonfunctional in the IU condition (as all infants have experience with whole-hand grasps). However, this was not the case. Infants without contrastive experience for each type of grasp do not appear to have functional understanding for the single grasp they do have experience with.

One of our aims was to discover whether ten-month-olds selectively attend to featural information for the purposes of action prediction. Although our results partly support this hypothesis, they are likely not the whole story. First, ten-month-olds as a group (ignoring motor skill) did not evaluate the functional consequences of precision and whole-hand grasps: Only particularly skilled infants demonstrated this understanding. However, infants at this same age from our previous study (Loucks & Sommerville, 2012a), as a group, do selectively attend to featural information. Although we did not assess infants' motor abilities in that study, it is likely that there was a similar degree of variability in motor skill in that sample. Second, we have also found that six-month-olds are at the beginning stages

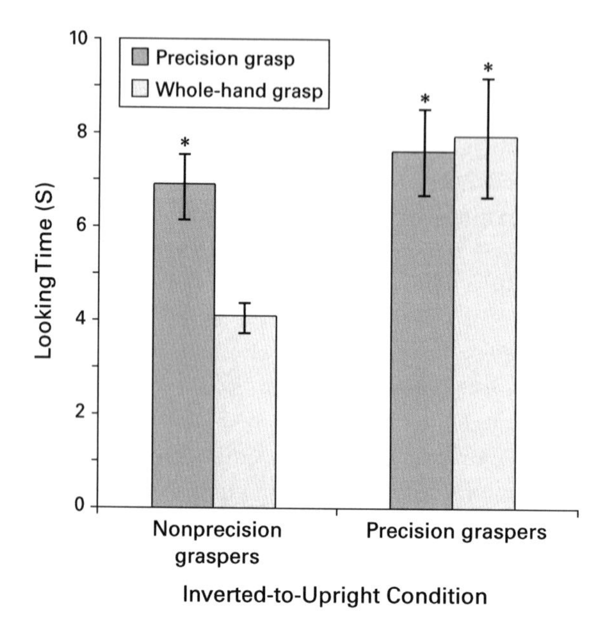

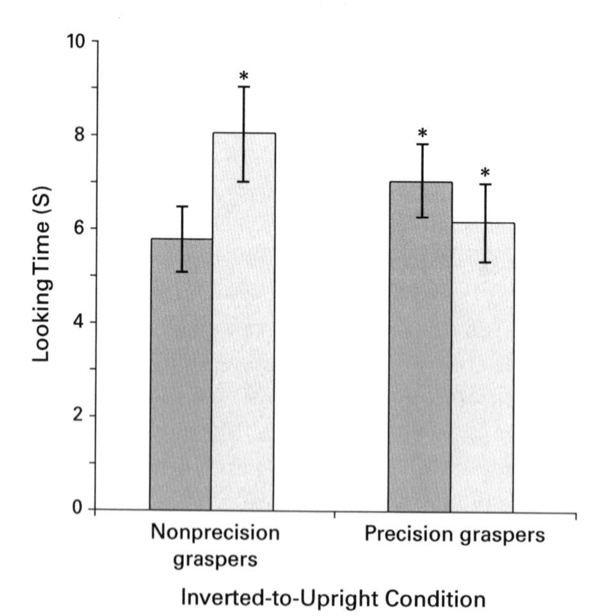

Figure 15.3
Mean looking times to the test events as a function of grasp type, action group, and condition. The asterisk indicates significant recovery from habituation. Error bars represent standard error.

of showing this selective attention bias, and they are likely considerably less skilled at precision grasping.

However, recent evidence indicates that infants' early experience with precision grasping influences their action prediction and perception of grasps. Daum, Prinz, and Aschersleben (2011) examined six-month-old infants' expectations that the preconfiguration of a hand into either a large or small hand aperture during a reaching movement would lead to a particular type of final grasp—either a whole-hand or precision grasp, respectively. They also assessed infants' grasping skills in a separate task and coded whether infants were able to perform thumb-opposite grasps (thumb-opposite group) or only palmar grasps (palmar group). Only the thumb-opposite group formed expectations in the perception task regarding the final grasp based on hand aperture size.

Although this study did not specifically examine infants' understanding of functional consequences, it nonetheless indicates that infants' production and perception of action are related early on—and in ways that would support the learning of the functional dimensions of grasps. In any case, whatever the original cause of the featural bias, evaluating the functional consequences of grasps may further strengthen this bias in infancy. As infants approaching one year of age begin to learn about more complex activities, such as tool use, selectively attending to featural information for the purposes of predicting future actions greatly increases the efficiency of infants' action perception—and likely accelerates their learning from others.

Flexibility in Attention to Action Dimensions

The actions that other people can perform are numerous. Not only are there a variety of different possible actions, but also many of these actions can be performed using various objects and can occur in a wide variety of different contexts. In order to cope with such variability, the action processing system should be flexible with respect to what perceptual dimensions are attended to and encoded.

So far we have demonstrated that (1) infants and adults selectively attend to featural information for simple object-directed actions, such as reaching and moving actions, and (2) this bias may support a functional understanding of grasps. Although this may be a pervasive bias in action perception, it is also possible that this bias is not rigid. Different actions might require selective attention to a different dimension of action for the purposes of predicting future actions or action outcomes. The perceptual information that might be relevant for action prediction might also vary as a function of context.

For instance, imagine someone dropping an object into a container. For such an action, attending to configural information may be especially beneficial—and potentially more beneficial than attending to featural information. In particular, the height at which something is dropped (configural information) can influence the outcome of the drop. When an object is dropped from relatively higher up, there is an increase in the possibility of targeting error—that is, the ultimate location of the drop will be more variable. Thus, if an observer were trying to predict the outcome of the drop, attention to configural information regarding drop height might be elevated in priority. This might be especially true in particular contexts. For instance, if an actor is trying to drop the object into a relatively wide container, then increasing the drop height is not so relevant to predicting the outcome: Even if the targeting is slightly off, the object will still, given the container's width, likely land in the container. However, if the actor is trying to drop the object into a relatively narrow container, then attending to drop height may be especially critical for prediction, as the increased variability in targeting may exceed the width of the container, and the object may miss.

Thus, we wondered if attention to action dimensions might be selective and flexible as a function of context and the observers' tendency to predict action. Research indicates that adults' and infants' action anticipation varies as a function of context: Neither age group anticipates the motion of inanimate objects (Falck-Ytter et al., 2006; Woodward & Cannon, this volume), and infants' goal anticipation is modulated as a function of the type of action (Gredebäck et al., 2009). However, these studies do not tell us whether observers attend differentially to perceptual information in action as a function of action prediction and context. To our knowledge, no research to date has examined this particular issue.

A perceptual system that is flexible according to context in the service of action prediction seems especially valuable for younger infants. Since infants have limited experience with human activities and minimal information-processing abilities, efficiency of processing is critical. If infants can guide their processing of action to focus only on *relevant* perceptual dimensions depending on the context, then they can utilize their limited observational skills to maximize their processing of others' action.

Thus, we recently carried out a set of experiments to determine whether adults' and younger infants' attention to action dimensions is modulated as a function of predictive processes and the context in which the action occurs (Loucks & Sommerville, under review). We were also interested in assessing whether perception of dropping actions might be related to observers' production of dropping. Perhaps sensitivity to configural information in others' dropping actions is based on observers' active experience with dropping. An observer who understands how drop height influences targeting may (1) perform better at precision dropping and (2) be more sensitive to drop height in others' action.

Because we were investigating a previously unexplored aspect of action prediction, we first carried out an experiment with adults. Adults participated in both a perception task and a dropping task. The stimuli for the perception task involved videos of three different actors successfully dropping a small beanbag into both a wide and a narrow container. Each of the actors dropped the beanbag from five different body-centered heights: Hip height, middle torso height, shoulder height, forehead height, and overhead height. Importantly, the only difference between the videos using the wide and narrow containers was the container itself. Still frames were taken from each of these videos for use in the perception task. Each still frame depicted an actor holding the beanbag over the container at one of the five heights.

During each trial of the perception task, adults watched a video of a dropping event and were then presented with a still frame, at which point they were asked to determine whether the frame was from the video or not. Still frames for half of the trials depicted the same height used in the video, while the other half depicted a different height. Half of the adults only viewed dropping into the wide container (wide condition), while the other half only viewed dropping into the narrow container (narrow condition). We hypothesized that adults would be especially accurate at detecting increases in height (upward changes) relative to decreases in height (downward changes) in the narrow condition, since dropping from higher up increases the chance of a miss. We hypothesized that there would be no such difference in the wide condition.

In the dropping task, adults attempted to drop the same beanbag from the same five body-centered heights into the same narrow container used in the videos. During each trial, after lifting the beanbag to a specified height, an experimenter positioned the container on the floor in front of the participant. Several floor positions were used in order to collect multiple trials for each height. The experimenter coded whether or not each drop was successful.

Considering first the results from the dropping task, analyses revealed the predicted pattern between drop height and accuracy. As drop height increased, dropping accuracy decreased; this was reflected in a significant linear trend to the data. Thus, increasing drop height does indeed increase the variability in specifically targeting a drop location. To our knowledge, this is the first empirical evidence of this phenomenon.

In analyzing the results from the perception task, we focused only on trials in which the height changes were one step upward or downward (e.g., from shoulder to forehead and shoulder to torso). This was a measure of how finely sensitive adults were at detecting height changes. The analysis confirmed the hypothesized pattern: Adults were significantly more accurate at detecting one-step upward changes relative to one-step downward changes in the narrow condition, but were equally accurate at detecting these same two height changes in the wide condition. These findings indicate that action perception in adults is predictive and flexible according to context. Adults appear to be attempting to predict the outcome of the dropping action; thus, they key on information that will be relevant for this prediction. However, adults increase attention to this information only when it is directly relevant to predicting the outcome. Such selective and flexible attention no doubt enhances the efficiency of action perception.

Finally, we also examined the relation between perception and production of dropping actions. As we hypothesized, there was a significant positive relation between overall performance on the perception task and overall performance on the dropping task. That is, adults who were more accurate at targeting their drops tended to be more accurate at detecting changes in drop height in another's dropping actions. Although this finding adds to an existing literature on the common coding of action production and perception, it is the first evidence to indicate a relation between production and perception for dropping actions and, also, the first evidence that motor production is related to the encoding of action dimensions in others. These results may indicate that adults' experience with dropping leads to an increase in attention to configural information for dropping actions. Motor experience may increase observers' awareness that targeting variability increases with increasing height, which then influences perception of this dimension more broadly.

Having established the basic dropping phenomena and having revealed that action perception is selective and flexible in adults, we next investigated whether ten-month-old infants' action perception also shares these attributes. We designed a habituation task that closely matched the perception task used with adults. The stimuli were three videos of precision dropping—a standard video, a higher height video, and a lower height video—filmed for involving both a narrow container and a wide container (resulting in six videos total). Different groups of infants participated in each of the container conditions. For both conditions, infants were first habituated to the standard video, in which the actor reached for and lifted an object (a purple elastic ball) over the transparent container and successfully dropped it in. Following habituation, infants were shown (1) the higher height video, in which the object was dropped from higher than in the standard video, and (2) the lower height video, in which the object was dropped from lower than in the standard video. Importantly, the height differences were matched between conditions, and the actor moved in sync with a metronome to equalize the duration of each movement. Thus, the only difference between conditions was the width of the container.

As predicted, we found an interaction between drop height and container width on infants' looking times. In the narrow container condition, infants looked significantly longer at the higher height over the lower height drop, but in the wide container condition, no significant preference was observed. These results indicate that by ten months of age, infants' attention to action dimensions, like that of adults, is modulated as a function of action prediction and the context in which action occurs. The fact that these attributes can be observed at a relatively early age suggests that they are basic and fundamental attributes of action perception.

While these results indicate that action perception is flexible, they do not speak to whether the featural bias extends to other actions. We did not compare infants' attention to featural changes for dropping actions, and thus we cannot say whether infants attend *more* to featural or configural information in this context. Adults' attention to configural information in dropping actions did seem elevated compared to their attention to configural changes in Loucks and Baldwin (2009) (average accuracy approximately 15 percent higher), but differences between the tasks make a proper comparison difficult. Future research should examine whether selective attention to featural information is ubiquitous in action perception.

Thus, attention during action perception is not rigid and fixed; rather, it is flexible early on in life and remains flexible throughout the lifespan. Such flexibility is an important asset for infants' analysis of human behavior. As there is a large amount of perceptual information that could be attended to for even a simple action, selective and flexible sampling of information allows infants to focus attention on what is most relevant, which helps infants to optimize their learning from others' action.

Conclusion

As we view the complex actions of other people, we selectively sample distinct dimensions of the action for further analysis. This chapter has focused on attention to three distinct dimensions of action: (1) Featural information regarding fine-motor detail, (2) configural information regarding global spatial relationships, and (3) temporal information regarding action speed. For many object-directed actions, adults selectively attend to featural information over the other two dimensions. However, this attentional bias may not be fixed; evidence indicates that attention to action dimensions can be modulated according to context in order to serve predictive processes.

Our research has demonstrated that infants do not begin with this featural bias. Infants first attend broadly to these multiple dimensions of action, and do not selectively attend to any one dimension over another. Through a process of perceptual tuning, infants attend less to configural and temporal information and maintain their level of attention to featural information; by ten months of age, infants appear to be attending to action in the same manner as adults. This achievement may go hand in hand with an ability to understand the functional consequences of different hand grasps for the purposes of action prediction. But as with adults, infants' attentional bias may not be fixed; their attention during action perception is similarly flexible in the service of action prediction.

Throughout this chapter, we have also shown that observers' production of action is related, from infancy through adulthood, to their perception of action. Perceptual tuning in action perception may be mediated by infants' developing grasping skills, and infants' understanding of functional consequences may be mediated by infants' developing use of multiple different types of grasps. We have also shown that adults' attention to specific perceptual dimensions in action may be mediated by their motor experience with specific types of actions. Such relationships are reminiscent of Piaget's theory that perception and production of action are tightly linked in development (Piaget & Inhelder, 1962) and support more contemporary theories of action perception and production (Prinz, 1997). The findings from this chapter thus further underscore the importance of infants' developing motor capacities in their cognitive development, and urge us to consider the role of infants' understanding of themselves in our study of their understanding of others.

Acknowledgments

The research presented in this chapter was supported by a National Research Service Award postdoctoral fellowship (F32HD058445-01A2) to the first author and a National Institute of Child Health and Human Development grant (1R03HD053616-01A1) to the second author. We are grateful to Jen Paur and other members of the Early Childhood Cognition Lab for their assistance with data collection and coding, and we would also like to thank all of the families who volunteered to participate in this research.

Note

1. Another form of temporal information in action is sequential information—the order in which multiple goal-directed actions are carried out. However, as we are primarily concerned with the perception of single, goal-directed actions, we have defined temporal information as the speed at which actions are carried out.

References

Brass, M., Bekkering, H., & Prinz, W. (2001). Movement observation affects movement execution in a simple response task. *Acta Psychologica, 106*(1–2), 3–22.

Cannon, E. N., Woodward, A. L., Gredebäck, G., von Hofsten, C., & Turek, C. (2012). Action production influences 12-month-old infants' attention to others' actions. *Developmental Science, 15*(1), 35–42.

Daum, M. M., Prinz, W., & Aschersleben, G. (2011). Perception and production of object-related grasping in 6-month-olds. *Journal of Experimental Child Psychology, 108*(4), 810–818.

Decety, J., Grézes, J., Costes, N., Perani, D., Jeannerod, M., Procyk, E., et al. (1997). Brain activity during observation of actions: Influence of action content and subject's strategy. *Brain: A Journal of Neurology, 120*(10), 1763–1777.

Dobkins, K. R. (2009). Does visual modularity increase over the course of development? *Optometry and Vision Science, 86*(6), E583–E588.

Falck-Ytter, T., Gredebäck, G., & Von Hofsten, C. (2006). Infants predict other people's action goals. *Nature Neuroscience, 9*(7), 878–879.

Flanagan, J. R., & Johansson, R. S. (2003). Action plans used in action observation. *Nature, 424*(6950), 769–771.

Freire, A., Lee, K., & Symons, L. A. (2000). The face-inversion effect as a deficit in the encoding of configural information: Direct evidence. *Perception, 29*(2), 159–170.

Goffaux, V., Hault, B., Vuong, Q. C., & Rossion, B. (2005). The respective role of low and high spatial frequencies in supporting configural and featural processing of faces. *Perception, 34*(1), 77–86.

Gredebäck, G., & Kochukhova, O. (2009). Goal anticipation during action observation is influenced by synonymous action capabilities, a puzzling developmental study. *Experimental Brain Research, 202*(2), 493–497.

Gredebäck, G., Stasiewicz, D., Falck-Ytter, T., Rosander, K., & von Hofsten, C. (2009). Action type and goal type modulate goal-directed gaze shifts in 14-month-old infants. *Developmental Psychology, 45*(4), 1190–1194.

Hannon, E. E., & Trehub, S. E. (2005a). Tuning in to musical rhythms: Infants learn more readily than adults. *Proceedings of the National Academy of Sciences of the United States of America, 102*(35), 12639–12643.

Hannon, E. E., & Trehub, S. E. (2005b). Metrical categories in infancy and adulthood. *Psychological Science, 16*(1), 48–55.

Hari, R., Forss, N., Avikainen, S., Kirveskari, E., Salenius, S., & Rizzolatti, G. (1998). Activation of human primary motor cortex during action observation: A neuromagnetic study. *Proceedings of the National Academy of Sciences of the United States of America, 95*(25), 15061–15065.

Kanakogi, Y., & Itakura, S. (2011). Developmental correspondence between action prediction and motor ability in early infancy. *Nature Communications, 2*, 341.

Kelly, D. J., Quinn, P. C., Slater, A. M., Lee, K., Ge, L., & Pascalis, O. (2007). The other-race effect develops during infancy: Evidence of perceptual narrowing. *Psychological Science, 18*(12), 1084–1089.

Knoblich, G., & Flach, R. (2001). Predicting the effects of actions: Interactions of perception and action. *Psychological Science, 12*(6), 467–472.

Kochukhova, O., & Gredebäck, G. (2010). Preverbal infants anticipate that food will be brought to the mouth: An eye tracking study of manual feeding and flying spoons. *Child Development, 81*(6), 1729–1738.

Kuhl, P. K., Stevens, E., Hayashi, A., Deguchi, T., Kiritani, S., & Iverson, P. (2006). Infants show a facilitation effect for native language phonetic perception between 6 and 12 months. *Developmental Science, 9*(2), F13–F21.

Kuhl, P. K., Tsao, F., & Liu, H. (2003). Foreign-language experience in infancy: Effects of short-term exposure and social interaction on phonetic learning. *Proceedings of the National Academy of Sciences of the United States of America, 100*(15), 9096–9101.

Leder, H., & Carbon, C. (2006). Face-specific configural processing of relational information. *British Journal of Psychology, 97*(1), 19–29.

Lewkowicz, D. J., & Ghazanfar, A. A. (2006). The decline of cross-species intersensory perception in human infants. *Proceedings of the National Academy of Sciences of the United States of America, 103*(17), 6771–6774.

Lewkowicz, D. J., & Ghazanfar, A. A. (2009). The emergence of multisensory systems through perceptual narrowing. *Trends in Cognitive Sciences, 13*(11), 470–478.

Lockman, J. J. (2000). A perception–action perspective on tool use development. *Child Development, 71*(1), 137–144.

Loucks, J. (2011). Configural information is processed differently in human action. *Perception, 40*(9), 1047–1062.

Loucks, J., & Baldwin, D. (2009). Sources of information for discriminating dynamic human actions. *Cognition, 111*(1), 84–97.

Loucks, J., & Sommerville, J. A. (2012a). Developmental changes in the discrimination of dynamic human actions in infancy. *Developmental Science, 15*(1), 123–130.

Loucks, J., & Sommerville, J. A. (2012b). The role of motor experience in understanding action function: The case of the precision grasp. *Child Development, 83*(3), 801–809.

Loucks, J., and Sommerville, J. A. (under review). Prediction of action outcomes modulates attention during action observation in adults and young infants.

Loucks, J., & Sommerville, J. A. (in preparation). Perceptual tuning in action perception: Is motor experience the cause?

Maurer, D., Le Grand, R., & Mondloch, C. J. (2002). The many faces of configural processing. *Trends in Cognitive Sciences, 6*(6), 255–260.

McCarty, M. E., Clifton, R. K., & Collard, R. R. (1999). Problem solving in infancy: The emergence of an action plan. *Developmental Psychology, 35*(4), 1091–1101.

McCarty, M. E., Clifton, R. K., & Collard, R. R. (2001). The beginnings of tool use by infants and toddlers. *Infancy, 2*(2), 233–256.

Mondloch, C. J., Le Grand, R., & Maurer, D. (2002). Configural face processing develops more slowly than featural face processing. *Perception, 31*(5), 553–566.

Pascalis, O., de Haan, M., & Nelson, C. A. (2002). Is face processing species-specific during the first year of life? *Science, 296*(5571), 1321–1323.

Pascalis, O., Scott, L. S., Kelly, D. J., Shannon, R. W., Nicholson, E., Coleman, M., et al. (2005). Plasticity of face processing in infancy. *Proceedings of the National Academy of Sciences of the United States of America, 102*(14), 5297–5300.

Paulus, M., Hunnius, S., & Bekkering, H. (2011). Can 14- to 20-month-old children learn that a tool serves multiple purposes? A developmental study on children's action goal prediction. *Vision Research, 51*(8), 955–960.

Paulus, M., Hunnius, S., van Wijngaarden, C., Vrins, S., van Rooij, I., & Bekkering, H. (2011). The role of frequency information and teleological reasoning in infants' and adults' action prediction. *Developmental Psychology, 47*(4), 976–983.

Piaget, J., & Inhelder, B. (1962). *The psychology of the child.* New York: Basic Books.

Prinz, W. (1997). Perception and action planning. *European Journal of Cognitive Psychology, 9*(2), 129–154.

Rizzolatti, G., & Craighero, L. (2004). The mirror-neuron system. *Annual Review of Neuroscience, 27*, 169–192.

Rosander, K., & von Hofsten, C. (2011). Predictive gaze shifts elicited during observed and performed actions in 10-month-old infants and adults. *Neuropsychologia, 49*(10), 2911–2917.

Rosenbaum, D. A., Vaughan, J., Barnes, H. J., & Jorgensen, M. J. (1992). Time course of movement planning: Selection of handgrips for object manipulation. *Journal of Experimental Psychology: Learning, Memory, and Cognition, 18*(5), 1058–1073.

Schank, R. C., & Abelson, R. P. (1977). *Scripts, plans, goals, and understanding: An inquiry into human knowledge structures.* Mahwah, NJ: Erlbaum.

Scott, L. S., & Monesson, A. (2009). The origin of biases in face perception. *Psychological Science, 20*(6), 676–680.

Scott, L. S., Pascalis, O., & Nelson, C. A. (2007). A domain-general theory of the development of perceptual discrimination. *Current Directions in Psychological Science, 16*(4), 197–201.

Sommerville, J. A., Hildebrand, E. A., & Crane, C. C. (2008). Experience matters: The impact of doing versus watching on infants' subsequent perception of tool-use events. *Developmental Psychology, 44*(5), 1249–1256.

Sommerville, J. A., & Woodward, A. L. (2005). Pulling out the intentional structure of action: The relation between action processing and action production in infancy. *Cognition, 95*(1), 1–30.

Sommerville, J. A., Woodward, A. L., & Needham, A. (2005). Action experience alters 3-month-old infants' perception of others' actions. *Cognition, 96*(1), B1–B11.

Southgate, V., & Csibra, G. (2009). Inferring the outcome of an ongoing novel action at 13 months. *Developmental Psychology, 45*(6), 1794–1798.

Werker, J. F., & Tees, R. C. (1984). Cross-language speech perception: Evidence for perceptual reorganization during the first year of life. *Infant Behavior and Development, 7*(1), 49–63.

White, B. L., Castle, P., & Held, R. (1964). Observations on the development of visually-directed reaching. *Child Development, 35*(2), 349–364.

Woodward, A. L. (1998). Infants selectively encode the goal object of an actor's reach. *Cognition, 69*(1), 1–34.

Woodward, A. L., & Guajardo, J. J. (2002). Infants' understanding of the point gesture as an object-directed action. *Cognitive Development, 17*(1), 1061–1084.

Yovel, G., & Kanwisher, N. (2004). Face perception: Domain specific, not process specific. *Neuron, 44*(5), 889–898.

16 Online Action Analysis: Infants' Anticipation of Others' Intentional Actions

Amanda Woodward and Erin N. Cannon

Abstract

The ability to understand others' actions as intentional is a critical foundation for human social functioning. Equally critical is the ability to recruit this knowledge rapidly in the course of social interactions to generate online predictions about others' actions. Recent experiments have recruited eye-tracking methods to investigate infants' visual anticipation of others' actions. In this chapter, we consider this newly emerging literature in the context of the larger existing body of work that has principally used visual habituation methods to investigate infants' offline action understanding. This older body of work has shown that infants have relatively rich and generative knowledge about others' intentional actions and that this knowledge is structured, at least in part, by infants' own action experience. We consider whether infants' online anticipation of others' actions recruits this body of knowledge and conclude that although there is initial evidence to indicate that it does, many questions are yet to be answered.

Fundamental to human experience is the perception that we live in a world of intentional agents. We see others' actions not as purely physical movements, but rather as movements that are structured by intentions. This social worldview is pervasive in adults' memory for, reasoning about, and communication of event information, and it has ontogenetic roots early in life. Studies reviewed in this chapter inform us that infants, well before their first birthdays, see others' actions as structured by intentions. In adults, this analysis of others' intentions occurs rapidly and automatically, and it can play a critical role in guiding online social reasoning, allowing interpretations of actions to be informed by what has come before and supporting predictions about others' future actions. The ability to rapidly analyze and anticipate others' intentional actions is essential for engagement in collaborative activities, competition with social partners, communication, and, more generally, the coordination of one's own actions with those of others. Even a simple social response, such as avoiding passersby on a crowded sidewalk, requires that we monitor behavioral indicators of others' intended paths (e.g., gaze direction and shifts in posture) so that we can respond quickly and appropriately to them. More extended social interactions, such as working with someone to prepare a meal, or exchanging information during a conversation, require that this rapid prospective reasoning operate iteratively, updating predictions as the interaction unfolds.

In this chapter we will consider the early ontogeny of this ability. Specifically, we will ask when and whether infants are able to use their understanding of others' intentions to (1) generate predictions about others' next actions and (2) do so in the timescale required to implement this analysis in interactions with others. Recent advances in the use of eye-tracking paradigms in studies with infants have provided initial evidence about infants' visual anticipation of others' actions. Using this approach, several recent studies have investigated infants' ability to use what they know about intentional action to generate predictions, in the moment, about others' subsequent actions. We will review emerging findings that bear on this question, consider the initial conclusions they support, and lay out the questions that will focus future investigations. To frame this review, we begin by briefly summarizing a body of findings that elucidates infants' understanding of others' intentional actions. We then turn to the question of whether and when infants recruit this knowledge in their online responses to others' actions.

Infants' Understanding of Intentional Action

A great deal of research has shown that the propensity to view others' actions as goal directed emerges early in life. In particular, experiments using looking-time paradigms have shown that infants represent others' actions as directed at goals and objects of attention rather than as purely physical movements. To illustrate, when infants are habituated to a goal-directed action (e.g., a person reaching toward and grasping a toy), they subsequently show a stronger novelty response (i.e., longer looking) for test events that alter the goal of the action than they do for test events that preserve the goal while varying the physical properties of the action (e.g., Bíró & Leslie, 2007; Brandone & Wellman, 2009; Luo, 2011, this volume; Woodward, 1998; Woodward et al., 2009). Infants respond selectively to goal changes involving simple, instrumental actions, like reaching, by three to six months (Bíró & Leslie, 2007; Woodward, 1998; Gerson

& Woodward, in press; Luo, 2011; this volume; Sommerville, Woodward, & Needham, 2005). Infants show this sensitivity to others' goals with their hands as well as with their eyes: On viewing an adult who produces a goal-directed action, such as reaching for a toy, seven-month-old infants selectively reproduce the adult's goal, reaching for the same toy (Mahajan & Woodward, 2009).

Critically, infants do not show selective attention to goals when the moving entity being observed is not readily identified as an agent (Hofer et al., 2005; Woodward, 1998) or when the action is ambiguous (Woodward, 1999). For example, when infants see an inanimate claw move toward and contact a toy, they do not respond by looking selectively longer at goal changes (Woodward, 1998) or reproducing the claw's goal (Gerson & Woodward, 2012). Similarly, when infants view human movements that seem accidental or ambiguous, such as contacting an object with the back of the hand, they do not respond as if the event were goal directed (Hamlin et al., 2008; Woodward, 1999). The movements of claws and inert hands toward objects lead infants to attend to the contacted object, just as they attend to the objects at which goal-directed actions are directed (see, e.g., Gerson & Woodward, 2012; Hamlin et al., 2008; Woodward, 1998; 1999); nevertheless, infants represent the movements of claws and inert hands differently from the ways they represent goal-directed actions. Thus, infants' responses to goal changes are not readily explained by lower-level factors such as the association between the agent and the object or the way the action draws attention to the object.

Infants' action knowledge also reflects the higher-order plans that organize assemblies of actions. One way that this is evident is in infants' reasoning about means-end actions, in which a person's actions on an intermediary or tool are directed at the attainment of a downstream goal. By nine-to-twelve months of age, infants understand the actions on the tool as directed toward the ultimate goal, rather than at the tool itself (Woodward & Sommerville, 2000; Sommerville & Woodward, 2005). In addition, infants integrate information about a person's different actions over time and across contexts. For example, they can use a person's prior focus of attention to interpret his or her subsequent actions (e.g., Phillips et al., 2002; Vaish & Woodward, 2010; Luo & Baillargeon, 2007), and they use information about a person's preference in one context to interpret his or her actions in a new context (Sommerville & Crane, 2009). Infants also engage in this kind of integrative reasoning when viewing the complementary, collaborative actions of two individuals. They understand that although the specific actions of the two people differ, their goal is the same

(Henderson & Woodward, 2011). These integrative, plan-level aspects of infant action knowledge support inferences about novel actions. For example, although infants do not spontaneously view the movements of a mechanical claw as being goal directed, they do so when they are shown a person using the claw who is coordinating visual attention with the claw's movements in order to act on objects (Hofer, Hauf, & Aschersleben, 2005; cf. Gerson & Woodward, 2012).

Both the action-level and plan-level aspects of infants' intentional action knowledge have been linked to developments in infants' own actions. Infants begin to show sensitivity to the intentional structure of specific actions at the same time that the actions are emerging in their own productive repertoires, and developments in the two kinds of abilities are correlated (Brune & Woodward, 2007; Woodward & Guajardo, 2002). Similarly, infants' understanding of the means-end structure of others' actions correlates with their own ability to produce means-end actions (Sommerville & Woodward, 2005). Moreover, interventions that alter infants' individual and means-end actions affect their understanding of those actions in others (Gerson & Woodward, in press; Sommerville et al., 2008; Sommerville et al., 2005), and active engagement in actions influences infants' subsequent action understanding in ways that simply watching others produce an action does not (Gerson & Woodward, 2012; Sommerville et al., 2008).

These findings indicate that the representations that structure infants' own actions play a role in supporting their understanding of others' actions. That is, these findings suggest that infants' action representations are *embodied* in the sense, described by Wilson (2002), of being "mental structures that originally evolved for perception or action [that are] co-opted and run ... decoupled from the physical inputs and outputs that were their original purpose, to assist in thinking and knowing" (p. 633). This conclusion is consistent with a number of recent theories about the embodied nature of intentional action knowledge, both in its mature state (e.g., Gallese & Goldman, 1998; Rizzolatti & Craighero, 2004) and during early development (e.g., Hauf, this volume; Meltzoff, 2007; Gerson & Woodward, 2010).

Even so, the question of whether embodied action knowledge comprises all of what infants know about intentional action is not fully resolved. For example, in some cases, infants seem able to reason about events that do not involve human actions at all as if they were goal directed (e.g., Luo, 2011; Luo & Choi, this volume; Gergely & Csibra, 2003; Hamlin et al., 2008). Whether these responses reflect the generalization of embodied action representations to novel events or instead reflect a separate, "disembodied" set of conceptual representations is an open question. This issue aside, it is nevertheless clear that significant aspects of the knowledge infants bring to bear in making sense of others' actions are closely linked to their own experiences as agents.

Do Infants Generate Online Action Predictions?

The evidence reviewed in the previous section shows that infants have relatively rich and generative knowledge about intentional action. Do infants employ this knowledge to form predictions about others' actions? The evidence summarized so far cannot address this question. Although looking-time methods yield evidence about infants' cognitive representations, they often do not provide clear evidence as to whether infants have generated a prediction. To illustrate, consider the finding that, having been habituated to an action repeatedly directed at one object, infants look longer when the action is directed at a new object than at the same object (e.g., Woodward, 1998). One interpretation of this result is that infants expected the actor to continue to act on the same object, and thus their longer looking during goal-change trials indicates surprise when this expectation is violated. However, it is also possible that infants detect the goal change in test events without having first formed the expectation that the agent would maintain the same goal. That is, infants' responses in looking-time procedures could reflect post hoc detection of goal changes rather than prior expectations. The same concern applies to studies that evaluate infants' ability to relate an agent's actions at one time, such as looking at an object, to that agent's subsequent actions, such as grasping that object versus a different object (e.g., Phillips et al., 2002). Infants might have generated a prediction early in the event, based on the agent's visual attention toward the object. Alternatively, infants might have detected the inconsistency between the two actions by comparing them after the fact. Thus, although the findings of looking-time studies are consistent with the conclusion that infants form action predictions, these findings in themselves do not provide conclusive evidence that infants do so.

A second issue is that, from looking-time data alone, it is not clear whether infants can generate predictions on the time scale required by online social interactions like collaboration, competition, and communication. In looking-time studies, infants have repeated opportunities and relatively long time intervals to encode, analyze, and respond to the actions that they view. In a typical experiment, an infant may view the same action repeated as many as fourteen times and have an open-ended trial length (generally at least several seconds long) to observe and respond to each example. But using intentional analysis to inform real-time social interaction would require that infants generate predictions rapidly in the course of observing an action.

For these reasons, a different kind of evidence is needed, both in terms of requiring clearer evidence for infants' generation of predictions and in terms of the timescale of the response. Measures of infants' anticipatory gaze shifts offer such a method. Even very young infants generate predictive eye movements in response to ongoing events. For example, Haith, Canfield, and their colleagues (Canfield et al., 1997; Haith et al., 1988) documented that infants as young as three-and-a-half months of age can learn a regular sequences of light movements and move their eyes in anticipation of the next light in the sequence. In this case, infants' anticipatory gaze shifts reflected their learning about the novel light pattern over a number of trials (Wentworth & Haith, 1998). Other studies have shown that infants are also able to launch anticipatory gaze shifts based on predictions derived from a priori physical knowledge. To illustrate, when four- to six-month-old infants observe an object passing behind an occluder, they spontaneously anticipate the reemergence of the object on the other side. Further, they do so from the first trials in which they saw this event, thus indicating that this anticipation does not depend on infants' learning within the session that the object will reemerge (Johnson et al., 2003; von Hofsten et al., 2007). Instead, infants' anticipatory responses reflect their knowledge about physical objects and their possible patterns of movement.

The bodies of work outlined in the previous paragraph suggest at least two kinds of mechanisms by which infants could anticipate others' actions. First, drawing on general-purpose learning, infants may come to anticipate regular patterns in others' movements. For example, infants may learn to expect that hands holding phones end up near the ear or that hands holding cups end up near the mouth. In fact, by six months of age, infants show just these kinds of expectations by looking systematically to the body parts associated with familiar objects like phones and cups when these objects are held by people (Gredebäck & Melinder, 2010; Hunnius & Bekkering, 2010; Kochukhova & Gredebäck, 2010). Learning about predictable movement patterns can happen in the laboratory as well. For example, Paulus and colleagues (2011) found that, after viewing an agent who repeatedly took a circuitous path toward a goal, infants visually anticipated that the agent would continue to take the same path even when a more efficient path was available. These findings show that infants are able to extract information about patterns in other's movements and that this information supports their online visual predictions. Even so, these findings do not clarify whether infants recruit their analysis of others' intentions to derive these expectations.

A second means by which infants could generate predictions about others' actions is via their conceptual knowledge about intentional action. That is, infants could use the knowledge described in the first part of this chapter to generate action predictions independent of particular patterns of movement. Southgate and colleagues (2007) reported an elegant demonstration of this kind of anticipation in two-year-old children. Children viewed events in which a protagonist repeatedly retrieved an object that she saw hidden by a puppet in each of two boxes. Following a final hiding event, the protagonist turned away and the puppet removed the toy, taking it offstage. Then, the protagonist turned back toward the boxes as if to approach them and retrieve the toy. Children's visual responses to the event were measured using eye-tracking. Even though the object was no longer hidden in either box, children looked predictively toward the box in which the protagonist had last seen the object hidden. Because the protagonist had previously retrieved the toy from each of the boxes an equal number of times, children's anticipatory responses could not have reflected learning about the prior movements of the protagonist. Instead, children must have generated predictions based on an analysis of the protagonist's prior goals and states of attention (see Senju, Southgate, Snape, Leonard, & Csibra 2011 for similar findings with 18-month-old infants).

Infants' Action Anticipation

What about younger infants? As described above, infants less than half the age of Southgate and colleagues' (2007) subjects see others' actions as structured by intentions. Do infants implement this action analysis in their online predictions about others' actions? To date, this question has been addressed by evaluating infants' visual anticipation of concrete, manual actions, such as reaching for and grasping objects or moving objects from one location to another. Although these are highly familiar actions, there is evidence, from both adults and infants, to suggest that anticipatory responses in these studies reflect more than simply movement-based expectations.

Consider the event depicted in figure 16.1. A person reaches for each of three balls one at time, moving each one across a table and placing it

Figure 16.1
Stimulus events used by Cannon and colleagues (2012).

into a bucket. When adults and twelve-month-old infants view events like these, they spontaneously look to the bucket before the ball arrives (Cannon et al., 2012; Eshuis et al., 2009; Falck-Ytter et al., 2006). Although this response is generally assessed in experiments with repeated trials, adults and twelve-month-old infants show anticipation from the earliest trials onward, suggesting that the response does not depend on learning that takes place over trials (Cannon et al., 2012; Falck-Ytter et al., 2006; Flanagan & Johansson, 2003). Moreover, both infants and adults show more robust anticipation of the balls' arrival when a person was seen to move the balls (as in figure 16.1) as opposed to when the balls traversed the same path to the bucket apparently on their own (Falck-Ytter et al., 2006). Thus, infants and adults show heightened anticipation for intentional actions compared to matched movements that did not involve intentional actions.

Infants' anticipation of the endpoints of actions is also influenced by the presence of a goal. Gredebäck and colleagues (2009) found that fourteen-month-old infants showed strong anticipation when actions culminated in a salient goal (e.g., putting an object into a container). However, when infants viewed similar actions for which the goal was less salient (e.g., transporting an object to an unmarked location) or arm movements that followed the same path but did not involve moving an object, they did not show robust anticipation. Eshuis and colleagues (2009) report similar findings for adults.

Several studies report that infants younger than twelve months of age fail to show reliable action anticipation. For example, six-month-old infants tested with "bucket" events like the one depicted in figure 16.1 failed to look to the bucket before the ball's arrival (Falck-Ytter et al., 2006; see also Gredebäck et al., 2009). However, Kanakogi and Itakura (2011) recently reported that infants as young as six months of age are able to anticipate a simpler action—a reach toward an object. When viewing a reaching hand, infants shifted their gaze to the object before the hand arrived, but they did not shift attention to the toy as rapidly when viewing a mechanical claw that moved toward the toy or an ambiguous hand gesture that was directed at the toy. Thus, younger infants seem to show similar patterns of selective anticipation for goal-directed actions. Even so, Kanakogi and Itakura's (2011) findings also indicate that infants' anticipation became more robust between ages six and ten months, and that there were large differences in the timing of anticipatory gaze shifts between ten-month-olds and adults. Given the timing of the events and the magnitude of infants' anticipatory responses, six-month-olds likely shifted gaze to the toy only when the hand was quite close to it, in contrast to older infants and adults who shifted attention to the toy much earlier.

Taken together, these findings indicate that infants have a special propensity to anticipate goal-directed actions. Like adults, infants (1) look ahead to the endpoints of others' actions, (2) do so from the first experimental trials onward, and (3) show this response most robustly for actions of agents that are directed toward goals. To go back to the distinction we raised earlier, we can ask whether these findings show that infants' anticipatory responses reflect something more than learned movement regularities. On the one hand, infants' anticipatory responses seem to be tuned to goal-directed actions and seem not to reflect learning about the experimental events during the testing session. On the other hand, it seems possible that these anticipatory responses could reflect learned regularities from everyday experience—for example, when a container is present, hands tend to move objects to it—in much the same way that infants have learned that phones are held to the ear and cups to the mouth. Therefore, although they are suggestive, these findings do not resolve with certainty the question of whether infants generate anticipatory responses based on an analysis of an agent's goals.

Goal-Based Action Prediction in Infants

The principal paradigm that has been used to investigate action anticipation in both adults and infants demonstrates rapid online anticipation of others' actions, but ultimately leaves open the question of the nature of the cognitive representations that underlie these anticipatory responses. In the studies described in the previous section, the goal and pattern of movement were confounded: Infants always saw the same pattern of movement directed to the same goal. Thus, the results do not clarify whether infants anticipated the goal per se. Infants may anticipate regularities in movement but fail to predict that an agent will maintain the same goal despite changes in movements.

To distinguish goal-based action prediction from movement anticipation, we adapted the logic used in our prior visual-habituation studies. Eleven-month-old infants viewed events in which a hand reached for and

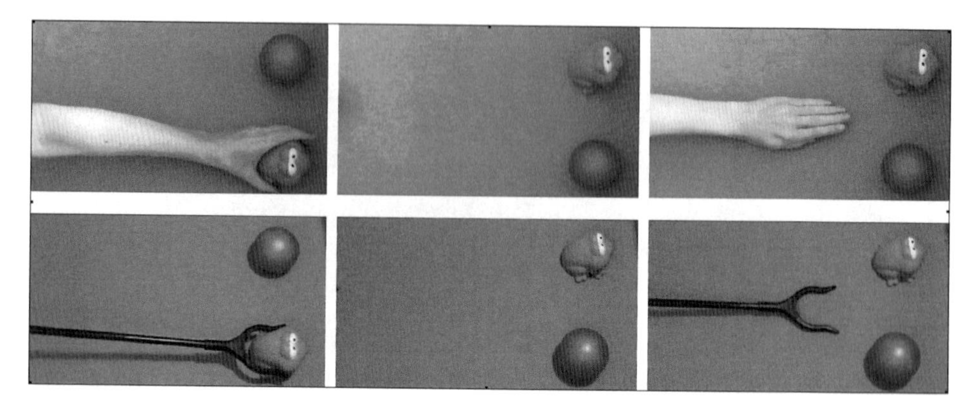

Figure 16.2
Events used in the hand (top row) and claw (bottom row) conditions in Cannon and Woodward (2012).

grasped one of two objects (see figure 16.2; Cannon & Woodward, 2012). Infants viewed three familiarization trials showing this event, each one two-and-a-half seconds in duration. Next, the positions of the objects were reversed and infants were given one trial to view the objects in their new locations. Then, in the test trial, infants saw the hand reach halfway across the stage and stop one-and-a-half second later, equidistant between the two objects. We evaluated whether infants launched an anticipatory saccade from the hand to one of the objects, and if so, whether they predicted that the hand would move to the same goal object or, instead, to the same location as it had previously. Infants generated predictive eye movements in most test trials, looking first to the hand region and then to one of the two toys. Critically, these saccades were systematically directed to the prior goal object rather than to the location that had been previously reached toward.

To evaluate whether this pattern of response was selective for goal-directed actions, we tested a second group of infants using events in which a mechanical claw moved toward and grasped the object (see figure 16.2). Prior studies have shown that infants do not spontaneously view events like these as being goal directed, although when additional cues to the goal-directedness of the action are present, infants can sometimes make use of them (Gerson & Woodward, 2012; Hofer et al., 2005; Woodward, 1998). Thus, the claw events provided a test of whether infants' attention would be drawn to the previously contacted object even when the events did not involve a goal-directed action. The findings indicated that infants' responses to the claw differed from responses to the hand. Infants in this condition showed an equal propensity to launch predictive eye movements during test trials as in the hand condition, but they systematically looked to the prior location rather than the prior goal-object.

Importantly, infants in the hand and claw conditions were equally attentive to the experimental events. They looked equally to the two objects and showed similar patterns of monitoring the movements of the claw and hand toward the object. Thus, the difference in their responses to hands and claws on test probes was not due to differences in overall attention to the events. These findings suggest that infants at this age generate at least two different kinds of predictions—goal-based predictions for the action of an agent and movement-based predictions for the motion of an inanimate object.

These findings indicate that by eleven months of age, conceptual analysis drives infants' predictions about others' actions. That is, the knowledge that is evident in looking-time studies seems to contribute to infants' rapid, online responses to others' actions. One obvious next question is whether younger infants generate such predictions. As reviewed earlier, infants demonstrate an understanding of actions as being goal directed as early as three-to-six months of age in looking-time procedures. One possibility is that infants' action knowledge is accessible to online prospective reasoning from the start. Infants at these ages recruit physical knowledge to predict the movements of inanimate objects, and so it seems possible that they would be similarly able to use social knowledge in this way. On the other hand, the recruiting of action knowledge to generate predictions may rely on later aspects of executive function, such as developments in working memory or attentional control, or on later aspects of social cognitive development. In this case, the findings of habituation experiments may reflect a more limited sensitivity, supporting the perception of action structure but not prospective social reasoning.

If there were a lag in the emergence of goal-based action prediction relative to earlier aspects of action knowledge, this could help to explain a developmental paradox: Looking-time studies tell us that younger (i.e., under twelve months of age) infants understand others' actions as goal-directed, and yet younger infants do not engage in many of the overt social behaviors that are generally taken to reflect an understanding others as intentional. For example, infants twelve months of age and older actively engage in collaborative and referential, communicative interactions with social partners, but these social behaviors are minimal or absent in younger infants. If younger infants possess limited abilities to generate goal-based action predictions, then this is exactly what would be expected. That is, younger infants may not be able to use what they know to generate the online anticipations needed to regulate these social interactions. Further studies are required to evaluate whether and how infants' action anticipation relates to their emerging competence in real-world social interactions.

A second question is whether infants' action anticipation goes beyond the understanding of isolated actions as being goal directed and reflects the knowledge of higher-order action plans that has been revealed in infant looking-time studies. In looking-time paradigms, infants integrate information about an agent's actions over time, using prior actions to interpret subsequent actions. This ability enables infants to interpret novel actions in the context of higher-order action plans. For example, this ability allows infants to understand the use of a novel tool as goal directed based on their seeing the tool use coordinated with an agent's other actions. Do infants engage in similar reasoning to modulate their online predictions about others actions? The findings of Southgate and colleagues (2007), described earlier, indicate that by two years of age, children generate predictions of this sort, such as predicting an agent's reaching actions based on her prior focus of attention. As yet, there have been no studies of this kind of predictive reasoning in younger infants.

Embodiment and Anticipation

As described earlier, prior research has documented a close relation between infants' emerging control of their own actions and their understanding of others' actions as measured in looking-time studies. Developments in infants' own actions correlate with developments in their action understanding, and training interventions that change infants' actions also influence their understanding of others' actions as goal directed. These findings suggest that infants' action representations are embodied, in the sense that they borrow structure from representations that guide action production. This possibility has particular significance for the question of whether and how infants anticipate others' goal-directed actions because action production is an inherently prospective process, requiring the generation and updating of predictions about one's actions as they occur (Rosenbaum et al., 2009; Wolpert et al., 2001). Actions are prospectively organized from early in infancy (von Hofsten, 2004), and this fact raises the possibility that embodied action representations could support infants' prospective attention to others' actions.

Indeed, several converging lines of evidence indicate that embodied action representations support action anticipation in adults. Visual attention to others' actions closely mirrors the anticipatory patterns that accompany one's own actions, suggesting that both kinds of anticipation may be driven by the same underlying mechanism (Flanagan & Johansson, 2003; Hauf, this volume). Further, when adults view a predictable action, electrophysiological activity associated with motor preparation occurs just before the observed action (Kilner et al., 2004). Recent studies from our laboratory have confirmed that motor processes play a functional role in adults' action anticipation. Specifically, we found that engaging in a concurrent action task disrupts anticipation of others' actions (Cannon & Woodward, 2008, in preparation). In these studies, adults viewed the events depicted in figure 16.1 while, in one condition, tapping four fingers on one hand in a scripted cascade pattern. This concurrent motor task significantly reduced adults' anticipation of the actions they viewed relative to baseline trials with no concurrent task and relative to a condition in which the concurrent task (a verbal working-memory task) was not motor in nature. Further, the finger-tapping task did not disrupt adults' anticipation of inanimate events (e.g., tracking a ball as it rolled behind an occluder), indicating a specific connection between action production and action anticipation.

If embodied action representations support infants' action anticipation, then developments that occur in infants' action control during the first years of life would be expected to have an effect on developments in their action anticipation. Falck-Ytter and colleagues' (2006) findings suggested this may be the case: Six-month-old infants tested in their procedure (anticipating actions in which balls are placed into a bucket) did not anticipate the action. The researchers speculated that, because infants at this age do not engage in containment actions, their lack of anticipation might reflect their lack of motor representation for that action. Gredebäck and Kochukhova (2010) conducted a more direct test of this possibility by assessing both action production and action anticipation in two-year-old children. They found that children's level of skill at placing pieces into a puzzle was correlated with their tendency to visually anticipate an adult's actions with the puzzle.

In a recent study (Cannon et al., 2012), we asked whether this relation could be traced to earlier in infancy by testing whether infants' own engagement in containment actions predicted their anticipation of others' actions with containers. Although most twelve-month-old infants are able to place objects into containers, there is variation in the extent to which they spontaneously do so. We assessed this individual variation by giving infants the opportunity to act on containers and small toys either before or after we assessed their action anticipation for the events depicted in figure 16.1. We found that when infants were given the opportunity to engage in containment actions prior to the observation task, their spontaneous level of activity when placing objects into containers predicted their subsequent tendency to anticipate the observed containment actions. The same relation did not hold when the action and perception tasks were given in the reverse order, suggesting that engaging in the actions had primed infants' subsequent anticipatory responses.

Kanakogi and Itakura (2011) report a similar correlation in six- to ten-month-old infants. As described earlier, they assessed infants' anticipation of events in which a hand moved toward and grasped one of two objects; they found that infants six months of age and older reliably anticipated the arrival of the hand at the object. They also assessed the quality of infants' own reaches by analyzing the extent to which infants reached toward a toy using just one hand. Infants tend to act bimanually earlier in development and later transition to more efficient, unimanual reaches. They found that this measure of reaching ability correlated with infants' visual anticipation of the reaching events independent of the effects of increasing age on visual anticipation. Further, on analogy with Kilner and colleagues (2004) findings in adults, Southgate and colleagues (2007) report that infants show electroencephalographic (EEG) activity associated with motor system activation when they view the initial stages of an event that includes a predictable, repeated reaching action (see also Nystrom, 2008; Marshall & Meltzoff, 2011).

Thus, findings from several laboratories suggest that for infants, as for adults, visual anticipation of others' actions draws on embodied action representations. However, further research is needed to fully evaluate this possibility: Unlike the studies with adults, all of the current evidence for infants' action anticipation derives from correlational studies, so it is not yet clear whether embodied representations play a functional role in infants' action anticipation. Our studies with adults have begun to move beyond documenting correlated activation to interventions that can test the causal effects of motor processes on action perception. It will be important to pursue this same strategy, using motor training or motor-interference manipulations, with infants.

A further open issue concerns the level at which embodied action representations support action anticipation. In these studies, as in much of the research reviewed earlier, the research design does not provide a clear test of whether the anticipatory responses were based on an analysis of action goals. Because these anticipation tasks confounded goal-based and movement-based regularities, it is not clear from these findings whether embodied representations contribute to movement-based or goal-based anticipations of others' actions. Either (or both) is possible. The prospective representations that guide action production reflect multiple levels of analysis—from the generation of specific movements to the higher-order plans that organize sequences of actions (Rosenbaum et al., 2009; Wolpert et al., 2001). This is true in infants as well as adults, as evidenced by the fact that infants, like adults, shape the first actions in a sequence with respect to the downstream goals toward which the sequence is directed (e.g., Claxton et al., 2003; von Hofsten, 2004). Thus there is, in principle, no reason why embodied action representations could not reflect both higher-order and lower-order aspects of action structure.

The findings from our finger-tapping studies (Cannon & Woodward, 2008, in preparation) suggest that, for adults, the relevant embodied representations reflect structure above the level of movement generation. For example, the interference effects declined as the finger-tapping sequence became automatized, suggesting it was the planning of components of motor performance, and not the generation of the finger movements per se, that mattered. In addition, concurrent motor activity interfered with the anticipation not only of specific body movements, but also with more abstract goal-directed events. Specifically, finger tapping interfered with the anticipation of tool-use events (e.g., a claw picking up a ball and placing it into a bucket) in which the presence of the agent could only be inferred; this interference effect was reduced by a manipulation that led participants to view the claw events as mechanical rather than generated by a person.

These findings with adults suggest that embodied action representations reflect action structure that is more abstract than specific motor movements, instead reflecting an analysis of movements as goal directed. However, further research is needed to evaluate the other levels at which embodied action representations may support action anticipation. For example, are the embodied representations that are involved in adults' anticipation of others' actions based on higher-order plans or prior states of attention?

The level at which embodied representations are involved in infants' action anticipation is an open question at this point. As we concluded earlier, aside from the findings of the goal-prediction study with eleven-month-olds (Cannon & Woodward, 2012), little is known about infants' ability to recruit goal information for the generation of action anticipations. More research is needed to evaluate when in development this ability emerges, as well as the extent to which infants can integrate action information to generate flexible online predictions. As this research moves forward, a second question will be whether and how embodied action representations are involved. One possibility is that we will find developmental change in the level at which embodied representations support infants' action anticipation. Over the course of their first two years, infants become able to form and implement progressively more abstract action plans that organize longer and more complex chains of actions. These developments in action control may have implications for infants' ability to anticipate actions structured by higher-order plans in others.

Conclusions

The development of infant eye-tracking paradigms has allowed researchers access to aspects of social cognitive development that were not easily studied in the past. In particular, these methods have given researchers a window into infants' rapid, online responses to others' intentional actions—an aspect of social information processing that is critical for engaging in well-structured social interactions. In recent years, a number of studies have employed these methods to investigate infants' prospective attention to others' actions. These studies have revealed that (1) infants, like adults, generate rapid anticipatory responses when viewing others' actions, and (2) this response is particularly robust for events that involve well-structured, goal-directed actions as compared to events that involve only the movements of objects or ambiguous human movements.

These findings raise a number of new questions. Our goal in this chapter has been to (1) articulate some of these questions, (2) consider the progress that has been made in addressing them, and (3) highlight the important issues that are still unresolved. The open questions center on two general issues. First, research over the past fifteen years has documented that infants possess rich and generative knowledge about intentional action. Is this knowledge recruited in infants' online action anticipation? Second, research with adults indicates that the visual anticipation of others' actions recruits representations involved in action production. That is, action anticipation is supported by embodied action representations. Is this also true of infants?

As we hope our review makes clear, the current literature offers preliminary affirmative answers to both of these questions. But a number of striking gaps remain in the empirical record. Most critically, these gaps concern the level of description of the representations that guide infants' action anticipation, and how this may change over early development. Infants represent others' actions not merely as movements through space, but as actions structured by goals, both at the level of individual actions and at the level of higher-order plans that structure groups of actions. Our findings in studies with eleven-month-olds indicate that this knowledge informs action predictions. However, as yet, we do not know when this connection is established in early development, and there is no evidence, in infants under the age of two years, for anticipation based on a plan-level analysis. It remains possible, therefore, that infants' action knowledge is not fully expressed in their online responses to others' actions until later points in development.

Further, there are reasons to suspect that the embodied nature of action knowledge may explain the engagement of this knowledge in infants' action anticipation, but, as yet, several links in the account are untested. There is evidence that connects both action-level and plan-level knowledge to developments in infants' own actions, suggesting that infants' embodied action knowledge reflects both levels of analysis. Further, evidence from studies with adults indicates that embodied action representations support the anticipation of goal-directed actions. But the only evidence that relates action production to action anticipation in infants is correlational, leaving open the question of whether embodied representations play a functional role in infants' action anticipation.

In pursuing these questions, an additional challenge is the likely possibility that the answers will change as early development unfolds. Infancy is a period of dramatic changes not only in social cognition, but also in motor competence. Prior research has made headway in uncovering relations between these domains of development, but considering these relations in the context of infants' active online responses to others' actions opens new questions and new possibilities.

Acknowledgments

The research we describe in this chapter was supported by grants to the first author from the National Science Foundation (DLS 0951489), the National Institute of Child Health and Human Development (P01-HD064653), and the Office of Naval Research, and by a grant from the Jacobs Foundation to the second author.

References

Bíró, S., & Leslie, A. M. (2007). Infants' perception of goal-directed actions: Development through cue-based bootstrapping. *Developmental Science, 10*, 379–398.

Brandone, A. C., & Wellman, H. M. (2009). You can't always get what you want: Infants understand failed goal-directed actions. *Psychological Science, 20*, 5–91.

Brune, C. W., & Woodward, A. L. (2007). Social cognition and social responsiveness in 10-month-old infants. *Journal of Cognition and Development, 8*, 133–158.

Canfield, R. L., Smith, E. G., Bresznyak, M. P., and Snow, M. P. (1997). *Informational processing through the first year of life: A longitudinal study using the visual expectation paradigm.* Monographs of the Society for Research in Child Development, serial no. 250.

Cannon, E. N., & Woodward, A. L. (2008). Action anticipation and interference: A test of prospective gaze. In B. C. Love, K. McRae, & V. M. Sloutsky (Eds.), *Proceedings of the 30th annual conference of the Cognitive Science Society.* Austin, TX: Cognitive Science Society.

Cannon, E., & Woodward, A. L. (2012). Infants generate goal-based action predictions. *Developmental Science, 15*, 292–298.

Cannon, E. and Woodward, A. L. (in preparation). Effects of action production on the online visual anticipation of others' actions.

Cannon, E., Woodward, A. L., Gredebäck, G., von Hofsten, C., & Turek, C. (2012). Action production influences 12-month-old infants' attention to others' actions. *Developmental Science, 15*, 5–42.

Claxton, L., Keen, R., & McCarty, M. E. (2003). Evidence of motor planning in infant reaching behavior. *Psychological Science, 14*, 354–356.

Eshuis, R., Coventry, K. R., & Vulchanova, M. (2009). Predictive eye movements are driven by goals, not by the mirror neuron system. *Psychological Science, 20*, 438–440.

Falck-Ytter, T., Gredebäck, G., & von Hofsten, C. (2006). Infants predict other peoples' action goals. *Nature Neuroscience, 9*, 878–879.

Flanagan, J. R., & Johansson, R. S. (2003). Action plans used in action observation. *Nature, 24*, 769–771.

Gallese, V., & Goldman, A. (1998). Mirror neurons and the simulation theory of mind reading. *Trends in Cognitive Sciences, 12*, 493–501.

Gergely, G., & Csibra, G. (2003). Teleological reasoning about actions: The one-year-old's naïve theory of rational action. *Trends in Cognitive Sciences, 7*, 287–292.

Gerson, S., & Woodward, A. (2012). A claw is like my hand: Comparison supports goal analysis in infants. *Cognition, 122,* 181–192.

Gerson, S., and Woodward, A. (in press). Learning from their own actions: The unique effect of producing actions on infants' action understanding. *Child Development.*

Gerson, S., & Woodward, A. L. (2010). Building intentional action knowledge with one's hands. In S. P. Johnson (Ed.), *Neo-constructivism.* Oxford: Oxford University Press.

Gredebäck, G., & Kochukhova, O. (2010). Goal anticipation during action observation is influenced by synonymous action capabilities, a puzzling developmental study. *Experimental Brain Research, 202,* 493–497.

Gredebäck, G., & Melinder, A. M. D. (2010). Infants understanding of everyday social interactions: A dual process account. *Cognition, 114,* 197–206.

Gredebäck, G., Stasiewicz, D., Falck-Ytter, T., Rosander, K., & von Hofsten, C. (2009). Action type and goal type modulate goal-directed gaze shifts in 14-month-old infants. *Developmental Psychology, 5,* 1190–1194.

Haith, M. M., Hazan, C., & Goodman, G. S. (1988). Expectation and anticipation of dynamic visual events by 3.5-month-old babies. *Child Development, 9,* 467–479.

Hamlin, J. K., Hallinan, E. V., & Woodward, A. L. (2008). Do as I do: 7-month-old infants selectively reproduce others' goals. *Developmental Science, 11,* 487–494.

Henderson, A. M. E., & Woodward, A. L. (2011). Let's work together: What do infants understand about collaborative goals? *Cognition, 121,* 12–21.

Hofer, T., Hauf, P., & Aschersleben, G. (2005). Infant's perception of goal-directed actions performed by a mechanical device. *Infant Behavior and Development, 28,* 466–480.

Hunnius, S., & Bekkering, H. (2010). The early development of object knowledge: A study on infants' visual anticipations during action observation. *Developmental Psychology, 6,* 446–454.

Johnson, S. P., Amso, D., & Slemmer, J. A. (2003). Development of object concepts in infancy: Evidence for early learning in an eye tracking paradigm. *Proceedings of the National Academy of Sciences of the United States of America, 100,* 10568–10573.

Kanakogi, Y., & Itakura, S. (2011). Developmental correspondence between action prediction and motor ability in early infancy. *Nature Communications, 2,* 341.

Kilner, J. M., Vargas, C., Duval, S., Blakemore, S. J., & Sirigo, A. (2004). Motor activation prior to observation of a predicted movement. *Nature Neuroscience, 12,* 1299–1301.

Kochukhova, O., & Gredebäck, G. (2010). Preverbal infants anticipate that food will be brought to the mouth: An eye tracking study of manual feeding and flying Spoons. *Child Development, 81*, 1729–1738.

Luo, Y. (2011). Three-month-old infants attribute goals to a non-human agent. *Developmental Science, 14*, 453–460.

Luo, Y., & Baillargeon, R. (2007). Do 12.5-month-old infants consider what objects others can see when interpreting their actions? *Cognition, 105*, 489–512.

Mahajan, N., & Woodward, A. L. (2009). Infants imitate human agents but not inanimate objects. *Infancy, 14*, 667–679.

Marshall, P. J., & Meltzoff, A. N. (2011). Neural mirroring systems: Exploring the EEG mu rhythm in human infancy. *Developmental Cognitive Neuroscience, 1*, 110–123.

Meltzoff, A. (2007). The "like me" framework for recognizing and becoming an intentional agent. *Acta Psychologica, 124*, 26–43.

Nystrom, P. (2008). The infant mirror neuron system studied with high density EEG. *Social Neuroscience, 3*, 334–347.

Paulus, M., Hunnius, S., van Wijngaarden, C., Vrins, S., van Rooij, I., & Bekkering, H. (2011). The role of frequency information and teleological reasoning in infants' and adults' action prediction. *Developmental Psychology, 7*, 976–983.

Phillips, A. T., Wellman, H. M., & Spelke, E. S. (2002). Infants' ability to connect gaze and emotional expression to intentional action. *Cognition, 5*, 3–78.

Rizzolatti, G., & Craighero, L. (2004). The mirror-neuron system. *Annual Review of Neuroscience, 27*, 169–192.

Rosenbaum, D. A., Vaughan, J., Meulenbroek, R. G. J., Jax, S., & Cohen, R. (2009). Smart moves: The psychology of everyday perceptual-motor acts. In E. Morsella, J. A. Bargh, & P. M. Gollwitzer (Eds.), *Oxford handbook of human action*. Oxford: Oxford University Press.

Senju, A., Southgate, V., Snape, C., Leonard, M., & Csibra, G. (2011). Do 18-month-old infants attribute mental states to others? A critical test. *Psychological Science, 22*, 878–880.

Sommerville, J. A., & Crane, C. C. (2009). Ten-month-old infants use prior information to identify an actor's goal. *Developmental Science, 12*, 314–325.

Sommerville, J. A., Hildebrand, E. A., & Crane, C. C. (2008). Experience matters: The impact of doing versus watching on infants' subsequent perception of tool use events. *Developmental Psychology, 4*, 1249–1256.

Sommerville, J. A., & Woodward, A. L. (2005). Pulling out the intentional structure of human action: The relation between action production and processing in infancy. *Cognition, 5*, 1–30.

Sommerville, J. A., Woodward, A. L., & Needham, A. (2005). Action experience alters 3-month-old infants' perception of others' actions. *Cognition, 6,* B1–B11.

Southgate, V., Johnson, M. H., El Karoui, I., & Csibra, G. (2010). Motor system activation reveals infants' on-line prediction of others' goals. *Psychological Science, 21,* 355–359.

Southgate, V., Senju, A., & Csibra, G. (2007). Action anticipation through attribution of false belief by 2-year-olds. *Psychological Science, 18,* 587–592.

Vaish, A., & Woodward, A. L. (2010). Infants use attention but not emotions to predict others' actions. *Infant Behavior and Development, 3,* 9–87.

von Hofsten, C. (2004). An action perspective on motor development. *Trends in Cognitive Sciences, 8,* 266–272.

von Hofsten, C., Kuchukhova, O., & Rosander, K. (2007). Predictive tracking over occlusions by 4-month-old infants. *Developmental Science, 10,* 625–640.

Wentworth, N., & Haith, M. M. (1998). Infants' acquisition of spatiotemporal expectations. *Developmental Psychology, 4,* 247–257.

Wilson, M. (2002). Six views of embodied cognition. *Psychonomic Bulletin & Review, 9,* 625–636.

Wolpert, D. M., Ghahramani, Z., & Flanagan, J. R. (2001). Perspectives and problems in motor learning. *Trends in Cognitive Sciences, 5,* 487–493.

Woodward, A. L. (1998). Infants selectively encode the goal object of an actor's reach. *Cognition, 9,* 1–34.

Woodward, A. L. (1999). Infants' ability to distinguish between purposeful and non-purposeful behaviors. *Infant Behavior and Development, 22,* 145–160.

Woodward, A. L., & Guajardo, J. J. (2002). Infants' understanding of the point gesture as an object-directed action. *Cognitive Development, 17,* 1061–1084.

Woodward, A. L., & Sommerville, J. A. (2000). Twelve-month-old infants interpret action in context. *Psychological Science, 11,* 73–76.

Woodward, A. L., Sommerville, J. A., Gerson, S., Henderson, A. M. E., & Buresh, J. S. (2009). The emergence of intention attribution in infancy. In B. Ross (Ed.), *The psychology of learning and motivation* (Vol. 51). New York: Academic Press.

Contributors

Dare Baldwin University of Oregon, USA

Lara Bardi University of Padova, Italy

H. Clark Barrett University of California, Los Angeles, USA

Erin N. Cannon University of Maryland, College Park, USA

You-jung Choi University of Missouri, USA

Willem E. Frankenhuis Central European University, Hungary, and Radboud University Nijmegen, the Netherlands

Tao Gao Massachusetts Institute of Technology, USA

Emily D. Grossman University of California, Irvine, USA

Antonia F. de C. Hamilton University of Nottingham, UK

Petra Hauf St. Francis Xavier University, Canada

Valerie A. Kuhlmeier Queen's University, Canada

Jeff Loucks University of Regina, Canada

Scott A. Love University of Glasgow, Scotland

Yuyan Luo University of Missouri, USA

Elena Mascalzoni University of Padova, Italy

Phil McAleer University of Glasgow, Scotland

Richard Ramsey Bangor University, UK

Lucia Regolin University of Padova, Italy

M. D. Rutherford McMaster University, Canada

Kara D. Sage Hamilton College, USA

Brian J. Scholl Yale University, USA

Maggie Shiffrar Rutgers University, Newark, USA

Francesca Simion University of Padova, Italy

Jessica Sommerville University of Washington, USA

James P. Thomas Rutgers University, Newark, USA

Nikolaus F. Troje Queen's University, Canada

Amanda Woodward University of Chicago, USA

Index

The letter *f* or *t* following a page number denotes a figure or table, respectively.